Advances in State and Parameter Estimation

This book deals with the basics of parameter estimation and state estimation as the fundamental building blocks of mathematical modelling activity in the broader field of control theory. All the methods are validated using MATLAB®-based implementations with realistically simulated data for general dynamic systems, as well as for aircraft parameter estimation. This book includes several illustrative examples and chapter-end exercises.

Features:

- Provides comprehensive coverage of all issues related to parameter and state estimation.
- Discusses advanced topics related to Kalman filter, stability analysis, image centroid tracking and neural networks for parameter estimation.
- Explores convergence and stability results for the discussed methods.
- Reviews the estimation of parameters in linear/nonlinear models, and distributed fitting.
- Includes MATLAB®-based illustrative examples, and exercises.

This book is aimed at researchers and graduate students in systems and control, signal processing, estimation theory, engineering mathematics, and aerospace engineering.

Advances in State and Parameter Estimation

Theory and Practice

Jitendra R. Raol, Parimala P., Reshma V. and
Sara M. George

CRC Press
Taylor & Francis Group
Boca Raton London New York

CRC Press is an imprint of the
Taylor & Francis Group, an **informa** business

First edition published 2025
by CRC Press
2385 NW Executive Center Drive, Suite 320, Boca Raton FL 33431

and by CRC Press
4 Park Square, Milton Park, Abingdon, Oxon, OX14 4RN

CRC Press is an imprint of Taylor & Francis Group, LLC

ISBN: 9781032654867 (hbk)
ISBN: 9781041026211 (pbk)
ISBN: 9781003620211 (ebk)

DOI: 10.1201/9781003620211

Typeset in Times
by Newgen Publishing UK

Contents

Preface xiii
Acknowledgements xv
About the Authors xvii

Chapter 1 Introduction 1

1.1 State Estimation 1
1.2 Parameter Estimation 1
1.3 Sensor Data Fusion State Vector and Measurement-Level Fusion 2
1.4 Highlights of the Chapters 2

Chapter 2 Least-Squares and Maximum-Likelihood Methods 5

2.1 Introduction 5
2.2 Principle of Least Squares 6
2.2.1 Properties of the Least-Squares Estimates 7
2.3 Generalized Least Squares 8
2.3.1 A Probabilistic Version of the LS 8
2.4 Nonlinear Least Squares 9
2.5 Equation-Error Method 11
2.6 Gaussian Least-Squares Differential Correction Method 12
2.7 Principle of Maximum Likelihood and Cramer–Rao's Lower Bound 14
2.7.1 Cramer–Rao's Lower Bound 15
2.7.2 The Maximum-Likelihood Estimate is Efficient 17
2.8 Maximum-Likelihood Estimation for a Dynamic System 18
2.8.1 Derivation of the Likelihood Function 18
2.8.2 Accuracy Aspects 20
2.9 Output-Error Method: Features and Numerical Aspects 21
2.9.1 Features and Numerical Aspects 23
Exercises 24

Chapter 3 Kalman Filtering Methods 27

3.1 Introduction 27
3.2 Kalman Filtering 28
3.2.1 Covariance Matrix Formulation 28
3.2.2 Discrete-Time Filtering Algorithm 29
3.2.3 Continuous-Time Kalman Filter 32
3.2.4 Interpretation and Features of KF 32
3.3 Kalman *U-D* Factorization Filter 33
3.3.1 Time Propagation 34
3.3.2 Measurement Update 35
3.4 Extended Kalman Filter 36
3.4.1 Time Propagation 38
3.4.2 Measurement Update 38
Exercises 43

Chapter 4 Filtering-cum-Data Fusion with State Delay and Missing Measurements................46

4.1 Introduction..46
4.2 Filtering-cum-Fusion Algorithms 47
4.2.1 Modified Gain Fusion Algorithm (GFA)..47
4.2.2 Modified H-Infinity *A Posteriori* Filtering Algorithm (HIPF) 48
4.2.3 Modified H-Infinity Global Filtering-cum-Fusion Algorithm (HIGF) 49
4.2.4 Modified H-Infinity *A Posteriori* Risk-Sensitive Filtering Algorithm (HIRF)..50
4.3 Nonlinear Observer Based on a Continuous-Time Data-Fusion Filter............51
4.3.1 Asymptotic Stability Analysis..52
4.4 Performance Evaluation of the Four Modified HI-Based Fusion Algorithms 53
4.5 Nonlinear Observers with State Delay and Missing Data...............................54
4.5.1 Nonlinear System and Observer Error Dynamics..............................58
4.5.2 Asymptotic Stability Analysis..59
4.6 Nonlinear Observers for Sensor Data Fusion 61
4.6.1 Observer for Measurement Data-Level Fusion 61
4.6.2 Observer for State-Vector-Level Fusion...63
4.6.3 Performance Evaluation 63
4.7 Nonlinear Observers for Data Fusion Based on the HI Norm 65
4.7.1 Nonlinear System and Observer Error Dynamics..............................65
4.7.2 Asymptotic Results 68
4.8 Nonlinear Observers for Data Fusion Based on HI Schemes 71
4.8.1 Measurement Data-Level Fusion 71
4.8.2 State-Vector-Level Fusion..72
4.8.3 Performance Evaluation 73

Chapter 5 Gaussian-Sum Extended Kalman Filter with Lyapunov Stability Analysis...............76

5.1 Introduction...76
5.2 Gaussian-Sum Filter (GSF)...77
5.2.1 Time Propagation/Evolution...78
5.2.2 Measurement/Data Update...78
5.2.3 Gaussian-Sum Estimate..79
5.3 GSF with Pruning 79
5.3.1 Algorithm for GSFP 80
5.4 Asymptotic Stability Analysis of GSF via an Observer.................................80
5.5 Aircraft Parameter Estimation 84
5.6 Results and Discussions...85
5.6.1 Effects of Pruning..88
Exercises...88

Chapter 6 Gaussian-Sum Information Filter with Lyapunov Stability Analysis 90

6.1 Introduction...90
6.2 GS Extended Information Filter (GDEIF) 91
6.2.1 Time Propagation/Evolution...91
6.2.2 Measurement/Data Update...92
6.2.3 Gaussian-Sum Estimates 93

6.3 Asymptotic Stability Analysis of GSEIF via an Observer 93
6.4 Aircraft Parameter Estimation 97
6.5 Results and Discussions 99
Exercises 101

Chapter 7 Image Centroid Tracking with Square-Root Kalman Filters 103

7.1 Introduction 103
7.1.1 Target Tracking and Fusion 103
7.1.2 Importance and Application of Tracking and Fusion 104
7.1.3 Image Processing for Tracking Applications 104
7.1.4 Filtering and Image Fusion 104
7.1.5 Filtering Algorithms for Tracking Applications 105
7.1.6 Merits of the Square-Root Information Filter 105
7.1.7 Image Fusion 105
7.2 Computation of an Image Centroid 106
7.2.1 Synthetic Image Generation 107
7.2.2 Image Centroid Tracking using KF 108
7.2.3 Performance with Target-Image Noise Variances for CTKF 109
7.2.4 Performance with Target-Image Noise Variances for CTUDF 109
7.3 Square-Root Information Filter 111
7.3.1 Information Filter 112
7.3.2 Square-Root Information Filter (SRIF) 113
7.3.3 Image Centroid Tracking using CTSRIF-Performance Evaluation 115
7.3.4 Image Centroid Fusion using the SRIF Algorithm 116
Exercises 116

Chapter 8 Image Centroid Tracking with Fuzzy-Logic-Augmented Filters 119

8.1 Introduction 119
8.1.1 Fuzzy-Logic System 119
8.2 Fuzzy-Logic System: Adaptive Neuro-Fuzzy Inference System – ANFIS 120
8.3 Fuzzy-Logic-Based KF 121
8.4 Fuzzy-Logic-Based SRIF 122
8.5 Performance Evaluation of the FL-Based CT Algorithms 122
8.5.1 Mathematical Model and Data Generation 122
8.5.2 Performance with ANFIS-CTA and FLKF-CTA 123
8.5.3 Performance Evaluation of FL-SRIF-Based CTA 125
8.5.4 Performance Evaluation of FL-SRIF-Based CTA for Image Fusion 125
Exercises 126

Chapter 9 Image Centroid Tracking-cum-Fusion using New Factorization Filtering Algorithms 128

9.1 Introduction 128
9.2 Eigenfactor Square-Root Filter SRVD: Time Propagation and Data Update 128
9.2.1 *V-D* Discrete-Time Measurement Update 128

9.2.2 *V-D* Time Propagation 129
9.3 Performance Evaluation of the SRIF Fusion 129
9.4 Performance Evaluation of the SRVD Algorithm 130
9.5 A New Filtering Algorithm Based on SRVD and SRIF 132
9.5.1 Time-Propagation Part 132
9.5.2 Measurement-Data Update 133
9.5.3 Performance Evaluation of the VDSRIF Algorithm 134
9.6 A New Filtering Algorithm Based on UDF and SRIF 135
9.6.1 Initial Part with Initial Conditions 136
9.6.2 Time-Propagation Part 136
9.6.3 Measurement-Data Update 136
9.6.4 Performance Evaluation of the UDSRIF Algorithm 137
9.7 Observer Based on the Fuzzy Nonlinear Function of Residuals 139
9.7.1 Linear System and Observer Error Dynamics 140
9.7.2 Asymptotic Stability Analysis 141
9.7.3 Performance Evaluation of an FL-Modulated Observer for Centroid Tracking 142
9.7.4 Overall Performance of the Square-Root-Type Image Centroid Tracking Algorithms 147
Exercises 148

Chapter 10 H-Infinity Fuzzification Filtering and Target Tracking 150

10.1 Introduction 150
10.2 Fuzzy-Infinity Filter Fuzzification (FLHIFF) – Scheme 1 151
10.3 FL-Based HI Fuzzified Filter (FLHIFF) – Scheme 2 151
10.4 Degree of Matching (DOM) – Scheme 3 154
10.5 Performance Evaluation 156
10.5.1 Target Tracking for Maneuvering Objects using HIF and FLHIFF 156
10.5.2 Target Tracking for Mild Maneuvering Objects using HIF and FLHIFF 157
10.5.3 Target Tracking for Evasive Maneuvering Objects using HIF and FLHIFF 157
10.5.4 Target Tracking for Nonmaneuvering Objects using AKFFDOM and AHIFFDOM 160
10.6 Adaptive H-Infinity Filter (AHIF) – Scheme 4 162
10.6.1 Target Tracking for Maneuvering Objects using HIF and AHIF 164
10.6.2 Target Tracking for Mild Maneuvering Objects using HIF and AHIF 165
10.6.3 Target Tracking for Evasive Maneuvering Objects using HIF and AHIF 166
10.7 Fuzzy-Logic-Based H-Infinity Filter (FLHIF) 167
Exercises 176
Appendix 10A: FL-Based Adaptive Tuning 177

Chapter 11 H-Infinity-Based Observer 180

11.1 Introduction 180
11.2 HI-Based Nonlinear Observer 180
11.3 Nonlinear FMF-Based Observer 183

11.4 Asymptotic Result for Observer Error Dynamics ... 184
11.5 Evaluation of the Nonlinear Observer ... 187
Exercises ... 190

Chapter 12 Deterministic Nonlinear Estimators–Observers and Stability Results ... 192

12.1 Introduction ... 192
12.2 Concept of Model Error and Cost-Function Formulation ... 192
12.3 Pontryagin's Conditions ... 193
12.4 Method of Invariant Embedding ... 195
12.4.1 Continuous-Time Estimator ... 197
12.4.2 Discrete-Time Estimator ... 199
12.5 Nonlinear Observers ... 202
12.5.1 Observer Structure with IE Gain ... 202
12.5.2 Asymptotic Stability with Lyapunov Method ... 203
12.5.3 Performance Evaluation ... 205
12.5.4 Observer with IEHI (IE and HI)-Based Gain ... 207
12.5.5 Asymptotic Stability Result ... 208
12.5.6 Performance Evaluation ... 210
12.6 New Continuous-Time HI Norm-Based Estimator ... 212
12.6.1 H-Infinity *A Posteriori* Filter ... 212
12.6.2 Derivation of the Continuous-Time HI Filter ... 213
12.6.3 Observer Based on the CTHF ... 217
12.6.4 Asymptotic Analysis of the Observer ... 218
12.6.5 Performance Evaluation of the CTHF ... 220
12.7 Data Fusion with IE- and HI-Based Estimators ... 221
12.7.1 Model Error Estimators ... 223
12.7.2 DF Scheme Utilizing ME Estimators ... 225
12.7.3 Performance Evaluation of the DF Scheme ... 226

Chapter 13 Hybrid Global H-Infinity Fusion Filter ... 234

13.1 Introduction ... 234
13.2 Global H-Infinity Fusion Filter – DGHIFF in Discrete Time ... 234
13.2.1 At the Local Sensors ... 235
13.2.2 At the Fusion Centre ... 235
13.3 Derivation of the Hybrid Global HI Fusion Filter – HGFF ... 236
13.4 Hybrid Global HI-Based Fusion Filter: HGFF Collected ... 239
13.4.1 At the Local Sensors ... 239
13.4.2 At the Fusion Centre ... 240
13.5 Observer Based on the Global Hybrid HI Fusion Filter (HGFF) ... 240
13.5.1 Analytical Results ... 241
13.6 Performance Evaluation of the New Hybrid Fusion Filter (HGFF) ... 244

Chapter 14 Neural Networks for Parameter Estimation with Lyapunov Stability ... 247

14.1 Introduction ... 247
14.2 Feedforward Neural Networks ... 248
14.2.1 FFNN-BPN (Back-Propagation) Training Algorithm ... 248
14.2.2 FFNN-RLS (Recursive Least-Squares) Training Algorithm ... 250

14.3 Lyapunov Stability Analysis 251
14.3.1 For FFNN-BPN 252
14.3.2 For FFNN-RLS 254
14.4 Aircraft Parameter Estimation with FFNNs 256
14.5 Recurrent Neural Networks 260
14.5.1 RNN-Weight-Bias Method 261
14.5.2 RNN-Gradient-Based Method 263
14.6 Lyapunov Stability Analysis 263
14.6.1 For RNN-Weight-Bias Method 264
14.6.2 For RNN-Gradient-Based Method 266
14.7 Aircraft Parameter Estimation with RNNs 267
14.8 Conclusion 271
Exercises 272

Chapter 15 Interactive Multiple Modelling for Target Tracking with New Algorithms 275

15.1 Introduction 275
15.2 Process of Target Tracking 276
15.2.1 Multiple Models for Maneuvering Targets 276
15.2.2 Process of Interaction 276
15.3 Performance Evaluation 278
15.3.1 With GSKF 280
15.3.2 With GSIF 281
15.4 Discussion of the Results 283
Exercises 284

Chapter 16 Machine Learning for Estimation 286

16.1 Introduction 286
16.2 Extended Luenberger Observer 287
16.2.1 Notations 288
16.3 Recurrent Neural Network (RNN)-Based State Estimator 289
16.3.1 Remarks 290
16.3.2 RNN-Based State Estimator Model 291
16.4 Hybrid Model-Based State Estimator 292
16.4.1 Feedforward Neural Network (FFNN) 292
16.4.2 Hybrid Model-Based Estimator 293
16.5 Linear Regression and Support Vector Machine (SVM) 294
16.6 Regression Trees and Their Ensembles 295
16.6.1 Regression Trees 295
16.6.2 Ensembles of Regression Trees 295
16.7 Flight Simulation Data Repeatability 295
16.8 Feature Reduction 295
16.9 Altitude Data 297
16.9.1 SVM Optimizable Regression 297
16.9.2 Ensembles of Regression Trees 297
16.9.3 Ensembles of Regression Trees with Cross Validation 297
16.10 Results and Discussions 298

16.10.1 Roll Results 298
16.10.2 Pitch Results 298
16.10.3 Altitude Plus Pitch Results 298
Exercises 299

Appendix A Mathematical Modelling and System Identification 301
Appendix B Sensors, Signals and Estimation-Error Propagation 313
Appendix C Neural Networks, Fuzzy Logic and Genetic Algorithms for Estimation 321
Index 329

Preface

Mathematical modelling of dynamic systems is a very important branch of control and system engineering disciplines. The study of parameter and state estimation is crucial in mathematical model building; and these math models are required to understand the behaviour of many dynamic systems: transportation vehicles, aircrafts, spacecrafts, missiles, chemical plants, environmental phenomena, usual electro-mechanical systems, and robots. These mathmathetical models are used for analysis, forecast, prediction, design and synthesis of several control systems in the broader field of systems engineering.

Parameter estimation is the process of using measurements of a dynamic system to develop mathematical models that adequately represent the system characteristics. The assumed mathematical model consists of a finite set of parameters, the values of which are estimated using estimation techniques. Fundamentally, the approach is based on the least-squares minimization of error between the model response and the actual system's response.

State estimation is the process of determining the states of a dynamic system, expressed in the state-space form, using the noisy-measurements from the system, the criterion is basically the same or similar as for parameter estimation; interchangeably the specific jargon of 'filtering' is used, whereas state estimation is the general and broader phrase used. A closely related crucial study of system identification is briefly discussed. Also, certain aspects of sensors, signals and the estimation-error propagation are dealt with.

The main idea behind modelling, parameter and state estimation of an engineering system or a (chemical) process is to improve its performance or design a control system to regulate/monitor/control the dynamic behaviour to achieve the desired functionality of the overall system. With the advent of high-speed, and large memory-digital computers, more complex and sophisticated techniques like filter error methods, Gaussian-sum filters and innovative methods based on artificial neural networks would find increasing use in parameter and state estimation problems.

This book presents various existing and some *novel and recent* parameter and state estimation techniques based on the following: least squares, nonlinear least squares, output error, several filtering methods, Gaussian-sum extended Kalman filter, Gaussian-sum extended information filter, fuzzy-logic-augmented H-infinity (HI) filters, several new factorization filtering algorithms, nonlinear observers, artificial neural networks; and many specific ramifications, based on the HI filter, to handle state delays and randomly missing measurements; more details can be found in the table of contents.

Main applications covered in this book are image centroid tracking and fusion, target tracking, sensor data fusion, robust-nonlinear system identification and aircraft parameter estimation; however, more applications can be easily made to many dynamic systems. Also, estimation techniques based on neural networks, fuzzy logic and machine learning are included.

Most of the novel approaches and methods presented, are supported by the Lyapunov stability analyzes and the convergence conditions for these algorithms are obtained, for the first time for some estimators; and in most cases the techniques are validated by numerically simulated results for which MATLAB-based scripts were written by the authors (MATLAB is a registered trademark of MathWorks, Inc.). Some of the codes/scripts (all nonproprietary nature) would be made available and can be downloaded from the book's website. What one needs is to have an access to MATLAB, control, signal processing and system identification toolboxes.

The emphasis in the present book is on description of the essential features, mathematical models, algorithmic steps, numerical simulation details and results to illustrate the efficiency and efficacy of the application of these methods to practical systems; and hence the theorem-proof approach is not followed.

Thus, the book should be useful to any general reader, senior undergraduates, postgraduates and doctoral students in science and engineering. Also, it should be useful to practicing project scientists/engineers and teachers pursuing parameter/state estimation activity in nonaero or aerospace fields.

Disclaimer: Although, great care has been taken in the preparation of the book and working out the illustrative examples, the readers should verify equations and the results before applying the algorithms to their own engineering problems; the practical applications of the methods presented in the book should be at their own risk, and it is not the responsibility of the authors, contributors or the publisher.

Acknowledgements

Jitendra: Primarily Gauss, Fisher, Winners, Kalman, and Kailath have enlivened the lives of millions of engineers and scientists all over the world by their original contributions in the area of parameter and state estimation.

I am very grateful to Dr. Mrs. Girija, G., Mr. V. Parameswaran and Dr. Jatinder Singh with whom I have carried out estimation work at the Flight Mechanics and Control Division (FMCD) of CSIR-NAL for three decades. Dr. S. Balakrishna (now in the USA) pioneered the research and practical application of estimation to flight test data analysis of aircraft in FMCD four decades ago. All the doctors from the CSIR-NAL health center-clinic have been instrumental in enhancing my family's health index for several years starting from 1975 until today. My wife Virmati, daughter Harshakumari Gohil, son Mayur, and our daughter-in-law Amina have been my lifeline for several years.

Parimala: I am very grateful to Dr. J. R. Raol for considering me worthy to be a joint author for the present book. I am also grateful to him for providing me with excellent guidance during my doctoral work, and for moral support and encouragement during my difficult times.

I would like to acknowledge the contributions of my M. Tech students Mrs. Roopa, H. R. and Mr. Manjunath for technical support by carrying out various experimental studies as a part of their theses work, which was also useful for SW generation in some chapters. My heartfelt thanks go to my family members (especially my son Karthik, and daughter Kirthana) and friends for their unbounded love, affection and support throughout this journey.

Reshma: I extend my deepest appreciation to my mentor Dr J. R. Raol, whose wisdom and guidance have been invaluable throughout this journey, thank you for pushing me to dig deeper and strive for excellence in every aspect of my doctoral work and this book project. I am garteful for the countless hours you dedicated to reviewing drafts, answering questions, and offering invaluable suggestions during my doctoral study. Your commitment to excellence and attention to detail have left an indelible mark on me. I want to thank my family, especially my daughter Aadhya (Putty), for their unwavering support and understanding throughout this journey. Your love and encouragement have been my greatest source of strength. I am indebted to my friends who provided valuable feedback and encouragement along the way.

Sara: I would like to thank my mentor, Dr. J. R. Raol, for his guidance, motivation and encouragement towards writing some chapters for the book, and during some phases of my doctoral study. I am grateful to Dr. Mrs. S. Sethu Selvi (Professor and Formar Head, EC Deptartment, MSRIT), my Ph.D. thesis supervisor, for her technical support and guidnace. I am grateful to my parents and sisters for showing faith in me and giving me liberty to choose what I desired. Thanks to my husband James for his unconditional love and understanding. I appreciate my daughter Joanna for her affection.

We are very grateful to Professors/Drs. N. Shivashankarappa, Department of Electronics and Communications, Atriya Institute of Technology (AIT), Bangalore; Mrs. S. Sethu. Selvi, Former Head, Department of Electronics and Communications, Ramaiah Institute of Technology (RIT), Bangalore; and Mrs. Swetha Amit, Business Development Head, Zosh Aerosapce Pvt. Ltd., Bangalore for technical discussions, and their help in certain chapters of the book.

We are also very grateful to Ms. Nora Konopka, Editorial Director of Engineering at CRC Press, and Dr. Gagandeep Singh, Senior Publisher at CRC Press, for their interest in our work and for their logistics support during the present book project.

About the Authors

Jitendra R. Raol received BE and ME degrees in electrical engineering from M. S. University (MSU) of Baroda, Vadodara, in 1971 and 1973, respectively, and a PhD (electrical & computer engineering) from McMaster University, Hamilton, Canada, in 1986; at both universities he was also a postgraduate research and teaching assistant. He joined the National Aeronautical Laboratory (NAL, Bangalore) in 1975. At CSIR-NAL he was involved with human pilot modelling activites in fix- and motion-based research flight simulators. He rejoined NAL in 1986 and retired in July 2007 as Scientist-G (and Head at the Flight Mechanics and Control Division at CSIR-NAL).

He visited Syria, Germany, the United Kingdom, Canada, China, the United States of America, and South Africa on deputation/fellowships to (i) work on research problems on system identification, neural networks, parameter estimation, multi-sensor data fusion and robotics; (ii) present technical papers at international conferences; and/or (iii) deliver guest lectures at some of these places. He has given several guest lectures at many Indian colleges and universities, and Honeywell (HTSL, Bangalore).

He was a Fellow of the IEE/IET (United Kingdom) and a senior member of the IEEE (United States). He is a life-fellow of the Aeronautical Society of India and a life member of the Systems Society of India. During his studies at the MSU, he received the Dean and Professor Suba Rao Memorial Prize and M. C. Ghia charitable fellowship. In 1976, he won the K. F. Antia Memorial Prize of the Institution of Engineers (India) for his research paper on nonlinear filtering with time delay. He was awarded a certificate of merit by the Institution of Engineers (India) for his paper on parameter estimation of unstable systems. He had received the best poster paper award from the National Conference on Sensor Technology (New Delhi) for a paper on sensor data fusion. He also received a gold medal and a certificate for a paper related to target tracking (from the Institute of Electronics and Telecommunications Engineers, India). He is also one of the five recipients of the CSIR (Council of Scientific and Industrial Research, India) prestigious technology shield for the year 2003 for the leadership and contributions to the development of integrated flight mechanics and control technology for aerospace vehicles in the country; the shield was associated with a plaque, a certificate, and a project-grant-prize of 300,0000 INRs for the project work. He was one of the five recipients of the Chellaram Foundation Diabetes Research Award-2018 for the best paper (presented at the 2nd International Diabetes Summit, March 2018, Pune, India; which carried a prize of 100,000 INRs). He received the Sir Thomas Ward Memorial Prize of the Institution of Engineers (India) in 2019 (jointly) for the paper on Image Centroid Tracking with Fuzzy Logic…, and it carried a gold medal and a certificate. He is featured on Stanford University's list (USA) as one of the top 2% scientists/researchers in the world for the year 2019, as well as 2021.

He has published nearly 160 research papers and numerous technical reports. He guest-edited two special issues of *Sadhana* (an engineering journal published by the Indian Academy of Sciences, Bangalore) on (i) advances in modelling, system identification and parameter estimation (jointly with the late Prof. Dr. Naresh Kumar Sinha) and (ii) multi-source, multi-sensor information fusion. He also guest-edited two special issues of the *Defense Science Journal* (New Delhi, India) on (i) mobile intelligent autonomous systems (jointly with Dr. Ajith K. Gopal, CSIR-SA), and (ii) aerospace avionics and allied technologies (jointly with Prof. A. Ramachandran, MSRIT).

He co-authored the (i) Control Series book *Modeling and Parameter Estimation of Dynamic Systems* (2004); and CRC Press (Taylor & Francis, Florida, USA) books, (ii) *Flight Mechanics Modeling and Analysis* (2009, and the revised edition in March 2023), (iii) *Nonlinear Filtering: Concepts and Engineering Applications* (2017), (iv) *Control Systems: Classical, Modern, and AI based Approaches* (2019) and (v) *Situation Assessment in Aviation: Bayesian Netwok and Fuzzy Logic Approach* (2024).

He also authored CRC Press books (i) *Multi-sensor Data Fusion with MATLAB* (2010), and (ii) *Data Fusion Mathematics–Theory and Practice* (2015). He edited (with Ajith K. Gopal, CSIR-SA) a CRC Press book *Mobile Intelligent Autonomous Systems* (2012).

He has served as a member/chairman of numerous advisory, technical project review and doctoral examination committees. He has also conducted sponsored research and worked on several projects from Industry as well as other R&D organizations to CSIR-NAL with substantial budgets. Under his formal technical guidance, 11 doctoral and 8 master research scholars have received their degrees successfully; he also guided jointly/informally one dozen research students for their doctoral studies. He is a reviewer of a dozen national/international journals, and has evaluated several M. Tech./Doctoral theses (from India and overseas). He had been with (i) MSRIT (M. S. Ramaiah Institute of Technology, Bengaluru) as emeritus professor for 5 years; (ii) the Govt. College of Engineering, Kunnur (Kerala) as a senior research advisor; and (iii) the Deptartment of Aerospace Engineering (IISc., Bangalore) as a consultant on modelling and parameter estimation for the type I diabetes-patients' data for a period of 3 months. He was also a consultant to a private company in Bangalore that provides software and algorithmic solutions to target tracking, navigation and data-fusion problems. He is a member of the governing body of the Centurion University of Technology and Management (CUTM), Andhra Predesh, India,

His main research interests have been and are data fusion, system identification, state/parameter estimation, flight mechanics–flight data analysis, H-infinity filtering, nonlinear filtering, artificial neural networks, fuzzy logic systems, genetic algorithms and soft technologies for robotics.

On the literary front, he has also authored a few books as the collection of his 320 (free-) verses on various facets closely related to science, philosophy, evolution, and life itself. He has also contributed 80 articles/stories and 1200 'bites' (long quotes) and 7 ebooks covering social, philosophical, science and human-life-related aspects and issues. He was featured as one of the top 100 downloaded English authors of the matrubharti.com for the years 2021–2024. His new area of study and research is data-systems analytics (DaSyA).

Parimala P. earned her PhD degree from Jain University, Karnataka, India in 2018 under the guidance of Dr. J. R. Raol, NAL, Bangalore. Her thesis was titled *Image Tracking and Fusion using Square Root Information Filter*. She earned her BE in Telecommunication Engineering and ME degree in Digital Communication Engineering from BMS College of Engineering, Bangalore University, Bangalore during the years 1996 and 1999, respectively. She has published several research papers in international and national journals and conferences. She has been working as Assistant Professor at Ramaiah Institute of Technology in the Department of Electronics and Telecommunication Engineering, since 1999. She has taught several UG and PG courses. Her research interests are image processing, antenna design, digital communication. She has guided several UG and PG projects, many of them received Best Project Awards at Institute Levels and MSRIT Alumni Association Funds. She is working with Industry collaborated projects. She is passionate to work with real-time projects, which are closely related to societal issues. She believes in providing technological solutions to societal issues and challenges. She is Co Pi for DST funded project on Navic system Design worth 52 Lakhs. She is a recipient of the Sir Thomas ward award for a research paper by the Institute of Engineers, 2019. She is serving as a Coordinator for COE-Center for Antennas and Radio Frequency Systems, RIT, Bangalore. She is working on a consultancy projects related to antenna design and validation and has completed consultancy projects worth 1.5 Lakh. She has obtained an Indian patent on 5G Antenna Design, 2022. She has authored a few chapters for two renowned textbooks, including one in CRC publications. She has co-authored a text book on antenna and microwave engineering. She was Sensor Council Track Session Chair of IEEE CONNECT 2021and I4C 2022 conference. She is currently serving as IEEE APS RIT student branch Faculty Advisor and IEEE MTTs RIT student branch Faculty Advisor. She is a member of several professional bodies like MIEEE, LMISTE and IETE, and has served as IEEE Sensor Council Chapter Treasurer for 2 years and now is an Execom member.

Reshma V. completed her doctoral studies at Jain University, Karnataka, India in 2018, focusing on the development of a 'Fuzzy Augmented H-I Filter for Target Tracking'. She obtained her M.Tech degree in Power Electronics from Visvesvaraya Technological University (VTU) in 2005. Throughout her career, she has made significant contributions to academia and research, publishing numerous papers in both international and national journals and conferences. She has been an Assistant Professor in the Department of Electronics and Communication Engineering at Ramaiah Institute of Technology since 2007. Her research interests span across artificial intelligence, embedded system design, and VLSI. She has actively supervised several undergraduate and postgraduate projects, many of which have received accolades such as Best Project Awards at institute level. Notably, her involvement in projects has secured funding from the Karnataka State Science and Technology Council (KSCST) and MSRIT Alumni Association. In addition to her academic endeavours, she collaborates with Industry on various projects. Notably, her guidance of a team working on GUI automation for Samsung SRIB PRISM earned them the Excellence Award in 2019. She is particularly drawn to projects addressing real-time societal issues, believing in the power of technology to provide solutions.

Sara M. George holds a B.Tech in Electronics and Communication from Mahatma Gandhi University, Kerala; M.Tech and PhD degrees from Visvesvaraya Technological University, Karnataka. She is currently an Assistant Professor at the Ramaiah Institute of Technology, Bangalore. Her research interests include nonlinear filtering, signal processing and embedded system design. She has published nearly 20 technical papers in various conferences and journals.

1 Introduction

Several dynamic systems abound in real-life environments: biological, mechanical, electrical, civil, chemical, aerospace, road traffic, mechatronics, and a variety of other systems. Understanding the dynamic behaviour of these practical systems is of primary interest to scientists and engineers. Mathematical modelling via parameter estimation (and system identification and state estimation) is one of the approaches that leads to deeper understanding of the system's characteristics. These parameters often describe the stability and control behaviour of the system. Estimation of these parameters from input–output data (empirical signals) of the system is thus an important step in the analysis of dynamic systems.

1.1 STATE ESTIMATION

Parameter and state estimation are very closely related aspects in the overall mathematical modelling of dynamic systems, and this is primarily accomplished via the fundamental process of system identification, the latter is dealt with in Appendix A. Almost all the real-life dynamic systems can be described by the so-called state space models, which have inputs, outputs, and internal states. Some states could be outputs also. These states are the internal variables of the dynamic system, which are often required to be estimated. These states are estimated by using input/output data in a state estimation algorithm/procedure. The most celebrated technique for the state estimation is the Kalman filtering (KF), for linear systems, which has been highly successful for last 7 to 8 decades.

1.2 PARAMETER ESTIMATION

Figure 1.1 shows a simple approach to parameter estimation. The parameters of the model are adjusted iteratively until such time that the responses of the model match closely with the measured outputs of the system under investigation in the sense specified by the chosen minimization criterion. It must be emphasized here that a good match is a necessary but not sufficient condition for achieving good estimates of the parameters.

In the present book, we are generally concerned with the estimation of the parameters of dynamic systems and the state estimation using KF and other novel algorithms. Often, the parameters and the states are jointly estimated using the so-called extended Kalman filtering (EKF) approach, in which the parameters to be estimated are considered as additional states in the state space model.

DOI: 10.1201/9781003620211-1

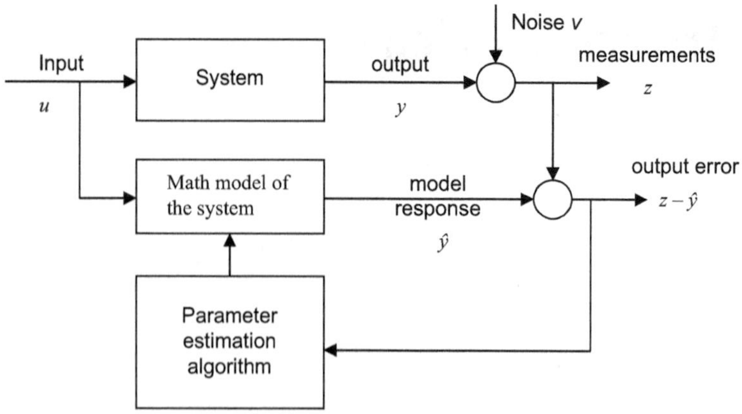

FIGURE 1.1 Parameter estimation procedure.

1.3 SENSOR DATA FUSION STATE VECTOR AND MEASUREMENT-LEVEL FUSION

Sensor data fusion (SDF, DF) is an additive, multiplicative and/or probabilistic numerical process that collects information from several sources and/or sensors and provides enhanced information about an object: its features and trajectories, the latter if an object is a moving one. In a state vector-level fusion (SVF), the estimated state trajectories are fused by using a weighted average fusion formula; and in a measurement-level fusion, the observations from several sensors are first fused and then the estimation procedure carried out, the latter one is supposed to be more accurate.

Some techniques presented in a few chapters of the book are also used for data-fusion applications.

1.4 HIGHLIGHTS OF THE CHAPTERS

Chapter 2 introduces the classical method of parameter estimation, the celebrated least-squares (LS) method invented by Gauss and independently studied by Legendre [1]; and deals with the widely used maximum-likelihood (ML-)based output error method (OEM). The principle of maximum likelihood and its related development are treated in sufficient details. In Chapter 3, we discuss the filtering methods, especially the KF algorithms and their applications.

Chapter 4 discusses the problem of state estimation with delayed states and randomly missing measurements. We also introduce new filtering-cum-fusion algorithms based on H-infinity (HI) concept, and nonlinear observers with state delay and missing data. Also, a nonlinear observer based on a continuous-time data-fusion filter is presented and asymptotic stability analysis is carried out.

Chapter 5 discusses the Gaussian sum extended KF (GSEKF). The asymptotic stability analysis of the GSEKF is established using the Lyapunov energy functional via an observer. An aircraft parameter estimation example is illustrated. Chapter 6 deals with the Gaussian sum extended information filter (GSEIF), its asymptotic stability analysis and application to aircraft parameter estimation.

In Chapter 7, we discuss the problem of image-centroid tracking using a square-root information filter. An illustrative example is also presented. Chapter 8 discusses the image-centroid tracking with a fuzzy logic-augmented filter, and their performance with simulated data. Chapter 9 presents the image-centroid tracking using several novel factorization filtering algorithms and their performance results.

In Chapter 10, we discuss the concept of the HI filter's fuzzification and its application to target tracking. Chapter 11 presents a HI-based nonlinear observer.

In Chapter 12, the deterministic nonlinear observers and their stability results are presented. A continuous-time HI norm-based estimator is derived from a discrete-time HI filter and its asymptotic stability results are derived. In Chapter 13, a continuous-time global HI-based fusion filter is derived from a global discrete-time HI fusion filter. It so happens that the newly derived filter turns out to be a hybrid HI-based global fusion filter. Its asymptotic results are also presented.

Chapter 14 discusses the aircraft parameter estimation with feedforward neural networks (FFNNs), and recurrent neural networks (RNNs). The asymptotic stability results are also derived for these two approaches.

Chapter 15 discusses the target-tracking applications with a few new algorithms. The performance evaluation results are also presented. In Chapter 16 we discuss the techniques of machine learning (ML) as applicable to state and parameter estimation.

Appendix A discusses the problem of system identification (SID). Appendix B deals with sensors, signals and error aspects in estimation. In Appendix C we discuss the fuzzy logic (FL) and genetic algorithms (GAs) for estimation.

There are several books on the problem of system identification, and state and parameter estimation [1–16]; however, the current book presents several novel and recent techniques and approaches with (MATLAB-based) illustrative examples where applicable, for state and parameter estimation: Gaussian sum EKF, Gaussian sum EIF, estimation with state delays and randomly missing data, image-centroid tracking with a square-root information filter (SRIF), HI-based target tracking and nonlinear observers, robust model error approach in the context of nonlinear system identification, aircraft parameter estimation applications using artificial neural networks (ANNs), hybrid global HI-based fusion filter, and machine-learning methods. Several new asymptotic results are derived based on the Lyapunov energy functional.

The results presented in the book have been generated with great care, but the use of these (and associated software) by the readers is their own responsibility. The present book would be useful to teachers, and researchers/engineers of all branches of engineering, and to students of upper grade UG, and PG levels, including doctoral research scholars.

REFERENCES

1. Mendel, J.M., *Discrete Techniques of Parameter Estimation: Equation Error Formulation*, Marcel Dekker, New York, 1976.
2. Ljung, L., *System Identification: Theory for the User*, Prentice-Hall, Englewood Cliffs, 1987.
3. Hsia, T.C., *System Identification – Least Squares Methods*, Lexington Books, Lexington, Massachusetts, Toronto, 1977.
4. Sorenson, H.W., *Parameter Estimation – Principles and Problems*, Marcel Dekker, New York and Basel, 1980
5. Graupe, D., *Identification of Systems*, Van Nostrand, Reinhold Company, New York, 1972.
6. Eykhoff, P., *System Identification: Parameter and State Estimation*, John Wiley, London, 1972.
7. Sinha, N.K., and Kuszta, B., *Modelling and Identification of Dynamic System*, Van Nostrand, New York, 1983.
8. Box, G.E.P., and Junkins J.L., *Time Series: Analysis, Forecasting and Controls*, Holden Day, San Francisco, 1970.
9. Dorsey, J., *Continuous and Discrete Control Systems – Modelling, Identification, Design and Implementation*, McGraw Hill, New York, 2002.
10. Goodwin, G.C., and Payne, R.L., *Dynamic System Identification: Experiment Design and Data Analysis*, Academic Press, New York, 1977.
11. Kailath, T., Sayad, A.H., and Hassibi, B., *Linear Estimation*, Prentice-Hall, New Jersey, 2000.
12. de Ridder, D., Tax, D.M.J., Lei, B., Xu, G., Feng, M., Zou, Y., and van der Heijden, F., *Classification, Parameter Estimation and State Estimation: An Engineering Approach Using MATLAB*, 2nd Edition, Wiley, USA, March 2017.

13. Boguslavskiy, J.A., and Borodovsky, M. (Editor), *Dynamic Systems Models: New Methods of Parameter and State Estimation*, Springers, New York, March 2016.
14. Englezos, P., and Kalogerakis, N., *Applied Parameter Estimation for Chemical Engineers*, CRC Press, Florida, September 5, 2019.
15. Wang, Z., Yi, D., Duan, X., Yao, J., and Gu, D., *Measurement Data Modelling and Parameter Estimation*, CRC Press, Florida, May 31, 2017.
16. Chen, H.-F., and Zhao, W., *Recursive Identification and Parameter Estimation*, CRC Press, Florida, October 12, 2017.

2 Least-Squares and Maximum-Likelihood Methods

2.1 INTRODUCTION

To address the parameter-estimation problem, we begin with the assumption that the data are contaminated by noise or measurement errors. We use these data in an identification-cum-estimation procedure (ICEP) to arrive at optimal estimates of the unknown parameters that best describe the behaviour of the data-system dynamics. This process of determining the unknown parameters of a postulated mathematical model from noisy input/output (IO) data is termed 'parameter estimation'. A closely related problem is that of 'state estimation' wherein the estimates of the so-called 'states' of the dynamic process/system (a power plant or an aircraft) are obtained by using the optimal linear or the nonlinear filtering theory; this is treated in Chapter 3.

Here, we discuss the least-squares/equation-error (LS/EE) techniques for parameter estimation of dynamic systems (including algebraic systems), in general, and the aerodynamic derivatives of aerospace vehicles from the flight data, in particular. In the first few sections, some basic concepts and techniques of the LS approach are discussed with a view to elucidating the more involved methods and procedures in the later chapters.

The LS is the most simple and perhaps, highly favoured approach to determine the system characteristics from its IO time histories. There are several methods that can be used to estimate system parameters [1]. These techniques differ from one another based on the optimal criterion used and the presence of process and measurement noise in the data. The maximum-likelihood method (MLM) invokes the probabilistic aspects of random variables (measurements, errors, etc.) and defines a process by which we obtain estimates of the parameters [2]. These parameters most likely produce the math-model responses, which closely match the observations. A likelihood function (akin to a probability density function, pdf) is defined when measurements are (collected and) used. This likelihood function is maximized to obtain the maximum-likelihood estimates of the parameters of the dynamic system. The equation-error method (EEM) is a special case of the maximum-likelihood estimator (MLE) for the data containing only process noise and no measurement noise. The output-error method is a MLE for data containing only measurement noise and no process noise. At times, one comes across statements in the literature mentioning that MLE is superior to EEM and output-error methods (OEM, see Figure 1.1). This falsely gives an impression that EEM and OEM are not MLEs.

The type of (linear or nonlinear) mathematical model, and the presence of process or measurement noise in data or both mainly drive the choice of the estimation method and the intended use of the obtained results. The EEM has a cost function that is linear in parameters. It is simple and easy to implement. The OEM is a more complex process and requires a nonlinear optimisation technique (Gauss–Newton method) to estimate model parameters. The iterative nature of the approach makes

DOI: 10.1201/9781003620211-2

it a little more computer intensive. The third approach is the filter-error method (FEM), which is the most general approach to the parameter-estimation problem accounting for both process and measurement noise; being a combination of KF and OEM, it is the most complex of the three techniques with high computational requirements. The OEM is perhaps the most widely used approach for aircraft parameter estimation. The Gaussian least-squares differential correction method (GLSDCM) is also an output-error method, but it is not based on the maximum-likelihood principle.

2.2 PRINCIPLE OF LEAST SQUARES

Since our approach is model-based, we need to define a mathematical model of the dynamic (or static) system. The measurement model is assumed to have the following form:

$$z = H\beta + v, \quad y = H\beta \tag{2.1}$$

where y is the $(m \times 1)$ vector of true outputs and z is the $(m \times 1)$ vector that denotes the measurements (affected by noise) of the unknown parameters (through H), β is the $(n \times 1)$ vector of the unknown parameters and v represents the measurement noise/errors, which are assumed to be zero mean and Gaussian. This model is called the measurement equation model, since it forms a relationship between the measurements and the parameters of a system.

It can be said that the estimation theory and the methods have a (measurement-)data-dependent nature, since the measurements used for estimation are invariably noisy. These noisy measurements are utilized in the estimation procedure/algorithm/software to improve upon the initial guesstimate of the parameters that characterize the signal or system. One of the objectives of the estimator is to produce the estimates of the signal (what it means is the predicted signal using the estimated parameters) with errors much less than the noise affecting the signal. In order to make this possible, the signal and the noise should have significantly differing characteristics, e.g. different frequency spectra, widely differing statistical properties (true signal being deterministic and the noise being of random nature). This means that the signal is characterized by a structure or a mathematical model ($H\beta$) and the noise (v), often or usually is assumed as zero mean and white process. In most cases, the measurement noise is also considered Gaussian. This Gaussian(-ness) assumption is supported by the central-limit theorem. We use discrete-time signals in carrying out analysis and generating computer-based numerical results in the illustrative examples.

The LS method is a special case of the well-known MLE for linear systems with Gaussian noise. In general, least-squares methods are applicable to both linear as well as nonlinear problems. They are applicable to multi-input multi-output (MIMO) dynamic systems. LS techniques can also be applied to the online identification problem. For this method, it is assumed that the system parameters do not rapidly change with time, thereby assuring almost stationarity of the plant or the process parameters. This may mean that the plant is assumed quasi-stationary during the measurement period. This should not be confused with the requirement of nonsteady IO data over the period for which the data is collected for parameter estimation. This means that during the measurement period there should be some activity. The LS method is a deterministic approach to the estimation problem. We choose an estimator of β that minimizes the sum of the squares of the error:

$$J \cong \frac{1}{2}\sum_{k=1}^{N} v_k^2 = \frac{1}{2}(z - H\beta)^T (z - H\beta) \tag{2.2}$$

Here J is a cost function and v, the residual errors at discrete-time index 'k'. Superscript T stands for the vector/matrix transposition. The minimization of J w.r.t. β yields

$$\partial J/\partial\beta = -(z - H\hat{\beta}_{LS})^T H = 0 \quad OR \quad H^T(z - H\hat{\beta}_{LS}) = 0 \tag{2.3}$$

$$H^T z - (H^T H)\hat{\beta}_{LS} = 0 \quad \text{or} \quad \hat{\beta}_{LS} = (H^T H)^{-1} H^T z \tag{2.4}$$

In eqn. (2.4), the *term before z* is a pseudo-inverse. Since, the matrix H and the vector (of measurements) z are known quantities, $\hat{\beta}_{LS}$, the least-squares estimate of β, can be readily obtained. The inverse will exist only if no column of H is a linear combination of other columns of H. It must be emphasized here that, in general, the number of measurements (of the so-called observables like y) should be more than the number of parameters to be estimated. This implies that, at least theoretically: *number of measurements = number of parameters* + 1. This applies to all the parameter-estimation techniques considered in this book. If this requirement were not met, then the measurement noise would not be smoothed out at all. If we ignore v in eqn. (2.1), we can obtain β using the pseudo-inverse of H, i.e. $(H^T H)^{-1} H^T$. This shows that the estimates can be obtained in a very simple way from the knowledge of only H. By evaluating the Hessian (second-order partial differentiation) of the cost function J, we can assert that the cost function will be minimum for the least-squares estimates.

2.2.1 Properties of the Least-Squares Estimates

The properties of LS are given below [1]:

(a) $\hat{\beta}_{LS}$ is a linear function of the data vector z (eqn. (2.4)), since H is a completely known quantity. H could contain input/output data of the system.

(b) The error in the estimator is a linear function of the measurement errors (v_k)

$$\tilde{\beta}_{LS} = \beta - \hat{\beta}_{LS} = \beta - (H^T H)^{-1} H^T (H\beta + v) = -(H^T H)^{-1} H^T v \tag{2.5}$$

Here $\tilde{\beta}_{LS}$ is the error in the estimation of β. If the measurement errors are large, then the error in parameter estimation is large.

(c) $\tilde{\beta}_{LS}$ is chosen such that the residual, defined by $r \cong (z - H\hat{\beta}_{LS})$ is perpendicular (in general orthogonal) to the columns of the observation matrix H. This is the principle of orthogonality. This property has a geometrical interpretation.

(d) If $E\{v\}$ is zero, then the LS estimate is unbiased. Let $\tilde{\beta}_{LS}$ be defined as earlier. Then, $E\{\tilde{\beta}_{LS}\} = -(H^T H)^{-1} H^T \; E\{v\} = 0$, since $E\{v\} = 0$. Here $E\{.\}$ stands for mathematical expectation of the quantity in braces. If, for all practical purposes, $z = y$, then $\hat{\beta}$ is a deterministic quantity and is then exactly equal to β. If the measurement errors cannot be neglected, i.e. $z \neq y$, then $\hat{\beta}$ is random. In this case, one can get $\hat{\beta}$ as an unbiased estimate of β. When the LSM leads to a biased estimate in the presence of measurement noise, then it can be used as a start-up procedure for other estimation methods like the GLSDCM and OEM.

(e) The covariance of the estimation error is given as

$$E\{\tilde{\beta}_{LS}\tilde{\beta}_{LS}^T\} \cong P = (H^T H)^{-1} H^T R H (H^T H)^{-1} \tag{2.6}$$

where R is the covariance matrix of v. If v is uncorrelated and its components have identical variances, then $R = \sigma^2 I$, where I is an identity matrix. Thus, we have

$$\operatorname{cov}\left(\tilde{\beta}_{LS}\right) = P = \sigma^2 (H^T H)^{-1} \tag{2.7}$$

Hence, the standard deviation of the parameter estimates can be obtained as $\sqrt{P_{ii}}$, ignoring the effect of cross terms of the matrix P. This will be true if the parameter-estimation errors like

$\tilde{\beta}_{ij}$ for $i \neq j$ are not highly correlated. Such a condition could prevail, if the parameters are not highly dependent on each other. If this is not true, then only ratios of certain parameters could be determined; such difficulties arise in closed-loop identification, e.g. data co-linearity.

(f) The residual has zero mean:

$$r \cong (z - H\hat{\beta}_{LS}) = \mathrm{H}\beta + v\text{-}\mathrm{H}\hat{\beta}_{LS} = H\tilde{\beta}_{LS} + v \tag{2.8}$$

$$E\{\mathrm{r}\} = H\,E\{\,\hat{\beta}_{LS}\,\} + E\{v\} = 0 + 0 = 0 \text{ for an unbiased LS estimate;} \tag{2.9}$$

If residual is not zero mean, then the mean of the residuals can be used to detect bias in the sensor data.

2.3 GENERALIZED LEAST SQUARES

The generalized least-squares (GLS) method is also known as the weighted least-squares method. The use of a weighting matrix in the least-squares criterion function gives the cost function for the GLS method:

$$J = (z - H\beta)^T W (z - H\beta) \tag{2.10}$$

Here, W is the weighting matrix, which is symmetric and positive definite and is used to control the influence of specific measurements upon the estimates of β. The solution would exist if the weighting matrix is positive definite. Let $W = SS^T$ and $S^{-1}WS^{-T} = I$; here, S is a lower triangular matrix and square root of W. We transform the observation vector z as follows:

$$z' = S^T z = S^T H\beta + S^T v = H'\beta + v' \tag{2.11}$$

Expanding the J, we get

$$\begin{aligned}(z - H\beta)^T W (z - H\beta) &= (z - H\beta)^T SS^T (z - H\beta) \\ &= (S^T z - S^T H\beta)^T (S^T z - S^T H\beta) \\ &= (z' - H'\beta)^T (z' - H'\beta)\end{aligned} \tag{2.12}$$

Due to similarity of the form of the above expression with the expression for LS, the previous results of Section 2.2 can be directly applied to the measurements z'.

We have seen that the error covariance provides a measure of the behaviour of the estimator. Thus, one can alternatively determine the estimator, which will minimize the error variances. If the weighting matrix W is equal to R^{-1}, then the GLS estimates are called Markov estimates [1].

2.3.1 A Probabilistic Version of the LS

Define the cost function as

$$J_{ms} = E\{(\beta - \hat{\beta})^T (\beta - \hat{\beta})\} \text{ where subscript } ms \text{ stands for mean square.} \tag{2.13}$$

Here 'E' stands for the mathematical expectation, which takes, in general, probabilistic weightage of the variables. Consider an arbitrary, linear and unbiased estimator $\hat{\beta}$ of β. Thus, we have $\hat{\beta} = Kz$, where K is $(n \times m)$ matrix that transforms the measurements (vector z) to the estimated parameters (vector β). Thus, we are seeking a linear estimator based on the measured data. Since $\hat{\beta}$ is required to be unbiased we have

$$E\{\hat{\beta}\} = E\{K(H\beta + v)\} = E\{KH\beta + Kv\} = KHE\{\beta\} + KE\{v\}$$

Since $E\{v\} = 0$, i.e. assuming zero mean noise, $E\{\hat{\beta}\} = KH\ E\{\beta\}$ and $KH = I$ for an unbiased estimate.

This gives a constraint on 'K', the so-called gain of the parameter estimator. Next, we recall that

$$\begin{aligned} J_{ms} &= E\{(\beta - \hat{\beta})^T(\beta - \hat{\beta})\} \\ &= E\{(\beta - Kz)^T(\beta - Kz)\} \\ &= E\{(\beta - KH\beta - Kv)^T(\beta - KH\beta - Kv)\} \\ &= E\{v^T K^T Kv\}; \quad \text{since} \quad KH = I \\ &= \text{Trace}\ E\{Kvv^T K^T\} \end{aligned} \tag{2.14}$$

and defining $R = E\{vv^T\}$, we get J_{ms} = Trace (KRK^T), where R is the covariance matrix of the measurement noise vector v. Thus, the gain matrix should be chosen such that it minimizes J_{ms} subject to the constraint $KH = I$. Such that the 'K' matrix is found to be [2]:

$$K = (H^T R^{-1} H)^{-1} H^T R^{-1} \tag{2.15}$$

With this value of K, the constraint will be satisfied. The error covariance matrix P is given by

$$P = (H^T R^{-1} H)^{-1} \tag{2.16}$$

It is easy to establish that the generalized LS method and linear minimum mean-squares method give identical results, if the weighting matrix W is chosen such that $W = R^{-1}$. Such estimates, which are unbiased, linear and minimize the mean-squares error, are called the best linear unbiased estimator (BLUE) [1], KF is such an estimator, Chapter 3. The matrix H, which determines the relationship between measurements and β will contain some variables that are known or measured. One important aspect about spacing of such measured variables in matrix H is that, if they are too close (due to fast sampling), then rows or columns of the matrix H will be correlated and similar, and might cause ill conditioning in matrix inversion or computation of parameter estimates. Matrix ill conditioning can be avoided by using the artifice: Let $H^T H$ be the matrix to be inverted, then use $(H^T H + \varepsilon I)$ with ε as a very small number, say 10^{-5} or 10^{-7} and I as the identity matrix of the same size $H^T H$. Alternatively, matrix factorisation and subsequent inversion can be used.

2.4 NONLINEAR LEAST SQUARES

Most real-life static/dynamic systems have nonlinear characteristics and for accurate modelling, these characteristics cannot be ignored. If the type of nonlinearity is known, then only certain unknown parameters need to be estimated. If the type of nonlinearity is unknown, then some approximated model should be fit to the data of the system. In this case, the parameters of the fitted model need to be estimated. In general, real-life practical systems are nonlinear and hence we apply the LS method to nonlinear models. Let such a process or system be described by

$$z = h(\beta) + v \tag{2.17}$$

where h is a known, the nonlinear vector-valued function of dimension m. With the LS criterion, we have

$$J = (z - h(\beta))^T (z - h(\beta)) \tag{2.18}$$

The minimization of J w.r.t. β results in

$$\partial J/\partial\beta = -2[z - h(\hat{\beta})]^T \, \partial h(\hat{\beta})/\partial\beta = 0 \tag{2.19}$$

We note that eqn. (2.19) is a system of nonlinear algebraic equations. For such a system, a closed form solution may not exist. This means that we may not be able to obtain $\hat{\beta}$ explicitly in terms of observation vector, without resorting to some approximation or numerical procedure. From eqn. (2.19) we get

$$[\partial h(\hat{\beta})/\partial\beta]^T (z - h(\hat{\beta})) = 0 \tag{2.20}$$

The second term in the above equation is the residual error and the form of the equation implies that the residual vector must be orthogonal to the columns of $\partial h/\partial\beta$, the principle of orthogonality. An iterative procedure to approximately solve the nonlinear least-squares (NLS) problem is described next. Assume some initial guess or estimate (called a guesstimate) β^* for β. We expand $h(\beta)$ about β^* via Taylor's series to obtain

$$z = h(\beta^*) + \{\partial h(\beta^*)/\partial\beta\}(\beta - \beta^*) + \textit{higher-order terms} + v$$

Retaining terms up to first order we get

$$(z - h(\beta^*)) = \{\partial h(\beta^*)/\partial\beta\}(\beta - \beta^*) + v \tag{2.21}$$

Comparing this with the measurement equation studied earlier and using the results of the previous sections we obtain

$$(\hat{\beta} - \beta^*) = (H^T H)^{-1} H^T \quad (z - h(\beta^*)) \; \hat{\beta} = \beta^* + (H^T H)^{-1} H^T (z - h(\beta^*)) \tag{2.22}$$

Here $H = \partial h(\beta^*)/\partial\beta$ at $\beta = \beta^*$. Thus, the algorithm to obtain $\hat{\beta}$ from eqn. (2.22) is given as follows: (i) Choose β^*, initial guesstimate; (ii) linearize h about β^* and obtain H matrix; (iii) compute residuals $(z - h(\beta^*))$ and then compute the $\hat{\beta}$; (iv) check for the orthogonality condition: $H^T (z - h(\beta))|_{\beta=\hat{\beta}}$ = orthogonality condition value = 0; (2.23) (v) if the above condition is not satisfied, then replace β^* by $\hat{\beta}$ and repeat the procedure; and (vi) terminate the iterations when the orthogonality condition is at least approximately satisfied. In addition, the residuals should be white as discussed below.

If the residuals $(z - h(\hat{\beta}))$ are not white, then a procedure called generalized least squares can also be adopted [1]. The main idea of the residual being white is that residual power spectral density is flat (w.r.t. frequency), and the corresponding autocorrelation is an impulse function. It means that the white process is uncorrelated at the instants of time other than *t=0*, and hence it cannot be predicted. It means that the white process has no model or rule that can be used for its prediction. It also means that if the residuals are white, complete information has been extracted from the signals used for parameter estimation and nothing; perhaps, more can be extracted from the signal.

If residuals are nonwhite, then a model (filter) can be fitted to these residuals using the LS method and parameters of the model/filter estimated:

$$\hat{\beta}_{rLS} = (X_r^T X_r)^{-1} X_r^T \tag{2.24}$$

Here, r is the residual time history and X_r is the matrix composed of values of r, and will depend on how the residuals are modelled. Once $\hat{\beta}_r$ is obtained by the LS method, it can be used to filter the original signal/data. These filtered data are used again to obtain the new set of parameters of the system and this process is repeated until $\hat{\beta}$ and $\hat{\beta}_r$ are converged. This is also called the GLS procedure (in the system identification literature) and it would provide more accurate estimates when the residual errors are autocorrelated, hence nonwhite.

2.5 EQUATION-ERROR METHOD

EEM is based on the principle of least squares. It minimizes a quadratic cost function of the error in the (state) equations to estimate the parameters. It is assumed that states, their derivatives, and control inputs are available or accurately measured. The EEM is relatively fast and simple, and applicable to linear as well as linear-in-parameter systems. If the system is described by the state equation

$$\dot{x} = Ax + Bu \qquad \text{with} \qquad x(0) = x_0 \tag{2.25}$$

the equation error can be written as

$$e(k) = \dot{x}_m - Ax_m - Bu_m \tag{2.26}$$

Here, x_m is the measured state, subscript m denoting 'measured'. Parameter estimates are obtained by minimizing the equation error w.r.t. β. The above equation can be written as

$$e(k) = \dot{x}_m - A_a x_{am} \tag{2.27}$$

where $A_a = [A\ B]$ and $x_{am} = \begin{bmatrix} x_m \\ u_m \end{bmatrix}$. In this case, the cost function is given by

$$J(\beta) = \frac{1}{2}\sum_{k=1}^{N}\left[\dot{x}_m(k) - A_a x_{am}(k)\right]^T \left[\dot{x}_m(k) - A_a x_{am}(k)\right] \tag{2.28}$$

The estimator is given as

$$\hat{A}_a = \dot{x}_m (x_{am}^T)(x_{am} x_{am}^T)^{-1} \tag{2.29}$$

We illustrate the above formulation as follows:

$$\text{Let} \begin{bmatrix} \dot{x}_1 \\ \dot{x}_2 \end{bmatrix} = \begin{bmatrix} a_{11} & a_{12} \\ a_{21} & a_{22} \end{bmatrix}\begin{bmatrix} x_1 \\ x_2 \end{bmatrix} + \begin{bmatrix} b_1 \\ b_2 \end{bmatrix} u$$

Then, if there are say two measurements, we have

$$x_{am} = \begin{bmatrix} x_{11m} & x_{12m} \\ x_{21m} & x_{22m} \\ u_{1m} & u_{2m} \end{bmatrix}_{3x2} ; \quad u_m = \begin{bmatrix} u_{1m} & u_{2m} \end{bmatrix}; \quad \dot{x}_m = \begin{bmatrix} \dot{x}_{11m} & \dot{x}_{12m} \\ \dot{x}_{21m} & \dot{x}_{22m} \end{bmatrix}$$

Then $\left[\hat{A}_a\right]_{2x3} = \left[\left[\hat{A}\right]_{2x2} \vdots \left[\hat{B}\right]_{2x1}\right] = \left[\dot{x}_m\right]_{2x2} \left[x_{am}^T\right]_{2x3} \left\{\left[x_{am}\right]_{3x2}\left[x_{am}^T\right]_{2x3}\right\}^{-1}$.

Application of the EEM to parameter estimation requires accurate measurements of the states and their derivatives. In addition, it can be applied to unstable systems because it does not involve any numerical integration of the dynamic system that would otherwise cause divergence. Utilization of measured states and state derivatives for estimation in the algorithm enables estimation of the parameters of even an unstable system directly. However, if the measurements are noisy, the method will give biased estimates. We would like to mention here that EE is amenable to be programmed in the structure of a recurrent neural network [1].

Example 2.1: The EE formulation for parameter estimation of an aircraft is illustrated with one such state equation here. Let the *Z*-force equation be

$$\dot{\alpha} = Z_u u + Z_\alpha \alpha + q + Z_{\delta_e} \delta_e \tag{2.30}$$

Then the coefficients of the equation are determined from the system of linear equations given by (eqn. (2.30) is multiplied in turn by u, α and δ_e):

$$\begin{aligned}
\sum \dot{\alpha} u &= Z_u \sum u^2 + Z_\alpha \sum \alpha u + \sum q u + Z_{\delta_e} \sum \delta_e u \\
\sum \dot{\alpha} \alpha &= Z_u \sum u \alpha + Z_\alpha \sum \alpha^2 + \sum q \alpha + Z_{\delta_e} \sum \delta_e \alpha \\
\sum \dot{\alpha} \delta_e &= Z_u \sum u \delta_e + Z_\alpha \sum \alpha \delta_e + \sum q \delta_e + Z_{\delta_e} \sum \delta_e^2
\end{aligned} \tag{2.31}$$

where $\sum$ is the summation over the data points $(k = 1,\ldots,N)$ of u, α, q and δ_e signals. Combining the terms, we get

$$\begin{bmatrix} \sum \dot{\alpha} u \\ \sum \dot{\alpha} \alpha \\ \sum \dot{\alpha} \delta_e \end{bmatrix} = \begin{bmatrix} \sum u^2 & \sum \alpha u & \sum q u & \sum \delta_e u \\ \sum u \alpha & \sum \alpha^2 & \sum q \alpha & \sum \delta_e u \\ \sum u \delta_e & \sum \alpha \delta_e & \sum q \delta_e & \sum \delta_e^2 \end{bmatrix} \begin{bmatrix} Z_u \\ Z_\alpha \\ 1 \\ Z_{\delta_e} \end{bmatrix}$$

The above formulation can be expressed in a compact form as

$Y = X\beta$; then the EE is formulated as $e = Y - X\beta$; keeping in mind that there will be modelling and estimation errors combined in '*e*'. It is presumed that measurements of $\dot{\alpha}$, u, α *and* δ_e are available. If the numerical values of $\dot{\alpha}, \alpha, u, q$ and δ_e are available, then the EE estimates of the parameters can be obtained by using the procedure outlined in eqn. (2.2) to eqn. (2.4).

2.6 GAUSSIAN LEAST-SQUARES DIFFERENTIAL CORRECTION METHOD

The method is based on the differential correction technique [1]. This algorithm can be used to estimate the initial conditions of states as well as parameters of a nonlinear dynamical model. It is batch iterative procedure and can be regarded as complementary to other nonlinear parameter-estimation procedures like the output-error method. One can use this technique to obtain the start-up values of the aerodynamic parameters for other methods. To describe the method used to estimate the parameters of a given model, let us assume a nonlinear system as

$$\dot{x} = f(x,t,C) \tag{2.32}$$

$$y = h(x,C,K) + v \tag{2.33}$$

Here, x is the $n \times 1$ state vector, y is the $m \times 1$ measurement vector and v is a random white Gaussian noise (WGN) process with covariance matrix R. The functions f and h are vector-valued nonlinear functions, generally assumed to be known. The unknown parameters in the state and measurement equations are represented by vectors C and K. Let x_0 be a vector of initial conditions at t_0. Then problem is to estimate the parameter vector

$$\hat{\beta} = \left[x_0^T C^T K^T\right]^T \tag{2.34}$$

It must be noted that the vector C appears in both state and measurement equations. Such situations often arise for aircraft parameter estimation. The iterative differential correction algorithm is applied to obtain the estimates from the noisy measured signals as

$$\hat{\beta}^{(i+1)} = \hat{\beta}^{(i)} + \left[\left(F^T W F\right)^{-1} F^T W \Delta y\right]^{(i)} \tag{2.35}$$

where

$$F = \left[\partial y/\partial x_0 \mid \partial y/\partial C \mid \partial y/\partial K\right] \tag{2.36}$$

We use ' ∂ ' to denote partial differentiation here. It can be noted that the above equations are a generalized version of eqn. (2.22). W is a suitable weighting matrix and Δy is a matrix of residuals of observables: $\Delta y = z(t_k) - y(t_k)$ where $k = 1,2,...,N$. The first sub-matrix in F is given as

$$\partial y(t_k)/\partial x(t_0) = \left[\partial h(x(t_k))/\partial x(t_k)\right]\left[\partial x(t_k)/\partial x(t_0)\right] \tag{2.37}$$

With

$$d/dt\left[\partial x(t)/\partial x(t_0)\right] = \left[\partial f(t,x(t))/\partial x(t)\right]\left[\partial x(t)/x(t_0)\right] \tag{2.38}$$

The transition matrix differential eqn. (2.38) can be solved with identity matrix as the initial condition. The second sub-matrix in F is

$$\partial y/\partial C = (\partial h/\partial x)\quad(\partial x/\partial C) + \partial h/\partial C \tag{2.39}$$

where $(\partial x(t)/\partial C)$ is the solution of

$$d/dt[\partial x/\partial C] = \partial f/\partial C + (\partial f/\partial x)(\partial x/\partial C) \tag{2.40}$$

The last sub-matrix in F is obtained as

$$\partial y/\partial K = \partial h/\partial K \tag{2.41}$$

Eqn. (2.41) is simpler than eqn. (2.39) and eqn. (2.40), since K is not involved in eqn. (2.32). The state integration is performed by fourth-order Runge-Kutta method (RK4). It is an iterative process.

In this case, the LS estimates obtained from the EEM can be used as initial parameters for GLSDC algorithm. In eqn. (2.35), if the matrix ill conditioning occurs, some factorization method can be used. It is a well-known fact that the quality of the measurement data significantly influences the accuracy of the parameter estimates. The technique can be employed to assess quickly the quality of the measurements (aircraft manoeuvres), polarities of signals, and to estimate bias and scale factor errors in the measurements.

2.7 PRINCIPLE OF MAXIMUM LIKELIHOOD AND CRAMER–RAO'S LOWER BOUND

The fundamental idea is to define a function of the data and the unknown parameters [3]. This function is called the likelihood function. The parameter estimates are then obtained as those values, which maximize the function. In fact, the likelihood function is the probability density of the observations (given the parameters!). Let $\beta_1, \beta_2, \ldots, \beta_r$ be unknown physical parameters of some system and $z_1, z_2, \ldots, z_n$ the measurements of the true (data) values $y_1, y_2, \ldots, y_n$. It is assumed that the true values are function of the unknown parameters: $y_i = f_i(\beta_1, \beta_2, \ldots, \beta_r)$. Let z be a random variable whose probability density $p(z, \beta)$ depends on unknown parameter 'β'. To estimate 'β' from measurements z, choose the value of 'β', which maximizes the likelihood function $L(z,\beta) = p(z, \beta)$. The MLM thus reduces the problem of parameter estimation to the maximization of a real function called the likelihood function. It is a function of the parameter 'β' and the experimental data z. The value of the likelihood function at 'β' and z is the pdf of the measurement evaluated at the given observations z and the parameter 'β'. This is to say that 'p' becomes 'L' when the measurements have been actually obtained and used in 'p'. Hence, the parameter 'β', which makes this function most probable to have yielded these measurements, is called the MLE. Next, presume that the true value y_i lies within a very small interval around measurement z_i and evaluate the related probability, probability that $y_i \in \left[z_i - \frac{1}{2}\,\delta z_i,\; z_i + \frac{1}{2}\,\delta z_i \right]$ is given as

$$\delta P_i = \int_{z_i - \frac{1}{2}\delta z_i}^{z_i + \frac{1}{2}\delta z_i} p(t).dt \approx p(z_i)\delta z_i;\ \text{for small } \delta z_i \tag{2.42}$$

The measurement errors are normally distributed and the probability is given by

$$\delta P_i = \frac{1}{\sqrt{2\pi}\,\sigma_i} \mathbf{exp}\left[-\frac{1}{2}\frac{(z_i - y_i)^2}{\sigma_i^2} \right] \delta z_i \quad \textit{where } \sigma_i^2 \text{ is the variance.} \tag{2.43}$$

The likelihood function is calculated for the statistically independent measurements, and this allows the joint probability density to be simply the product of the probabilities of the individual measurements, and is given by

$$\delta P = \prod_{i=1}^{n} p(z_i)\delta z_i = \frac{1}{(2\pi)^{n/2}\sigma_1 \ldots \sigma_n} \mathbf{exp}\left[\sum_{i=1}^{n} -\frac{1}{2}\frac{(z_i - y_i)^2}{\sigma_i^2} \delta z_1 \ldots \delta z_n \right] \tag{2.44}$$

The likelihood function is then given as

$$p(z \mid \beta) = p(z_1, \ldots, z_n \mid \beta_1, \ldots, \beta_r)$$

$$= \frac{1}{(2\pi)^{n/2}\sigma_1 \ldots \sigma_n} \mathbf{exp}\left[\sum_{i=1}^{n} -1/2 \frac{(z_i - y_i(\beta))^2}{\sigma_i^2}\right] \tag{2.45}$$

The parameter $\hat{\beta}$ that maximizes this likelihood function is called the maximum-likelihood parameter estimate of 'β'.

2.7.1 Cramer–Rao's Lower Bound

In this section, we derive certain theoretical properties of the maximum-likelihood estimator. The main point in any estimator is the error made in the estimates relative to the true parameters. However, these true parameters are unknown in the real case. Therefore, we only get some statistical indicators for the errors made. The Cramer–Rao lower bound (CRLB) is one such useful and, perhaps [3], the best measure for such errors. The likelihood function can also be defined as

$$L(z|\beta) = \log p(z|\beta) \tag{2.46}$$

since the function and its logarithm will have maximum at the same argument. The maximization yields the likelihood differential equation:

$$\frac{\partial}{\partial \beta} L(z \mid \beta) = L'(z \mid \hat{\beta}) = \frac{p'}{p}(z \mid \hat{\beta}) = 0 \tag{2.47}$$

This equation is nonlinear in $\hat{\beta}$ and a first-order approximation by Taylor's series expansion, can be used to obtain the estimate $\hat{\beta}$:

$$L'(z \mid \beta_0 + \Delta\beta) = L'(z \mid \beta_0) + L''(z \mid \beta_0)\Delta\beta = 0 \tag{2.48}$$

which gives increment in 'β' as

$$\Delta\beta = \frac{L'(z \mid \beta_0)}{-L''(z \mid \beta_0)} = -\left(L''(z/\beta_0)\right)^{-1} L'(z/\beta_0) \tag{2.49}$$

Eqn. (2.49) tells us that if we get the right-hand-side term computed, then we already have obtained $\Delta\beta$, the increment/change in parameter vector. This expression is based on computation of likelihood related partials, which can be evaluated when the details of the dynamical systems are known, as will be seen later in the chapter. The expected value of the denominator in eqn. (2.49) is defined as the information matrix (in general sense):

$$I_m(\beta) = E\{-L''(z \mid \beta)\} \tag{2.50}$$

The other form of $I_m(\beta)$ is derived next. Since, by the definition of the probability of a random variable: $\int_{-\infty}^{\infty} p(z \mid \beta)dz = 1$. We take first differentiation on both sides to obtain

$$\int_{-\infty}^{\infty} p'(z \mid \beta)dz = \int_{-\infty}^{\infty} [L'(z \mid \beta)\, p(z \mid \beta)dz = 0 \tag{2.51}$$

The second differentiation yields

$$\int_{-\infty}^{\infty} p''(z \mid \beta)dz = \int_{-\infty}^{\infty} [L''(z \mid \beta)\, p(z \mid \beta) + L'(z \mid \beta)^2\, p(z \mid \beta)]dz = 0 \tag{2.52}$$

From eqn. (2.52) we get

$$I_m(\beta) = E\{-L''(z \mid \beta)\} = E\{L'(z \mid \beta)^2\} \tag{2.53}$$

From the definition of information matrix, we can say that if there is large information content in the data, then $|L''|$ tends to be large, and the uncertainty in estimate $\hat{\beta}$ is small. The so-called Cramer–Rao inequality (information inequality) provides a lower bound to the variance of an unbiased estimator, as will be seen in the following. Let $\beta_e(z)$ be any estimator of 'β' based on the measurement z, and then $\bar{\beta}_e(z) = E\{\beta_e\ (z)\}$ is the expectation of the estimate (since it depends on the random signal z). Its variance is given as

$$\sigma^2_{\beta e} = E\{(\beta_e\ (z) - \bar{\beta}_e)^2\} \tag{2.54}$$

The bias in the estimator is defined as

$$E\{\beta_e - \beta\} = \int_{-\infty}^{\infty} \beta_e(z) p(z \mid \beta)dz - \beta = b(\beta) \tag{2.55}$$

If $b(\beta) = 0$, then it is called an unbiased estimator. We have thus

$$\int_{-\infty}^{\infty} \beta_e(z) p(z \mid \beta)dz = \beta + b(\beta) \tag{2.56}$$

Differentiating both the sides w.r.t. 'β' we get

$$\int_{-\infty}^{\infty} \beta_e(z) p'(z \mid \beta)dz = \int_{-\infty}^{\infty} \beta_e(z) L'(z \mid \beta) p(z \mid \beta)dz = 1 + b'(\beta) \tag{2.57}$$

since β_e is function of only z. In addition, we have $\int_{-\infty}^{\infty} p(z \mid \beta)dz = 1$ and differentiating both sides we get [3]

$$\int_{-\infty}^{\infty} p'(z \mid \beta)dz = \int_{-\infty}^{\infty} L'(z \mid \beta) p(z \mid \beta)dz = 0 \tag{2.58}$$

Multiplying eqn. (2.58) by $(-\bar{\beta}_e)$ and adding to the previous we get

$$\int_{-\infty}^{\infty} [\beta_e\ (z) - \bar{\beta}_e]\, L'(z \mid \beta) p(z \mid \beta)dz = 1 + b'(\beta)$$

$$\int_{-\infty}^{\infty} [\beta_e\ (z) - \bar{\beta}_e]\, \sqrt{p(z \mid \beta)}.L'(z/\beta)\, \sqrt{p(z \mid \beta)}dz = 1 + b'(\beta) \tag{2.59}$$

Now, we apply the following well-known Schwarz inequality to eqn. (2.64):

$$\left[\int f(z).g(z)dz\right]^2 \leq \int f^2(z)dz.\int g^2(z)dz \tag{2.60}$$

to get (the equality applies if $f(z) = kg(z)$):

$$[1+b'(\beta)]^2 \leq \int_{-\infty}^{\infty} [\beta_e(z)-\bar{\beta}_e]^2\, p(z\,|\,\beta)dz.\int_{-\infty}^{\infty} L'(z\,|\,\beta)^2\, p(z\,|\,\beta)dz \tag{2.61}$$

Using the definition of $I_m(\beta)$ and $\sigma^2_{\beta e}$, we get

$$[1+b'(\beta)]^2 \leq \sigma^2_{\beta e}\, I_m(\beta) \quad \text{or} \quad \sigma^2_{\beta e} \geq [1+b'(\beta)]^2\left(I_m(\beta)\right)^{-1} \tag{2.62}$$

This is called Cramer–Rao's inequality. For unbiased estimator, $b'(\beta) = 0$, and hence $\sigma^2_{\beta e} \geq I_m^{-1}(\beta)$. The equality sign holds if $\beta_e(z)-\bar{\beta}_e = kL'(z\,|\,\beta)$. For unbiased, efficient estimator we thus have

$$\sigma^2_{\beta e} = I_m^{-1}(\beta) \tag{2.63}$$

We emphasize here that the inverse of the information matrix is the covariance matrix and hence in eqn. (2.63), we have the theoretical expression for the variance of the estimator. The information matrix can be computed from the likelihood function or related data.

The foregoing development signifies that the variance in the estimator, for an efficient estimator, would be at least equal to the predicted variance, whereas for other cases, it could be greater but not lesser than the predicted value. Hence, the predicted value provides the lower bound. Thus, the ML estimate is also the minimum variance estimator.

2.7.2 The Maximum-Likelihood Estimate is Efficient

We assume that it is unbiased, then for efficiency we have to show that

$$\beta_e(z)-\bar{\beta}_e \overset{?}{=} kL'(z\,|\,\beta) \tag{2.64}$$

The likelihood equation is

$$L'(z\,|\,\beta)\,|_{\beta=\hat{\beta}(z)} = 0 \tag{2.65}$$

Substituting for the ML estimate: $\beta_e(z) = \hat{\beta}(z)$ and since it is unbiased ($\bar{\beta}_e = 0$), we get

$$\beta_e(z)-\bar{\beta}_e\,|_{\beta=\hat{\beta}(z)} = \hat{\beta}(z)-\beta\,|_{\beta=\hat{\beta}(z)} = 0 \tag{2.66}$$

Thus $0 = kL'(z|\beta)|_{\beta=\hat{\beta}(z)} = k \times 0$.

Hence, the equality is established and the ML estimator is proved efficient. This is a very important property of the MLE. As such, these results are quite general since we have not yet utilized the details of any dynamic system.

2.8 MAXIMUM-LIKELIHOOD ESTIMATION FOR A DYNAMIC SYSTEM

A linear dynamical system can be described as

$$\dot{x}(t) = Ax(t) + Bu(t) \tag{2.67}$$

$$y(t) = Hx(t) \tag{2.68}$$

$$z(k) = y(k) + v(k) \tag{2.69}$$

We emphasize here that in many applications, the actual systems are of continuous time. However, the measurements obtained are discrete time. The following assumptions are made on the measurement noise *v(k)*:

$$E\{v(k)\} = 0; \qquad E\{v(k)v^T(l)\} = R\delta_{kl} \tag{2.70}$$

It is assumed that the measurement noise is zero mean and white Gaussian with *R* as the covariance matrix of this noise, this allows us to use the Gaussian probability concept for deriving the MLE [4–6]. The assumption of whiteness of the measurement noise is quite standard and very useful in engineering practice. Strictly speaking, the assumption would not hold good. However, so long as the bandwidth of the noise spectrum is much larger than the system's bandwidth, the noise can be seen as practically 'white'.

2.8.1 Derivation of the Likelihood Function

If *z* is some real-valued Gaussian random variable, then its probability density is given by

$$p(z) = \frac{1}{\sqrt{2\pi}\,\sigma}\,\mathbf{exp}\left[-\frac{1}{2}\frac{(z-m)^2}{\sigma^2}\right] \tag{2.71}$$

where $m = E(z)$ and $\sigma^2 = E\{(z-m)^2\}$. For n – random variables $z_1, z_2, \ldots, z_n$ we have

$$p(z_1, z_2, \ldots, z_n) = \frac{1}{(2\pi)^{n/2}\sqrt{|R|}}\,\mathbf{exp}\left[-\frac{1}{2}(z-m)^T R^{-1}(z-m)\right] \tag{2.72}$$

Here $z^T = (z_1, z_2, \ldots, z_n)$; $\quad m^T = (m_1, m_2, \ldots, m_n)$, this is a vector of mean values.

and

$$R = \begin{bmatrix} r_{11} & \cdot & \cdot & r_{1n} \\ \cdot & \cdot & \cdot & \cdot \\ \cdot & \cdot & \cdot & \cdot \\ r_{1n} & \cdot & \cdot & r_{nn} \end{bmatrix} \text{ is the covariance matrix} \tag{2.73}$$

with $r_{ij} = E\{(z_i - m_i)(z_j - m_j)\} = \sigma_i \sigma_j \rho_{ij}$; ρ_{ij} = correlation coefficients, (ρ_{ii} = 1). Applying the foregoing development to measurements *z(k)*, and assuming that the measurement errors are Gaussian, we obtain

$$p(z(k) \mid \beta, r) = \frac{1}{(2\pi)^{m/2}\sqrt{|R|}}\,\mathbf{exp}\left[-\frac{1}{2}[z(k) - y(k)]^T R^{-1}[z(k) - y(k)]\right] \tag{2.74}$$

since in this case $m = E\{z\} = E\{v + y\} = E\{v\} + E\{y\}$ and $E\{v\} = 0$; we have the likelihood function as

$$p(z(1),...,z(N) \mid \beta, R) = \prod_{k=1}^{N} p(z(k) \mid \beta, R)$$
$$= \left((2\pi)^m |R|\right)^{-N/2} \mathbf{exp}\left[-\frac{1}{2}\sum_{k=1}^{N}[z(k)-y(k)]^T R^{-1}[z(k)-y(k)]\right] \quad (2.75)$$

The parameter vector 'β' is obtained by maximizing the likelihood function, eqn. (2.79) with respect to 'β' by minimizing the negative (log) likelihood function:

$$L = -\mathbf{log}\, p(z \mid \beta, R) = \frac{1}{2}\sum_{k=1}^{N}[z(k)-y(k)]^T R^{-1}[z(k)-y(k)]] + N/2 \log |R| + const \quad (2.76)$$

Based on this, the following two cases of minimization arise:

(i) if R is known then the cost function

$$CF = \sum_{k=1}^{N}[z(k)-y(k)]^T R^{-1}[z(k)-y(k)] \Rightarrow \text{minimum} \quad (2.77)$$

since the second term in eqn. (2.80) is constant

(ii) if R is unknown then we can minimize the function with respect to R and obtain

$$\partial L/\partial(R^{-1}) = 0$$

$$\text{to get} \quad \hat{R} = 1/N\sum_{k=1}^{N}[z(k)-\hat{y}(k)]\,[z(k)-\hat{y}(k)]^T \quad (2.78)$$

When R is substituted in the likelihood function the first term becomes $\frac{mN}{2}$ = constant, and we get $CF = |R| \rightarrow$ minimum. Minimization of CF in (i) w.r.t. 'β' results in

$$\partial L/\partial\beta = -\sum_{k}(\partial y(\beta)/\partial\beta)^T\ R^{-1}(z - y(\beta)) = 0 \quad (2.79)$$

This set is again a system of nonlinear equations and calls for an iterative solution. In the present case, we obtain an iterative solution by the so-called quasi-linearization method (also known as modified Newton–Raphson or Gauss–Newton method), i.e. we expand

$$y(\beta) = y(\beta_0 + \Delta\beta) \quad (2.80)$$

as

$$y(\beta) = y(\beta_0) + \frac{\partial y(\beta)}{\partial\beta}\Delta\beta \quad (2.81)$$

The quasi-linearization is an approximation method for obtaining solutions to nonlinear differential or difference equations with multi-point boundary conditions. A version of the quasi-linearization is used in obtaining practical workable solution in output error method; we have

$$-\sum_k \left[\frac{\partial y(\beta)}{\partial \beta}\right]^T R^{-1}\left[(z - y(\beta_0)) - \frac{\partial y(\beta)}{\partial \beta_0}\Delta\beta\right] = 0 \tag{2.82}$$

$$\left[\sum_k \left[\frac{\partial y(\beta)}{\partial \beta}\right]^T R^{-1}\frac{\partial y(\beta)}{\partial \beta}\right]\Delta\beta = \sum_k \left[\frac{\partial y(\beta)}{\partial \beta}\right]^T R^{-1}(z - y) \tag{2.83}$$

Next, we have

$$\Delta\beta = \left\{\sum_k \left[\frac{\partial y(\beta)}{\partial \beta}\right]^T R^{-1}\frac{\partial y(\beta)}{\partial \beta}\right\}^{-1}\left\{\sum_k \left[\frac{\partial y(\beta)}{\partial \beta}\right]^T R^{-1}(z - y)\right\} \tag{2.84}$$

The ML estimate is obtained as

$$\hat{\beta}_{new} = \hat{\beta}_{old} + \Delta\beta \tag{2.85}$$

2.8.2 Accuracy Aspects

Determining accuracy of the estimated parameters is an essential part of the parameter-estimation process. The absence of true parameter values for comparison makes the task of determining this very difficult. Cramer–Rao's bound is one of the primary criteria for evaluating the accuracy of the estimated parameters. MLE gives the measure of parameter accuracy without any extra computation, as can be seen from the following development. For a single parameter case, for unbiased estimate $\hat{\beta}(z)$ of β: $\sigma^2_{\hat{\beta}} \geq I_m^{-1}(\beta)$, where the information matrix is

$$I_m(\beta) = E\left\{-\frac{\partial^2 \log p(z \mid \beta)}{\partial \beta^2}\right\} = E\left\{\left(\frac{\partial \log p(z \mid \beta)}{\partial \beta}\right)^2\right\} \tag{2.86}$$

For several parameters, the Cramer–Rao inequality is given as $\sigma^2_{\hat{\beta}_i} \geq (I_m^{-1})_{ii}$; where the information matrix is

$$(I_m)_{ij} = E\left\{-\frac{\partial^2 \log p(z \mid \beta)}{\partial \beta_i \partial \beta_j}\right\} = \left\{\left(\frac{\partial \log p(z \mid \beta)}{\partial \beta_i}\right) \cdot \left(\frac{\partial \log p(z \mid \beta)}{\partial \beta_j}\right)\right\} \tag{2.87}$$

For efficient estimation, the equality holds, and the covariance matrix of the estimation errors is $P = I_m^{-1}$; the standard deviation of the individual parameters is given by $\sigma_{\hat{\beta}_i} = \sqrt{P_{ii}} = \sqrt{P(i,i)}$; and correlation coefficients are

$$\rho_{\hat{\beta}_i,\hat{\beta}_j} = P_{ij} / \sqrt{P_{ii} P_{jj}} \tag{2.88}$$

For MLM, we have

$$\log p(z|\beta) = -\frac{1}{2}\sum_{k=1}^{N}[z(k)-y(k)]^T R^{-1}[z(k)-y(k)] + \text{const} \tag{2.89}$$

The information matrix can be obtained as follows. Differentiate both sides w.r.t. β_i to get

$$\left(\frac{\partial \log p(z/\beta)}{\partial \beta_i}\right) = \sum_k \left(\frac{\partial y}{\partial \beta_i}\right)^T R^{-1}(z-y) \tag{2.90}$$

Again, differentiate both sides w.r.t. β_j to get

$$\left(\frac{\partial^2 \log p(z\,|\,\beta)}{\partial \beta_i\, \partial \beta_j}\right) = \sum_k \left(\frac{\partial^2 y}{\partial \beta_i\, \partial \beta_j}\right)^T R^{-1}(z-y) - \sum_k \left(\frac{\partial y}{\partial \beta_i}\right)^T R^{-1}\frac{\partial y}{\partial \beta_j} \tag{2.91}$$

Taking the expectation of the above equation, we get

$$(I_m)_{ij} = E\left\{-\frac{\partial^2 \log p(z\,|\,\beta)}{\partial \beta_i \partial \beta_j}\right\} = \sum_{k=1}^{N}\left(\frac{\partial y(k)}{\partial \beta_i}\right)^T R^{-1}\frac{\partial y(k)}{\partial \beta_j} \tag{2.92}$$

Since $E\{z-y\} = 0$; the measurement error has zero mean. We recall here from the previous section that the increment in parameter estimate $\Delta\beta$ is given by

$$\Delta\beta = \left[\sum_k \left(\frac{\partial y}{\partial \beta}\right)^T R^{-1}\frac{\partial y}{\partial \beta}\right]^{-1} \sum_k \left(\frac{\partial y}{\partial \beta}\right)^T R^{-1}(z-y) \tag{2.93}$$

Comparing with the expression for the information matrix in eqn. (2.96), we conclude that MLE gives a measure of accuracy without any extra computation. Several criteria are used to judge the 'goodness' of the estimator/estimates: CRLB of the estimates, correlation coefficients among the estimates, determinant of the covariance matrix of the residuals, plausibility of the estimates based on physical understanding of the dynamic system, comparison of the estimates with those of nearly similar systems or estimates independently obtained by other methods (analytical or other parameter-estimation methods), and model predictive capability.

2.9 OUTPUT-ERROR METHOD: FEATURES AND NUMERICAL ASPECTS

The OEM is based on the assumption that only the observations contain measurement noise and there is no noise in the state equations. The mathematical model of a linear system consists of the vector x representing the system states, vector y representing the computed system response (model output), vector z representing the measured variables and u representing the control input vector. The matrices A, B and H contain the parameters to be estimated. The OEM assumes that the measurement vector z is corrupted with noise, which is zero mean and has a Gaussian distribution with covariance R, i.e., $v \sim N(0,R)$. The aim is to minimize the error between the measured and model outputs by adjusting the unknown parameters contained in matrices A, B and H. Let the parameter vector to be estimated be represented by Θ, where Θ = [elements of A, B, H, initial condition of x]. Then, estimate of Θ is obtained by minimizing the cost function

$$J = \frac{1}{2}\sum_{k=1}^{N}[z(k)-y(k)]^T R^{-1}[z(k)-y(k)] + \frac{N}{2}\ln|R| \tag{2.94}$$

This cost function is like the WLS criterion with weighting matrix as W and with one extra term. The estimate of R can be obtained from

$$\hat{R} = \frac{1}{N}\sum_{k=1}^{N}[z(k) - \hat{y}(k)][z(k) - \hat{y}(k)]^T \tag{2.95}$$

once the predicted measurements are computed. Following the development of the previous Section 2.4, the estimate of Θ at the $(i+1)th$ iteration is obtained as

$$\Theta(i+1) = \Theta(i) + \left[\nabla_{\Theta}^2 J(\Theta)\right]^{-1}\left[\nabla_{\Theta} J(\Theta)\right] \tag{2.96}$$

where the first and the second gradients are defined as

$$\nabla_{\Theta} J(\Theta) = \sum_{k=1}^{N}\left[\frac{\partial y}{\partial \Theta}(k)\right]^T R^{-1}[z(k) - y(k)] \tag{2.97}$$

$$\nabla_{\Theta}^2 J(\Theta) = \sum_{k=1}^{N}\left[\frac{\partial y}{\partial \Theta}(k)\right]^T R^{-1}\left[\frac{\partial y}{\partial \Theta}(k)\right] \tag{2.98}$$

Eqn. (2.98) is a Gauss–Newton approximation of the second gradient. This approximation helps to speed up the convergence without causing significant error in the estimate of Θ. Starting with a set of suitable initial parameter values, the model response is computed with the input used for obtaining measurement data. The estimated response and the measured response are compared and the response error is used to compute the cost function. Eqns. (2.97) and (2.98) are used to obtain the first and second gradients of the cost function and then eqn. (2.96) is used to update the model parameter values. The updated parameter values are once again used in the mathematical model to compute the new estimated response and the new response error. This updating procedure continues until convergence is achieved.

The Gauss–Newton approximation for the second gradient is also called the Fisher information matrix provides a measure of relative accuracy of the estimated parameters. The diagonal elements of the inverse of the information matrix give the individual covariances, and the square root of these elements is a measure of the standard deviations called the Cramer–Rao bounds (CRB):

$$\text{Fisher information matrix} = \nabla_{\Theta}^2 J(\Theta) \tag{2.99}$$

$$\text{Standard deviation of estimated parameters} = \text{CRB}(\Theta) = diag\left[\sqrt{[\nabla_{\Theta}^2 \text{J}(\Theta)]^{-1}}\right] \tag{2.100}$$

The OEM also can be applied with equal ease to any nonlinear system, in principle,

$$\dot{x}(t) = f[x(t), u(t), \Theta] \quad \text{with initial } x(0) = x_0 \tag{2.101}$$

$$y(t) = h[x(t), u(t), \Theta] \tag{2.102}$$

$$z(k) = y(k) + v(k) \tag{2.103}$$

Wherein, f and h are general nonlinear functions, and the initial values x_0 of the state variables need to be estimated along with the parameter vector Θ. It is evident that the estimation of parameters with OEM would require computation of the state vector x, model output vector y and sensitivity coefficients $\partial y/\partial\Theta$. The sensitivity coefficients for a linear system can be obtained analytically by partial differentiation of the system equations. However, for a nonlinear system, each time the model structure changes, partial differentiation of the system equations needs to be carried out to obtain $\partial y/\partial\Theta$. A better approach is to approximate the sensitivity coefficients by finite differences. In this procedure, the parameters in Θ are perturbed one at a time and the corresponding perturbed model response y_p is computed. The sensitivity coefficient is then given by

$$\partial y/\partial\Theta_j = (y_p - y)/\Delta\Theta_j \tag{2.104}$$

The use of finite differencing in calculating $\partial y/\partial\Theta$ results in a program code that is more flexible and user friendly.

2.9.1 Features and Numerical Aspects

The MLM is very popular because of its several interesting features: (i) maximum-likelihood estimates are consistent, asymptotically unbiased and efficient; (ii) it is more general and can handle both measurement and process noise (of course it then incorporates Kalman filter into it, leading to filter error method); (iii) if process noise is absent and measurement noise covariance is known, it reduces to the output-error method; (iv) in case measurement noise is absent, it reduces to the equation-error method, if all the states are measured; (v) it is found to yield realistic values of the variances of the parameter estimates; and (vi) it can be used to estimate the covariance of the measurement noise, in fact, it gives the covariance of residuals.

The computation of the coefficients of parameter vector Θ requires the following: (i) initial values of the coefficients in Θ; (ii) current values of variables y at each discrete time point k; (iii) sensitivity matrix $(\partial y/\partial\Theta)_{ij} = \partial y_i/\partial\Theta_j$; and (iv) current state values are computed by numerical integration of the system state equations, which can be done by, say, the fourth-order Runge–Kutta method.

The sensitivity coefficients are approximately computed by using numerical differences. Assuming a small perturbation $\delta\Theta$ in the parameter Θ, the perturbed states x_p are computed which in turn are used to obtain the perturbed output variable y_p. The choice of the step size for evaluating the numerical difference is typically given as $\partial\Theta_j \rightarrow 10^{-7} * \Theta_j$. The gradient $\partial y/\partial\Theta_j$ may be computed using forward differencing. On comparing the optimization methods for ML estimation, it is found that the quasi-linearization method, which is equivalent to the modified Newton–Raphson method that neglects the computation of the second gradient of the error, is found to be 300–400 times faster than Powell's or Rosenbrock's method. It is also found to be about 150 times faster than the quasi-Newton method. The method also provides direct information on the accuracy of parameter estimates. However, it could have convergence problems with systems, which have discontinuous nonlinearities. The time-history match is a necessary but not the sufficient condition. It is quite possible that the response match would be good but some parameters could be unrealistic, e.g. unexpected sign behaviour. There could be one or more reasons for this kind of behaviour: deficient model used for the estimation or not all the modes of the system might have been sufficiently excited. One way to circumvent this problem is to add *a priori* information about the parameter in question. One more approach is to fix such parameters at some *a priori* value, which could have been determined by some other means or available independently from other sources from the system. The OEM/MLE method is so general that it can also be used for estimation of zero shifts in measured input/output data.

In this chapter, we discussed various LS methods and illustrated their performance using simple and realistically simulated data examples. A more involved example of data compatibility for aircraft was also illustrated. The GLS method does not consider the statistics of measurement errors. If there is a good knowledge of these statistics, then they can be used and it leads to minimum variance estimates. As we will see in Chapter 3, the KF is a method to obtain minimum variance estimates of states of a dynamic system described in state-space form. It can handle noisy measurements as well as partially account for discrepancies in a state model by using the so-called process noise. Thus, there is a direct relationship between sequential unbiased minimum variance algorithm and discrete KF. ML/OEM estimation of aircraft has been treated in [4–9]; the main reason for the success of this technique is that it has many nice theoretical properties and, it being an iterative process, generally gives reasonably accurate results for practical real data. The iterations refine the estimates. Another reason for its success is that it gives theoretical lower bounds on the variance of the estimates based on the Fisher information matrix. Thus, one can judge the accuracy of the estimates and obtain uncertainty bounds on the parameters. It can also be applied to nonlinear problems with equal ease.

EXERCISES

EXERCISE 2.1

One way of obtaining the least-squares estimate of (β) is shown in eqns. (2.2)–(2.4). Use the algebraic approach of eqn. (2.1) to derive a similar form. One extra term will appear. Compare this term with that of eqn. (2.5).

EXERCISE 2.2

Explain the significance of the property of the covariance of the parameter-estimation error (see eqns. (2.6), (2.7)). In order to keep estimation errors low, what should be done in the first place?

EXERCISE 2.3

Reconsider Example 2.1 and check the response of the motor speed, S beyond 1 sec. Are the responses for $\alpha \geq 0.1$ linear or nonlinear for this apparently linear system? What is the fallacy?

EXERCISE 2.4

Consider $z = mx + v$, where v is measurement noise with covariance matrix R. Derive the formula for covariance of ($z - \hat{y}$). Here, $y = mx$.

EXERCISE 2.5

Consider the generalized least-squares problem. Derive the expression for $P = \text{Cov}(\beta - \hat{\beta})$.

EXERCISE 2.6

Reconsider the probabilistic version of the least-squares method. Can we not directly obtain K from $KH = I$? If done so, what is the difference between this expression and the one in eqn. (2.15)? What assumptions you will have to make on H to obtain K from $KH = I$? What assumption will you have to make on R for both expressions to be the same?

EXERCISE 2.7

What are the three numerical methods to obtain partials of nonlinear function $h(\beta)$ w.r.t.β?

Exercise 2.8

Consider $z = H\beta + v$ and $v = X_v\beta_v + e$, where v is correlated noise, e is assumed to be white noise, and the second equation is the model of the correlated noise v. Combine these two equations and obtain expressions for the least-squares estimates of β and β_v.

Exercise 2.9

Based on Exercise 2.8, can you tell how one can generate a correlated process using white noise as the input process? (Hint: the second equation in Exercise 2.8 can be regarded as a low-pass filter).

Exercise 2.10

Derive the expression for time to double the amplitude, if σ is the positive real root of a first-order system. If σ is positive, then the system output will tend to increase when time elapses.

Exercise 2.11

Let the spring mass system be described by $m\ddot{y} + d\dot{y} + Ky = w(t)$. Obtain the state-space model in the form $\dot{x} = Ax + Bu$ and obtain $\frac{\partial \dot{x}}{\partial K}, \frac{\partial \dot{x}}{\partial d}$

Exercise 2.12

Comment on the differences and similarities between the GLSDCM and the output-error method, since both the methods use output-error criterion and are applicable to dynamic systems.

Exercise 2.13

Consider the equations $\dot{x}(t) = Ax(t) + Bu(t)$ and $y(t) = Cx(t) + Du(t)$. Assume that β_1 = unknown initial values of the state variables and β_2 = unknown parameters in matrices A, B, C and D. Postulate y as a function of β_1, β_2 and u. Let $\beta = [\beta_1^T, \beta_2^T]^T$. Obtain expressions for $\frac{\partial y}{\partial \beta}, \frac{\partial \dot{x}}{\partial \beta_1}$ and $\frac{\partial \dot{x}}{\partial \beta_2}$.

(Hint: Study GLSDCM)

Exercise 2.14

Consider Cramer–Rao's inequality and comment on this if there is a bias in the estimate.

Exercise 2.15

Comment on the relationship between ML and the LS methods,

Exercise 2.16

Compare and contrast the second-order gradient for ML estimation with the covariance matrix of estimation error.

REFERENCES

1. Raol, J.R., Girija, G., and Singh, J., *Modelling and Parameter Estimation of Dynamic Systems. IEE/IET Control Series*, Vol. 65, London, UK, Aug. 2004.
2. Astrom, K.J., Maximum likelihood and prediction error methods, *Automatica*, Vol. 16, pp. 551–574, 1980.
3. Plaetschke, E., Maximum likelihood estimation, Lectures presented at FMCD, NAL as a part of the IFM, DLR-FMCD, NAL collaborative programme, FMCD, NAL, Bangalore, Nov. 1987.
4. Mehra, R.K., Stepner, D.E., and Tyler, J.S., Maximum likelihood identification of aircraft stability and control derivatives, *Journal of Aircraft, AIAA, USA*, Vol. 11, No. 2, pp. 81–89, Feb. 1974.
5. Iliff, K.W., Parameter estimation for flight vehicles, *Journal of Guidance, Control and Dynamics, AIAA, USA*, Vol. 12, No. 5, pp. 609–622, Sept-Oct. 1989.
6. Maine, R.E., and Iliff, K.W., *Application of parameter estimation to aircraft stability and control-the output error approach, NASA RP-1168*, 1986.
7. Spall, J.C., and Garner, J.P., Parameter identification for state-space models with nuisance parameters, *IEEE Transactions on Aerospace and Electronic Systems*, Vol. 26, No. 6, pp. 992–998, Nov. 1990.
8. Chu, Q.P., Mulder, J.A., and Van Woerkom, P.T.L.M., Modified recursive maximum likelihood adaptive filter for nonlinear aircraft flight path reconstruction, *Journal of Guidance, Control and Dynamics*, Vol. 19, No. 6, pp. 1285–1295, 1996.
9. Girija, G., and Jategaonkar, R.V., Some results of ATTAS flight data analysis using maximum likelihood parameter estimation method, DLR-FB 91-04, IFM/Germany, 1991.

3 Kalman Filtering Methods

3.1 INTRODUCTION

In the area of signal processing, we discover analog and digital filtering concepts and methods. The real-life systems give rise to signals, which are invariably contaminated with so-called random noise. This noise could arise due to measurement errors from the sensors, instruments, data-transmission channels or human error. In most cases, the errors are random in nature and can be described best by a probabilistic model. A usual characteristic of such a random noise that affects the signal is Gaussian (normally distributed) noise with zero mean and a finite variance. This variance measures the power strength of the noise and it is often compared to the power of the signal that is influenced by the random noise. This leads to a measure called signal-to-noise ratio (SNR). Often the noise is assumed a white process. The aim is then to maximize SNR by filtering out the noise from the signal/data of the dynamic system. In the present chapter, we use the model-based approach and especially the approach based on the state-space model of a dynamical system (see Chapter 2). Therefore, our major goal is to get the best estimate or prediction of the signal, which is buried, in the random noise. This noise could be white or time-correlated (nonwhite). It could be coloured noise, i.e., output of a linear lumped parameter system excited by white noise.

In this chapter, we mainly study the problem of filtering and prediction using Kalman filtering (KF) methods [1–3]. KF has evolved to a very high state-of-the-art method for the state estimation of dynamic systems, which could be described by difference, or differential equations, especially in state-space form. The main reason for the success of KF is that it has an appealing state-space formulation and it gives algorithms that can be easily implemented on digital computers. In fact, KF is a numerical algorithm, which also has a tremendous real-time/online application because of its recursive formulation as against one-shot/batch-processing methods. For linear systems, it is an optimal state observer. The extended KF, EKF, can be used for nonlinear systems and the innovations computed from this approach are likely to be white Gaussian if we assure that the measurements are frequent. The most attractive feature of EKF is that it is one pass and therefore computationally less demanding. However, experience with the use of EKF for parameter estimation reveals that the estimated parameter values are very sensitive to the initial values of the measurement noise and the state-error covariance matrices. If the filter is not properly tuned, i.e. if the *a priori* values of the noise covariance matrices are not chosen appropriately, EKF can produce unsatisfactory results. Most of the applications with EKF reported in the literature relate to state estimation rather than parameter estimation.

There are three different ways to account for process noise in a linear system; all these formulations use the modified Gauss–Newton optimization to estimate the system parameters and the noise statistics. The major difference is the way by which the covariance matrices are estimated.

DOI: 10.1201/9781003620211-3

3.2 KALMAN FILTERING

A dynamic system is assumed as

$$x(k+1) = \varphi x(k) + Bu(k) + Gw(k) \tag{3.1}$$

$$z(k) = Hx(k) + Du(k) + v(k) \tag{3.2}$$

Here, x is the $n \times 1$ state vector; u is $p \times 1$ deterministic control input to the system; z is $m \times 1$ measurement vector; w is a white Gaussian noise (WGN) sequence with zero mean and covariance matrix Q. It is also called process noise with associated matrix G; v is a WGN sequence with zero mean and covariance matrix R. It is also called measurement noise; φ is the $n \times n$ transition matrix that takes states from k to $k + 1$; B is the input gain/magnitude vector/matrix; H is the $m \times n$ measurement model/sensor dynamics matrix; D is the $m \times p$ feedforward/direct control input matrix. Often D is dropped from the Kalman filter development.

We emphasize here that, although most dynamic systems are continuous time, the KF is an extremely popular filtering method and is best discussed using the discrete-time model. In addition, in the following, it will be seen that the solution of KF requires the Riccati equation to be handled, which is easier to manage in discrete form rather than in continuous-time form. One can convert the continuous-time system to the discrete-time model and use the discrete-time Kalman filtering algorithm, which can be easily implemented in a digital computer. Also, the fact that even the continuous-time filtering algorithm would be required to be implemented on a digital computer, it seems that both the approaches will lead to some approximations. We feel that understanding and implementing a discrete-time Kalman filter is easier.

The problem of state estimation using KF can be formulated as follows: given the model of the dynamic system, statistics of the noise processes and the noisy measurement data, and the input, determine the best estimate of the state, x, of the system. Since it is assumed that the dynamic system is known, it means that the (form and) numerical values of the elements of φ, B and H are accurately known. If some of these elements are not known, then these can be considered as additional unknown states and appended to the otherwise state vector x yielding the extended state vector. In most circumstances this will lead to a nonlinear dynamic system for which EKF can be used. Life would have been much easier or even trivial if the noise processes are not present, the dynamics of the system are accurately known and accurate information about the state initial values $x(0)$ is available. Then simple integration (analytical or numerical) of eqn. (3.1) would solve the (filtering) problem. The reality is not so simple. Initial conditions are often not known accurately, the system/plant dynamics are not always accurately known and the state and/or measurement noises are always present. And hence a filtering approach is needed for state estimation. The process noise accounts for modelling errors as well as an artefact to do filter tuning for trajectory matching. Since our aim is to obtain an estimate of the state of the dynamic system, we need to have measurements of the state; often these are available indirectly as eqn. (3.2) through the measurement model.

The mathematical models assumed are Gauss–Markov, since the noise processes assumed are Gaussian and the system described in eqn. (4.1) is linear. This model state is a Markov process or chain, mainly the model being the state equation of first order. This model is general and is readily amenable to recursive processing of the data. In addition, it is generally assumed that the system (in fact the representation of eqns. (3.1) and (3.2)) is controllable and observable.

3.2.1 Covariance Matrix Formulation

Consider the homogeneous state equation

$$\dot{x}(t) = A(t)x(t) \tag{3.3}$$

Then the state vector x evolves according to

$$x(t) = \varphi(t,t_0)x(t_0) \tag{3.4}$$

Here, $x(t_0)$ is the initial state at time t_0. For conformity with the discrete system, we rewrite eqn. (3.4) as

$$x(k+1) = \varphi(k,k+1)x(k) \tag{3.5}$$

The matrix φ is a state-transition matrix. It takes state from $x(k)$ at time k to $x(k + 1)$ at time $k + 1$ and so on. The equation for covariance matrix propagation can be easily derived based on its definition and eqn. (3.5).

Let $P(k) = E\{\underline{x}^T(k)\ \underline{x}^T(k)\}$ be the covariance matrix of $\underline{x}(k)$ at time index k, where $\underline{x}(k) = \hat{x}(k) - x(k)$ at k. It reflects the errors in the estimate of x at k. We want to know how the error propagates at other times. We have from eqn. (3.5):

$$\tilde{x}(k+1) = \beta\ \hat{x}(k) \tag{3.6}$$

Here, $\tilde{x}$ is a predicted estimate of x, considering $u = 0$, with no loss of generality. Then, we have, after adding a process noise term in eqn. (3.5):

$$\begin{aligned}\tilde{P}(k+1) &= E\{(\tilde{x}(k+1) - x(k+1))(\tilde{x}(k+1) - x(k+1)^T\} \\ &= E\{(\varphi\hat{x}(k) - \varphi x(k) - Gw(k))(\varphi\hat{x}(k) - \varphi x(k) - Gw(k))^T\} \\ &= E\{(\varphi\hat{x}(k) - \varphi\hat{x}(k))(\varphi\hat{x}(k) - \varphi x(k))^T\} + E\{Gw(k)w^T(k)G^T\}\end{aligned}$$

Here, we assume that state error and process noise are uncorrelated and hence the cross terms are neglected. Finally, we get

$$\tilde{P}(k+1) = \varphi\hat{P}(k)\varphi^T + GQG^T \tag{3.7}$$

Eqn. (3.7) is the state-error covariance propagation equation, i.e. the state-error variance at time k is modified by the process noise matrix and the new state-error variance is available at time $k + 1$. The transition matrix φ plays an important role.

3.2.2 Discrete-Time Filtering Algorithm

We presume that the state estimate at k is evolved to $k + 1$ using eqn. (3.6). Now at this stage a new measurement is available. This measurement contains information regarding the state as per eqn. (3.2). Therefore, intuitively, the idea is to incorporate the measurement into the data (filtering) process and obtain an improved/refined estimate of the state. We assume that the matrix H and *a priori* covariance matrix R are given or known.

3.2.2.1 Measurement/Data Update Algorithm

Given: H, R and measurements z

Assume: $\tilde{x}(k) \rightarrow$ *a priori* estimate of state at time k, i.e. before the measurement data is incorporated.

$\hat{x}(k) \rightarrow$ updated estimate of state at time k, i.e. after the measurement data is incorporated.

$\tilde{P} \rightarrow$ *a priori* covariance matrix of state-estimation err or (this was derived earlier).

Then measurement update algorithm is given as

$$\hat{x}(k) = \tilde{x}(k) + K[z(k) - H\ \tilde{x}(k)] \quad \text{(state estimate/filtered estimate)} \tag{3.8}$$

$$\hat{P}(k) = (I - KH)\tilde{P}(k)\,(\text{covariance update}) \tag{3.9}$$

The filtering eqns. (3.8) and (3.9) are based on the following development.

Our requirement is that we want the unbiased recursive form of the estimator (filter), with minimum errors in the estimates as measured by P. Let such a recursive form be given as

$$\hat{x}(k) = K_1\tilde{x}(k) + K_2 z(k) \tag{3.10}$$

The expression in eqn. (3.10) is a fair weighted combination of the *a priori* estimate (obtained) by eqn. (3.6) and the new measurement. The gains K_1 and K_2 are to be optimally chosen for the above requirement of unbiased estimate.

Let $\underline{x}(k) = \hat{x}(k) - x(k)$; $\underline{x}^*(k) = \tilde{x}(k) - x(k)$ be the errors in the state estimates.

Then, we have

$$\underline{x}(k) = [K_1\tilde{x} + K_2 z(k)] - x(k) = K_1\tilde{x} + K_2 Hx(k) + K_2 v(k) - x(k)$$

Using simplified measurement eqn. (3.2)

$$\begin{aligned}\underline{x}(k) &= K_1\,[\underline{x}^*(k) + x(k)] + K_2 Hx(k) + K_2 v(k) - x(k)\\ &= [K_1 + K_2 H - 1]x(k) + K_2 v(k) + K_1\underline{x}^*(k)\end{aligned}$$

Since $E\{v(k)\} = 0$ and if $E\{\underline{x}^*(k)\} = 0$ (unbiased *a priori* estimate), then

$$E\{\underline{x}(k)\} = E\{(K_1 + K_2 H - 1)\, x(k)\}$$

Thus, to obtain an unbiased estimate after the measurement is incorporated, we must have $E\{\underline{x}(k)\} = 0$, and hence

$$K_1 = 1 - K_2 H \tag{3.11}$$

Substituting the above equation in eqn. (3.10), we get

$$\begin{aligned}\hat{x}(k) &= (1 - K_2\mathrm{H})\ \tilde{x}(k) + K_2 z(k)\\ &= \tilde{x}(k) + K_2[z(k) - H\ \tilde{x}(k)]\end{aligned} \tag{3.12}$$

For further development, we change K_2 to K as the Kalman (filter) gain. Essentially, eqn. (3.12) is the measurement data update algorithm, but we need to determine the expression for gain K. The structure of the filter has now been well defined:

Current estimate = previous estimate + gain × (error in measurement prediction)

The term $[z(k) - H\ \tilde{x}(k)]$ is called the measurement prediction error or the residual of the measurement. It is also called innovation. The above form is common to many recursive algorithms. Next, we formulate P to determine the covariance of the state error after the measurement is incorporated.

$$\begin{aligned}\hat{P} &= E\{\underline{x}(k)\underline{x}^T(k)\} = E\{(\hat{x}(k)-x(k))\,(\hat{x}(k)-x(k))^T\}\\ &= E\{(\tilde{x}(k)-x(k)+K[Hx(k)+v(k)-H\tilde{x}(k)])\,(.)^T\}\\ &= E\{[(I-KH)\underline{x}^* + Kv(k)]\,[\underline{x}^{*T}(I-KH)^T + v(k)K^T]\}\end{aligned}$$

$$\hat{P} = (I-KH)\tilde{P}(I-KH)^T + KRK^T \tag{3.13}$$

In the above, '.' means that the second term within the parenthesis is the same as the first term. Next, we optimally choose K so that the error covariance matrix $\hat{P}$ is minimized in terms of some norm. Let the cost function $J = E\{\underline{x}^T(k)\underline{x}(k)\}$ be minimized with respect to the gain matrix K. This is equivalent to

$$\begin{aligned}J &= trace\{\hat{P}\}\\ &= trace\{(I-KH)\tilde{P}(I-KH)^T + KRK^T\}\end{aligned} \tag{3.14}$$

$$\frac{\partial J}{\partial K} = -2\,(I-KH)\tilde{P}H^T + 2KR = 0, \text{ the null matrix; } \begin{aligned}KR &= \tilde{P}H^T - KH\tilde{P}H^T\\ KR + KH\tilde{P}H^T &= \tilde{P}H^T\end{aligned}$$

$$K = \tilde{P}H^T(H\tilde{P}H^T + R)^{-1} \tag{3.15}$$

Substituting the expression of K into eqn. (3.13) and simplifying, we get:

$$\hat{P} = (I-KH)\tilde{P} \tag{3.16}$$

Finally, the Kalman filter equations are put collectively in the following form:

3.2.2.1.1 State Propagation

$$\text{State estimate: } \tilde{x}(k+1) = \varphi\,\hat{x}(k) \tag{3.17}$$

$$\text{Covariance } (a\ priori)\text{: } \tilde{P}(k+1) = \varphi\hat{P}(k)\varphi^T + GQG^T \tag{3.18}$$

3.2.2.1.2 Measurement Update

$$\text{Residual: } r(k+1) = z(k+1) - H\,\tilde{x}(k+1) \tag{3.19}$$

$$\text{Kalman gain: } K = \tilde{P}H^T(H\tilde{P}H^T + R)^{-1} \tag{3.20}$$

$$\text{Filtered estimate: } \hat{x}(k+1) = \tilde{x}(k+1) + K\,r(k+1) \tag{3.21}$$

$$\text{Covariance}(posteriori)\text{: } \hat{P} = (I-KH)\tilde{P} \tag{3.22}$$

Although K and P would vary as the filter is running, the time index is dropped for simplicity. However, Q and R are assumed pre-determined and constant.

We note here that $K = \tilde{P}H^T S^{-1}$ with $S = H\tilde{P}H^T + R$. This matrix S is the covariance matrix of residuals. The actual residuals can be computed from eqn. (3.19) and they can be compared with standard deviations obtained by taking the square root of the diagonal elements of S. This process of checking and tuning the filter to bring the computed residuals within the bound of at least two standard deviations is an important filter tuning exercise for correct solution of the problem. This process of tuning in conjunction with eqn. (3.18) is called the covariance-matching concept for adaptive estimation in Kalman filtering algorithm.

3.2.3 Continuous-Time Kalman Filter

Although the discrete-time filtering algorithm is widely preferred for digital implementation, we give continuous-time filtering algorithm here. Let us define the continuous-time model of the dynamical system as

$$\dot{x}(t) = Ax(t) + w(t) \tag{3.23}$$

$$z(t) = Hx(t) + v(t) \tag{3.24}$$

We have the following assumptions:

(i) The noise processes $w(t)$ and $v(t)$ are uncorrelated Gaussian random processes with spectral density matrices $Q(t)$ and $R(t)$, respectively.
(ii) $E\{x(0)\} = \hat{x}_0;\ E\{(\hat{x}_0 - x(0))\,(\hat{x}_0 - x(0))^T\} = P_0$
(iii) We have very accurate knowledge of A, H, Q and R.

Then, continuous-time KF is given as

$$\dot{\hat{x}}(t) = A\hat{x}(t) + K(t)[z(t) - H\hat{x}(t)] \quad \text{(state evolution)} \tag{3.25}$$

$$\dot{P}(t) = AP(t) + P(t)A^T + Q(t) - KRK^T; P(0) = P_0 \tag{3.26}$$

$$K = PH^T R^{-1} \quad \text{(Kalman gain)} \tag{3.27}$$

Eqn. (3.26) is called the matrix Riccati equation, which needs to be solved for obtaining P, which is used in computation of the Kalman gain. The comparison of eqns. (3.26) and (3.27) with eqns. (3.18) and (3.20) shows that the computations for continuous-time Kalman filter are more involved due to the continuous-time matrix Riccati equation. One simple route is to assume that a steady state is reached, thereby considering $\dot{P} = 0$, and solving eqn. (3.26) by an appropriate method.

3.2.4 Interpretation and Features of KF

Insight into KF's functioning can be easily obtained by considering the continuous-time Kalman filter gain eqn. (3.27). Let K for the scalar system be given as $K = c\ (\sigma_{\underline{x}}^2/\sigma_v^2)$. Here, $H = c$, $P = \sigma_{\underline{x}}^2$ and $R = \sigma_v^2$. The state eqn. (3.25) simplifies to $\dot{\hat{x}}(t) = a\hat{x}(t) + K[z(t) - c\hat{x}(t)]$.

If the measurement uncertainty is large, represented by σ_v^2, then Kalman gain will be low for fixed value of $\sigma_{\underline{x}}^2$. Then the filter does not put more emphasis on measurement and the state estimate will be based only on the previous estimate. Similarly, if $\sigma_{\underline{x}}^2$ is low, then also K will be low. This is intuitively appealing for the state update. If $\sigma_{\underline{x}}^2$ is large, then K will be large and more emphasis will be put on the measurement, assuming relatively low σ_v^2. Hence, based on the relative value of the scalar ratio $\sigma_{\underline{x}}^2/\sigma_v^2$, the Kalman gain adapts to the value, which is intuitively appealing. This is just achieved by the optimization of the cost function, without invoking this appealing feature, in the first place. For the discrete-time filter, we have the Kalman gain as

$K = \tilde{P}H^T(H\tilde{P}H^T + R)^{-1}$. For the scalar case, we have

$K = \sigma_{\underline{x}}^2\, c\,(c^2\sigma_{\underline{x}}^2 + \sigma_v^2)^{-1} = \dfrac{\sigma_{\underline{x}}^2\, c}{(c^2\sigma_{\underline{x}}^2 + \sigma_v^2)}$. We presume that $c = 1$, then

$$K = \frac{\sigma_{\underline{x}}^2}{(\sigma_{\underline{x}}^2 + \sigma_v^2)}$$

For constant process noise variance, increase in σ_v^2 signifies decrease in K and hence the filter puts more weightage on the previous state estimate and less on the new measurement. Similarly, for constant σ_v^2, increase in $\sigma_{\underline{x}}^2$ will cause K to increase, and more emphasis will be put on the measurement. Thus, in KF, the filter shifts its emphasis based on the information content/uncertainties in the measurement data. Ironically, this mechanization points to a major limitation of the Kalman filter, i.e. filter tuning of the parameters Q and R. However, it can be seen from the foregoing, that it is only the ratio of Q and R that matters. For matrices, the ratio will be in the form of individual norms of matrices Q and R or any other measure can be used. We need to evaluate the performance of the filter to see if proper tuning has been achieved or not and whether the estimates make sense. Several possibilities exist: (a) to check the whiteness of the measurement residuals; and (b) to see if the computed covariances match the theoretical covariances obtained from the covariance equations of the filter. The test (a) signifies that due to the fact the measurement residual is white, no information is left out from being utilized in the filter. The white process is an unpredictable process. The test (b) signifies that the computed covariances from the data match the filter predictions (theoretical estimates of the covariances), and hence proper tuning has been achieved. These tests are valid for all type of KF versions, be it extended KF or the factorization filtering algorithm.

Some features of the Kalman filter are as follows: (a) it is a finite-dimensional linear filter; (b) it can be considered as a system driven by residuals and producing the state estimates; (c) it obtains unbiased, and minimum variance estimates of the state; (d) it obtains theoretical estimates of the state-error covariance at each instant of time; (e) it is a recursive filter and incorporates the data as they are received. Uniform sampling of data is not a great need for this filter; and (f) it can be easily adapted to a real-time estimation of states. The only restriction is the computation of P and K, which would be time consuming. Often parallel Kalman filtering equations can be used. For linear systems, Kalman gain K and covariances can be pre-computed, since these computations do not depend upon the measurement data. Other features are as follows: (a) it can be extended to nonlinear systems; (b) with this modification, it can be used for joint state and parameter estimation; (c) it is also applicable to continuous time, time varying linear and nonlinear systems; and (d) it can be modified to handle the correlated process noise.

3.3 KALMAN *U-D* FACTORIZATION FILTER

The KF could diverge due to one or more of the following reasons [4]: (i) modelling errors (due to nonlinear system), (ii) wrong *a priori* statistics (P, Q, R), and (iii) finite word length implementation of the filter. For handling (i), properly tuned EKF should be used, and if feasible, accurate mathematical model of the system should be used, since KF utilizes the mathematical model of the underlying system itself. For handling (ii), proper tuning should be done, reliable estimates of Q and R or 'ratio' (in the sense of the norms) of Q and R should be determined. Adaptive tuning methods should be used. For (iii), factorization filtering methods should be used, or the filter should be implemented on a computer with large word length, as is often done.

In KF, eqn. (3.22) is especially ill-conditioned. Due to round off errors in computation and their propagation, the covariance matrix P could render to be nonpositive definite, whereas theoretically it should be a positive definite. In addition, matrix P should be symmetric, but during computation, it could lose this property. All these will lead the KF filter to diverge, meaning thereby, that the residuals will grow in size and the filter estimate will not converge in the sense of mean square to the true state. This is not the problem with the KF but its implementation on a finite word length computing machine. These effects are circumvented or greatly reduced by implementing KF in its factorized form. These algorithms do not process covariance matrix P in its original form, but process its square root. Such factorization implicitly preserves the symmetry and ensures the non-negativity of the covariance matrix P. There are several such algorithms available in the filtering literature. One such algorithm, which is widely used, called the *U-D* factorization (UDF) algorithm is given here. Here, U and D, are the matrix factors of the covariance matrix P of the KF, where U is a unit upper triangular matrix and D is a diagonal matrix. The UDF has several merits: (a) it is numerically reliable, accurate and stable, (b) it is a square-root-type algorithm, but does not involve square-rooting operations, (c) the algorithm is most efficiently and simply mechanized by processing vector measurements (observables) one component at a time, and (d) for linear systems, the UDF is algebraically equivalent to the KF.

The major advantage from *U-D* comes from the fact that the square-root-type algorithms process square roots of the covariance matrices and hence, they essentially use half the word length normally required by the conventional KF. In the *U-D* filter, the covariance update formulae of the conventional KF and the estimation recursion are reformulated, so that the covariance matrix does not appear explicitly. Specifically, we use recursions for U and D factors of covariance matrix $P = UDU^T$. Computing and updating with triangular matrices involve fewer arithmetic operations and thus greatly reduce the problem of round off errors, which might cause ill-conditioning and subsequent divergence of the algorithm, especially if the filter is implemented on a finite-word-length machine. This is more so for real-time implementation on board computers where the word length could be small, e.g. 16 or 32 bit. The filter algorithm for linear system is given in two parts:

3.3.1 Time Propagation

We have for the covariance update,

$$\tilde{P}(k+1 \mid k) = \varphi \hat{P}(k) \varphi^T + GQG^T \tag{3.28}$$

Given $\hat{P} = \hat{U}\hat{D}\hat{U}^T$ and Q as the process noise covariance matrix, the time update factors $\tilde{U}$ and $\tilde{D}$ are obtained through modified Gram–Schmidt orthogonalization process. We define $V = \left[\varphi\hat{U} \mid G \right]$ and $\bar{D} = \text{diag}[\hat{D}, Q]$, and $V^T = [v_1, v_2, \ldots, v_n]$. P is reformulated as $\tilde{P} = \tilde{V}\tilde{D}\tilde{V}^T$. The U and D factors of $\tilde{V}\tilde{D}\tilde{V}^T$ are computed as below.

For $j = 1, \ldots, n$ the following equations are recursively evaluated.

$$\tilde{D}_j = \langle v_j, v_j \rangle_{\bar{D}} \tag{3.29}$$

$$\tilde{U}_{ij} = (1/\tilde{D}_j) \langle v_i, v_j \rangle_{\bar{D}} \qquad i = 1, \ldots, j-1 \tag{3.30}$$

$$v_i = v_i - \tilde{U}_{ij} v_j \tag{3.31}$$

Here, $\langle v_i, v_j \rangle_{\bar{D}} = v_i^T \bar{D} v_j$ is the weighted inner product between v_i and v_j. Therefore, the time-propagation algorithm directly and efficiently produces the required U, D factors, taking the effect of previous U, D factors and the process noise. Thus, it also preserves the symmetry of the (original) P matrix.

3.3.2 Measurement Update

The measurement update in KF combines *a priori* estimates $\tilde{x}$ and error covariance $\tilde{P}$ with a scalar observation $z = cx + v$ to construct an updated estimate and covariance given as

$$K = \tilde{P}c^T/s; \hat{x} = \tilde{x} + K(z - c\tilde{x}); s = c\tilde{P}c^T + R; \hat{P} = \tilde{P} - Kc\tilde{P}; \tag{3.32}$$

Here, $\tilde{P} = \tilde{U}\tilde{D}\tilde{U}^T$; c is the measurement matrix, R is the measurement noise covariance, and z is the vector of noisy measurements. Kalman gain K, and updated covariance factors $\hat{U}$ and $\hat{D}$ can be obtained from the following equations:

$$g = \tilde{U}^T c^T; g^T = (g_1,\ldots,g_n); w = \tilde{D}g; \hat{d}_1 = \tilde{d}_1 R/s_1; s_1 = R + w_1 g_1 \tag{3.33}$$

For $j = 2,\ldots,n$ the following equations are evaluated:

$$s_j = s_{j-1} + w_j g_j; \hat{d}_j = \tilde{d}_j s_{j-1}/s_j;$$

$$\hat{u}_j = \tilde{u}_j + \lambda_j K_j, \qquad \lambda_j = -g_j/s_{j-1}$$

$K_{j+1} = K_j + w_j \tilde{u}_j$; $\tilde{U} = [\tilde{u}_1,\ldots,\tilde{u}_n]$; then, Kalman gain is given by

$$K = K_{n+1}/s_n \tag{3.34}$$

Here, $\tilde{d}$ is the predicted diagonal element, and $\hat{d}_j$ is the updated diagonal element of the D matrix.

The time propagation and measurement update of the state vector is like KF and hence, it is not repeated here. We also note that the measurement update/data processing can be done sequentially, meaning thereby that each observable can be processed in turn, and the state estimate updated. This avoids the matrix inversion in the Kalman gain formulation.

Example 3.1: Simulate data of a target moving with constant acceleration and acted on by an uncorrelated noise, which perturbs the constant acceleration motion. Add measurement noise with standard deviation of 1 to generate measurements of position and velocity. Estimate the states of the system using *U-D* factorization-based linear KF (UDKF) and the noisy position and velocity measurements. Evaluate the filter's performance using the standard procedure.

Solution 3.1: The target data (position and velocity) are generated using the state and measurement eqns. (3.1) and (3.2) by adding random process noise with $\sigma = 0.001$ (standard deviation), and measurement noise with $\sigma = 1$. The state vector, x consists of target position (x_p), velocity (x_v) and acceleration (x_a), $x = [x_p, x_v, x_a]$.

For this case, the state-transition matrix $\varphi = \begin{bmatrix} 1 & \Delta t & \Delta t^2/2 \\ 0 & 1 & \Delta t \\ 0 & 0 & 1 \end{bmatrix}; G = \begin{bmatrix} \Delta t^2/2 \\ \Delta t \\ 1 \end{bmatrix}; H = \begin{bmatrix} 1 & 0 \\ 0 & 1 \end{bmatrix}$

Using the program **Genmeas.m** in the folder **Ch3UDex1**, both the position and velocity measurements are generated for duration of 100 secs. The sampling interval of $\Delta t = 0.25$ sec is chosen for simulation. The initial condition of the states used for the simulation is $x_0 = [200, 10, 0.5]$. For use in UDKF, the state model is formulated with the three states and the measurement model is formulated using noisy measurements of position and velocity. The state-estimation programs are contained in the folder **Ch3UDex1.** The initial conditions for the filter are chosen as $\hat{x}_0 = [190.0, 8.0, 0.4]$. The

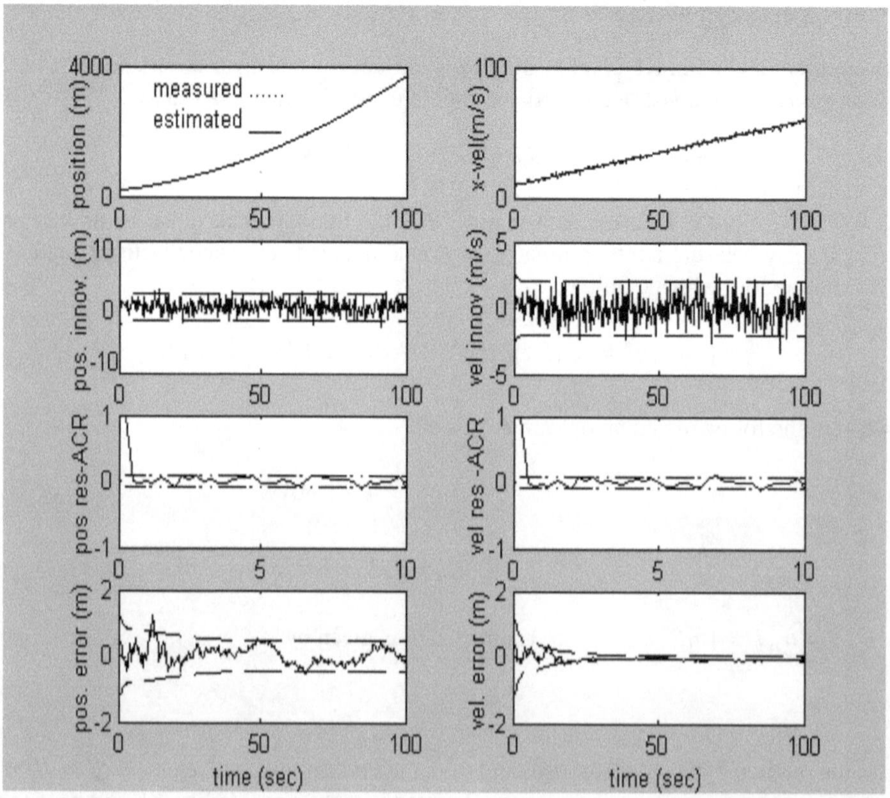

FIGURE 3.1 Measurements, innovations, autocorrelation of residuals and state errors. For the ACR plot, the X axis (time axis) is actually equivalent to the number of lags, e.g. 10 sec = 40 lags × 0.25 secs. (Raol, J.R., Girija, G., and J. Singh. *Modelling and Parameter Estimation of Dynamic Systems*. IEE/IET Control Series Vol. 65, London, UK, Aug. 2004.)

initial state-error covariance is chosen to reflect the difference in the true x_0 and $\hat{x}_0$. Figure 3.1 shows the estimated position and velocity measurements compared with the measured values, and the position and velocity innovations along with their theoretical bounds ($2\sqrt{S_{ii}(k)}$, S = innovation covariance), the autocorrelation function (ACR) of residuals with their bounds ($\pm 1.96/\sqrt{N}$, N = number of data points, $N = 400$) and the position and velocity state errors along with the $\pm 2\sqrt{P_{ii}(k)}$ bounds. The filter performance is very good as is evident from Figure 3.1 where all the estimated quantities fall within their theoretical bounds. For this example, the residual mean = [0.0656 and –0.0358] and PFE (percentage fit error) of the predicted measurements w.r.t. the true measurements =[0.0310, 0.4009].

3.4 EXTENDED KALMAN FILTER

Real-life dynamic systems are nonlinear and the estimation of states of such systems is often required. The nonlinear system can be expressed as

$$\dot{x}(t) = f[x(t), u(t), \Theta] \tag{3.35}$$

$$y(t) = h[x(t), u(t), \Theta] \tag{3.36}$$

$$z(k) = y(k) + v(k) \tag{3.37}$$

Here, f and h are general nonlinear vector-valued functions, and Θ is the vector of unknown parameters given by

$$\Theta = [x_0, b_u, b_y, \beta] \tag{3.38}$$

Here, x_0 represents values of the state variables at time $t = 0$; b_u represents the bias in control inputs (nuisance parameters); b_y represents the bias in model response y (nuisance parameters); β parameters in the mathematical model that defines the system characteristics.

Comparing eqns. (3.35) and (3.36) with eqns. (3.1) and (3.2), we see that the linear KF recursions eqns. (3.17)–(3.22) cannot be directly used for the state estimation of the nonlinear systems. One can, however, linearize the nonlinear functions f and h and then apply the KF recursions with proper modification to these linearized problems. The linearization of f and h could be around the pre-supposed nominal states, e.g. in the orbit-estimation problem, the nominal trajectory could be the circular orbit of the satellite to be launched. When the satellite is launched, it will acquire a certain orbit, which will be the actual orbit but affected by noisy measurements. Therefore, there will be three trajectories: nominal, estimated and the true trajectory. Often, EKF is preferred since the linearization will be around the previous/current best state estimates, which are more likely to represent the truth, rather than the linearization around the nominal states, leading to linearized KF (LKF). Hence, in this section, EKF is considered, which has application to aircraft parameter estimation as well. In EKF, the estimated state would converge to the true states for relatively large initial state errors, whereas this may not be true for the LKF. EKF is a sub-optimal solution to a nonlinear filtering problem. The nonlinear functions f and h are linearized about each new estimated/filtered state trajectory as soon as it becomes available. Simultaneous estimation of states and parameters is achieved by augmenting the state vector with unknown parameters (as additional states) and using the filtering algorithm with the augmented nonlinear model. The new augmented state vector is

$$x_a^T = \left[x^T \ \Theta^T\right] \tag{3.39}$$

$$\dot{x} = \begin{bmatrix} f(x_a, u, t) \\ 0 \end{bmatrix} + \begin{bmatrix} G \\ 0 \end{bmatrix} w(t) \tag{3.40}$$

$$\dot{x} = f_a(x_a, u, t) + G_a w(t) \tag{3.41}$$

$$y(t) = h_a\left(x_a, u, t\right) \tag{3.42}$$

$$z_m\left(k\right) = y\left(k\right) + u\left(k\right), \ k = 1, \ldots, N \tag{3.43}$$

Here,

$$f_a^T(t) = \left[f^T \ 0^T\right]; G_a^T = \left[G^T \ 0^T\right] \tag{3.44}$$

The estimation algorithm is obtained by linearizing eqns. (3.35) and (3.36) around the prior/current best estimate of the state at each time and then applying KF algorithm to the linearized model. The linearized system matrices are defined as

$$A(k) = \left.\frac{\delta f_a}{\delta x_a}\right|_{x_a = \hat{x}_a(k),\, u = u(k)} \tag{3.45}$$

$$H(k) = \left. \frac{\delta h_a}{\delta x_a} \right|_{x_a = \tilde{x}_a(k),\ u = u(k)} \tag{3.46}$$

and the state-transition matrix is given by

$$\varphi(k) = \exp\left[-A(k)\ \Delta t\right] \text{where } \Delta t = t_{k+1} - t_k \tag{3.47}$$

For the sake of clarity and completeness, the filtering algorithm is given in two parts: (i) time propagation, and (ii) measurement update. We notice the time-varying nature of A, H and φ, since they are evaluated at the current state estimate, which varies with time k.

3.4.1 Time Propagation

The current estimate is used to predict the next state, so that the states are propagated from the present state to the next time instant. The predicted state is given by

$$\tilde{x}_a(k+1) = \hat{x}_a(k) + \int_{t_k}^{t_{k+1}} f_a[\hat{x}_a(t), u(k), t]dt \tag{3.48}$$

In the absence of knowledge of process noise, eqn. (4.48) gives the predicted estimate of the state based on the initial/current estimate. The covariance matrix for state error (here state is x_a) propagates from instant k to $k + 1$ as

$$\tilde{P}(k+1) = \varphi(k)\hat{P}(k)\varphi^T(k) + G_a(k)QG_a^T(k) \tag{3.49}$$

Here, $\tilde{P}(k+1)$ is the predicted covariance matrix for the instant $k+1$, G_a is the process noise related coefficient matrix, and Q is the process noise covariance matrix.

3.4.2 Measurement Update

The EKF updates the predicted estimates by incorporating the measurements as and when they become available as follows:

$$\hat{x}_a(k+1) = \tilde{x}_a(k+1) + K(k+1)\left\{z_m(k+1) - h_a[\tilde{x}_a(k+1), u(k+1), t]\right\} \tag{3.50}$$

Here, K is the Kalman gain matrix. The covariance matrix is updated using the Kalman gain and the linearized measurement matrix from the predicted covariance matrix $\tilde{P}(k+1)$. The Kalman gain expression is given as

$$K(k+1) = \tilde{P}(k+1)H^T(k+1)[H(k+1)\tilde{P}(k+1)H^T(k+1) + R]^{-1} \tag{3.51}$$

A posteriori covariance matrix expression is given as

$$\hat{P}(k+1) = \left[I - K(k+1)H(k+1)\right]\tilde{P}(k+1) \tag{3.52}$$

The EKF is computationally more complex than simple KF. The major cost is due to the linearizations at every instant of time. For moderately nonlinear functions, the EKF would give reasonably accurate state estimates. If nonlinearities were severe, then repeated linearizations around newly estimated states, especially during measurement update, can be made. This yields so-called iterative EKF. In addition, a procedure called forward-backward filtering can be used. In this procedure, the EKF is used, in the first pass, as forward filtering. Then the EKF is run backward from the final point t_f to the initial time t_0, utilizing the same measurements. This process refines the estimates, but then it cannot be used in real-time applications. The *U-D* factorization filter can also be conveniently used in the EKF mode, since eqns. (3.51) and (3.52) can be put in the factorization form and processed.

We note from eqn. (3.48) that state (estimate) propagation is achieved by integration of the nonlinear function f_a between times t_k and t_{k+1}, thereby maintaining the effect of the nonlinearity of f. Also, in eqn. (3.50), nonlinear function h_a is used for predicting the measurements. These two features essentially give credence to the filter and hence to extended KF.

The EKF can be used for parameter estimation of linearized/nonlinear systems. However, since the covariance matrices are approximations, computed based on linearized nonlinear functions f and h, there is no guarantee of stability and performance, prior to experimental data analysis. However, in practice, the approach seems to work well if linearization is accurate and proper tuning of the filter is achieved. Although EKF is a nonlinear filtering solution, the modelling errors could prevail and these might degrade the performance of the algorithm. The proper tuning using Q matrix to have good matching of the states should be done. One major demerit of EKF is that it is computationally demanding and not easily amenable to parallelization of the algorithm, since the computations of the covariances are coupled with the filter computations.

Example 3.2: Simulate data of a second-order system with the following state and measurement matrices:

$$\begin{bmatrix} \dot{x}_1 \\ \dot{x}_2 \end{bmatrix} = \begin{bmatrix} a_{11} & a_{22} \\ a_{33} & a_{44} \end{bmatrix} \begin{bmatrix} x_1 \\ x_2 \end{bmatrix} + \begin{bmatrix} b_1 \\ b_2 \end{bmatrix} u = \begin{bmatrix} 0.06 & -2.0 \\ 2.8 & 0.08 \end{bmatrix} \begin{bmatrix} x_1 \\ x_2 \end{bmatrix} + \begin{bmatrix} -0.6 \\ 1.5 \end{bmatrix} u$$

$$\begin{bmatrix} z_1 \\ z_2 \end{bmatrix} = \begin{bmatrix} 1 & 0 \\ 0 & 1 \end{bmatrix} \begin{bmatrix} x_1 \\ x_2 \end{bmatrix} + v$$

Use a doublet signal as the input to the dynamic system (with the sampling interval of 0.05 sec). Use EUDF to estimate the states and parameters of the system using measurements of z_1 and z_2. Study the effect of measurement noise on the estimation results. Evaluate the performance of the filter using the standard procedure.

Solution 3.2: Simulated data of 10 secs duration is generated using the above equations (folder **Ch3EUDFex2sim**) with a sampling time of 0.05 secs. State noise with $\sigma = 0.001$ is added to generate the states. Measurements have SNR = 10. For state and parameter estimation, the state model is formulated with the two states x_1, x_2 and the six parameters of the A and B matrices in the above equations as augmented states in EUDF, eqn. (3.39). This results in a state model with eight states – two pertaining to the states x_1 and x_2 and six pertaining to the parameters $a_{11}, a_{12}, a_{21}, a_{22}, b_1, b_2$. The EUDF parameter-estimation programs are contained in the folder **Ch3EUDFex2.** The initial states/parameters for the KF are assumed 50% away from their true values. The initial state-error covariance matrix is chosen to reflect this uncertainty. The values of the process and measurement noise covariances are kept fixed at the values used in the simulation.

Figure 3.2a shows the estimated measurements compared with the noisy measurements. The figure also shows that the innovations pertaining to the two measurements fall within their theoretical

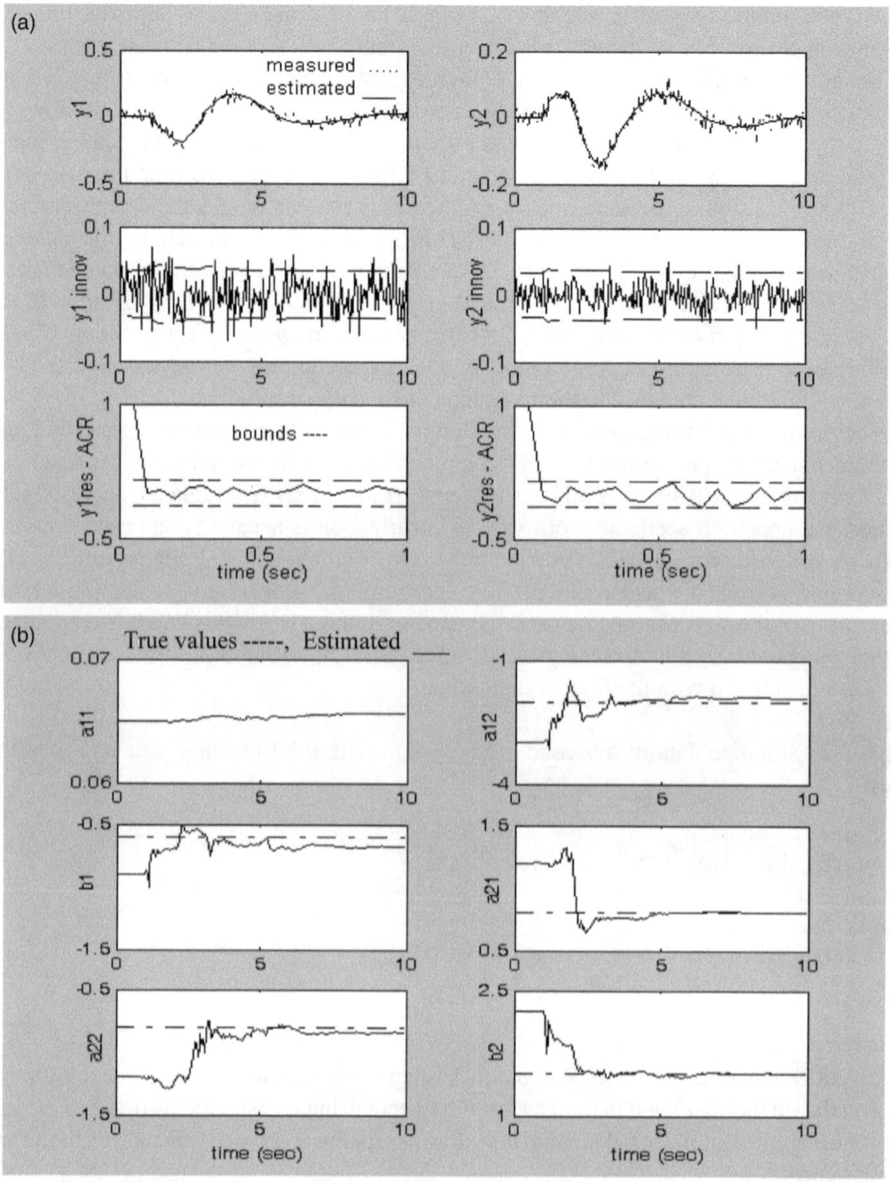

FIGURE 3.2 (a) Measurements, innovations and autocorrelation of residuals; (b) convergence of parameter estimates. (Raol, J.R., Girija, G., and J. Singh. *Modelling and Parameter Estimation of Dynamic Systems*. IEE/IET Control Series Vol. 65, London, UK, Aug. 2004.)

bounds and that the autocorrelation of the residuals fall within their theoretical bounds as well. Figure 3.2b shows the convergence of the parameters. In the presence of noise in the data, the parameters converge very close to their true values. Table 3.1 lists the estimated parameters along with their standard deviations; $\sigma = \sqrt{P_{ii}(k)}$. The estimated parameters and the standard deviations in Table 3.1 are those at the last data point (200 for this case). The parameter estimates are very close to the true values when there is no measurement noise in the data. In this case, very small value of R is used in the filter computation. However, it should be noted that process noise is present in the data. Some of the estimated parameters show slight deviations from the true values when there is noise in the data. However, the PEEN is less than 5%, which is acceptable when there is noise in the data.

TABLE 3.1
Parameter Estimates Using EUDF, Example 3.2

Parameters	True	Estimated (No Noise)	Estimated (SNR = 10)
a_{11}	0.06	0.0662 (0.0149)	0.0656 (0.0050)
a_{12}	-2.0	-2.0003 (0.0450)	-1.9057 (0.0956)
a_{21}	0.8	0.8005 (0.0202)	0.8029 (0.0892)
a_{22}	-0.8	-0.8038 (0.0340)	-0.8431 (0.0345)
b_1	-0.6	-0.5986 (0.0353)	-0.6766 (0.0548)
b_2	1.5	1.5078 (0.0356)	1.5047 (0.0734)
PEEN%	–	0.3833	4.5952

TABLE 3.2
Estimated Parameters of a LTA for Example 3.3

Parameter	True Values	Estimated (No Noise)	Estimated (SNR = 10)
C_{x0}	-0.0540	-0.05680 (0.0039)	-0.0592 (0.0085)
$C_{x\alpha}$	0.2330	0.2529 (0.0235)	0.2543 (0.0262)
$C_{x\alpha^2}$	3.6089	3.5751 (0.0619)	3.7058 (0.1131)
C_{z0}	-0.1200	-0.1206 (0.0046)	-0.1249 (0.0166)
$C_{z\alpha}$	-5.6800	-5.6759 (0.0196)	-5.7247 (0.0783)
$C_{z\delta}$	-0.4070	-0.4067 (0.0108)	-0.5049 (0.0477)
C_{m0}	0.0550	0.0581 (0.0049)	0.0576 (0.0081)
$C_{m\alpha}$	-0.7290	-0.7466 (0.0334)	-0.7092 (0.0433)
$C_{m\alpha^2}$	-1.7150	-1.6935 (0.0831)	-1.7843 (0.1097)
C_{mq}	-16.3	-16.2660 (0.3857)	-15.3075 (0.7980)
$C_{m\delta}$	-1.9400	-1.9397 (0.0110)	-1.8873 (0.0450)
PEEN(%)	–	0.4424	5.6329

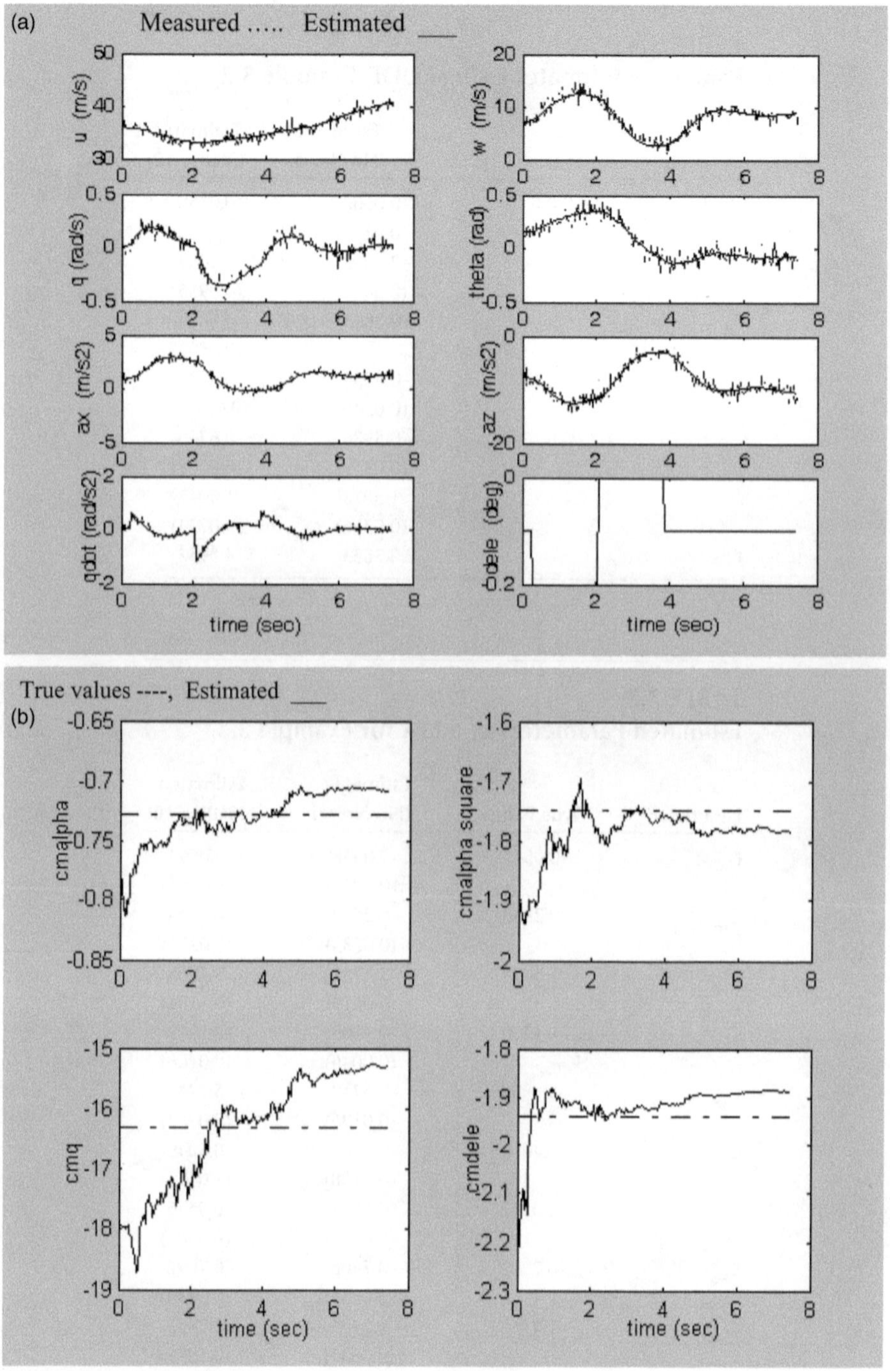

FIGURE 3.3 (a) Time history match; (b) parameter convergence: pitching moment derivatives. (Raol, J.R., Girija, G., and J. Singh. *Modelling and Parameter Estimation of Dynamic Systems*. IEE/IET Control Series Vol. 65, London, UK, Aug. 2004.)

Example 3.3: Use the simulated short period data of a LTA with process noise to estimate the non-dimensional longitudinal parameters of the aircraft using the KF method. Use the 4DOF longitudinal body axis model for estimation.

Solution 3.3: The data are generated with a sampling interval of 0.03 sec by giving a doublet input to the elevator. Random noise with $\sigma = 0.001$ is added to the states u, w, q, θ. The states with additive process noise are used to generate measurements (data set 1) of u, w, q, θ, $a_x, a_z, \dot{q}$. Random noise is added to these measurements to generate noisy data with SNR = 10 (data set 2). Both the sets of data are used for parameter estimation using UDKF. For estimating the parameters using UDKF, the parameters are modelled as augmented states in the state model, eqn. (3.39). For this case there are four states and 11 parameters so that the state model has 15 states. Seven measurements u, w, q, θ, $a_x, a_z, \dot{q}$ are used and all 11 parameters are estimated using the programs in the folder **Ch3EUDFex3**. The process and measurement noise covariances are kept fixed at the values used in the simulation of the data. The initial states and parameters for the Kalman filter are assumed 10% away from their true values. The initial state-error covariance matrix is chosen to reflect this uncertainty.

The estimated values of the parameters are compared with the true values (aerodynamic derivatives) in Table 3.2, which also shows the PEEN. The estimates are close to the true values even when there is noise in the data.

Figure 3.3a shows the estimated measurements compared with the noisy measurements. The convergence of the pitching-moment-related derivatives: $C_{m\alpha}, C_{m\alpha^2}, C_{mq}, C_{m\delta}$ is shown in Figure 3.3b; even in the presence of noise in the data, the parameters converge close to their true values. Some deviation is observed for the C_{mq} estimate.

Further applications of KF/EKF can be found in [5–7].

EXERCISES

Exercise 3.1

Let $z = y + v$. Obtain a variance of $z - \hat{z}$. We assume that v is the zero-mean white noise process and z is the vector of measurements.

Exercise 3.2

The transition matrix is defined as $\varphi = e^{A\Delta t}$ where A is the state-space system matrix and Δt the sampling interval. Obtain the state-transition matrix for $A = \begin{bmatrix} 0 & 1 \\ 0 & -a \end{bmatrix}$, if Δt is small; $a\Delta t$ is small. Use Taylor's series expansion for obtaining φ.

Exercise 3.3

Let the scalar discrete-time system be given by

$$x(k+1) = \varphi\, x(k) + bu + gw$$
$$z(k) = cx(k) + v$$

Here, u is the deterministic (control) input to the system and w is the process noise, which is assumed white and Gaussian. Obtain the complete set of Kalman filter equations. What happens to the 'u' term in the covariance update equation?

Exercise 3.4

Let $\dot{x} = Ax + w$ and let the elements of matrix A be unknown. Formulate the state-space model for the joint state and parameter estimation to be used in the EKF.

Exercise 3.5

Assume that the measurement noise is coloured (nonwhite) and is given by

$$\dot{v} = A_2 v + w_2$$

Then, append this equation to the state-space model of a linear system and obtain composite model suitable for the KF. Comment on the structure of the composite system model.

Exercise 3.6

What is the distinction between residual error, prediction error and filtering error in the context of the state/parameter estimation?

Exercise 3.7

What is the purpose/advantage of partitioning the KF algorithm into time propagation and measurement update parts? See eqns. (3.17) to (3.22).

Exercise 3.8

We have seen that the covariance matrix of the innovations (i.e. residuals) is $S = HPH^T + R$. We can also compute the residuals empirically from $r(k+1) = z(k+1) - H\tilde{x}(k+1)$. This gives Cov (rr^T). Explain the significance of both these computations. The matrix S is computed by the Kalman filter algorithm. (Hint: both the computations are for the same random variable r).

Exercise 3.9

Derive the explicit expression for P, the state covariance matrix of the Kalman filter, taking a scalar problem and comment on the effect of measurement noise variance on $\hat{P}$.

Exercise 3.10

Establish the following relationship:

Variance of (x) = mean squared value of (x) – square of mean value of (x), for a random variable x.

Exercise 3.11

Under what condition is the RMS value of a signal equal to the standard deviation of the signal?

Exercise 3.12

Why is the U-D filtering algorithm of square root type without involving the square-rooting operation in propagation of the covariance-related computations?

Exercise 3.13

Substitute eqn.(3.15) for the Kalman filter in eqn. (3.13) for the covariance matrix update, and obtain the compact form of $\hat{P}$ as in eqn. (3.16), by using only simple algebraic manipulations and no approximations.

Exercise 3.14

Why is the residual process in the KF called the 'innovations' process? (Hint: by innovations, it is meant that some new information is obtained/used).

Exercise 3.15

Derive recursive expressions for the determination of the average value and variance of a variable x.

REFERENCES

1. Kalman, R.E., A new approach to linear filtering and prediction problems, *Transactions of ASME, Series D, Journal of Basic Engineering*, Vol. 8, pp. 35–45, March 1960.
2. Maybeck, P.S., *Stochastic Models, Estimation and Control*, Vol. 1, Academic Press, New York, USA, 1979.
3. Gelb, A. (Ed.), *Applied Optimal Estimation*, MIT Press, Massachussetts, 1974.
4. Bierman, G.J., *Factorisation Methods for Discrete Sequential Estimation*, Academic Press, New York, 1977.
5. Raol, J.R., and Girija, G., *Evaluation of adaptive Kalman filtering methods for target tracking applications*, Paper No. AIAA-2001-4106, USA, Aug. 2001.
6. Girija, G., and Raol, J.R., PC based flight path reconstruction using UD factorization filtering algorithms, *Defense Science Journal*, Vol. 43, pp. 429–447, 1993.
7. Jategaonkar, R.V., and Plaetschke, E., Algorithms for aircraft parameter estimation accounting for process and measurement noise, *Journal of Aircraft*, Vol. 26, No. 4, pp. 360–372, April 1989.

4 Filtering-cum-Data Fusion with State Delay and Missing Measurements*

4.1 INTRODUCTION

The process of multisensory data fusion (MSDF) is very important for many civilian and military applications: (a) target tracking, (b) automatic target recognition (ATR) and (c) wireless sensor networks (WSNs) [1]. The idea is to ascertain the status and an identity of the object under investigation and observation wherefore the measurements are available from more than one sensor. Then more complete information about the status of the object is obtained by fusing the information from these sensors in an appropriate way. However, sometimes in a data-communication channel a few or many measurements might be missing (from one or more sensors). Then, it becomes very important to study and evaluate the performance of the data-processing algorithms in the presence (and despite) of missing measurements (missing data, MD), i.e. in the absence of certain signal data during certain time periods, however the random (measurement) noises present for these channels are ascertained [2]. Also, there might be some latency present in certain states of the dynamic system that models a WSN. Hence, it is important to consider the data processing and filtering algorithms in a synergistic manner.

The data might be missing due to the following: (i) a failure of a senor, and/or (ii) there being a problem in a communication channel, and the received data might be only the noise and the signal is missing. Hence, it is very important to handle the situation of missing data in a filter in some formal and optimal way. Time delay can be encountered in many real-time systems, due to the latency time of certain channels. Time delay influences the overall system stability and performance. If these aspects of the system's state delay and missing data are not handled appropriately in a tracking filter, then one might lose the track, and the performance of the tracking filter would also be very poor.

Four filtering-cum-fusion algorithms are presented, which are modified to include state delays and randomly missing measurement data; also a nonlinear observer based on the gain of the continuous-time data-fusion filter is presented with its asymptotic result [3]. A nonlinear observer that would handle state delay as well as randomly missing measurements is also proposed; and the asymptotic condition for the observer error dynamics is derived. The performance is illustrated with implementation in MATLAB. The nonlinear observer/s for the nonlinear continuous-time system with state delay and randomly missing measurements in the measurement data-level (MLF) and state-vector-level fusion (SVF) modes are presented. Also, it is ascertained that the SVF formula for this observer is similar to that for KF/EKF.

An H-infinity (HI)-based nonlinear observer that would handle state delay as well as randomly missing data leading to a sub-optimal estimator-cum-observer is proposed. Also, the asymptotic condition for the observer error dynamics based on the Lyapunov energy (LE) functional is derived. Many of the presented results and observations (inferences) here are novel in the area of observer theory and multisensory data fusion [3].

---

* This chapter is contributed by Jitendra R. Raol and N. Shankarappa.

DOI: 10.1201/9781003620211-4

4.2 FILTERING-CUM-FUSION ALGORITHMS

Four existing filtering algorithms are modified to include the state delay and randomly missing measurements to obtain new filtering-cum-fusion algorithms (FFAs): (a) the gain fusion or the so-called data-sharing filter (GFA), (b) H-infinity *a posteriori* filter (HIPF), (c) H-infinity *a posteriori* risk sensitive filter (HIRF) and (d) H-infinity global fusion filter (HIGF). The performance metrics are evaluated using MATLAB-based implementations of these algorithms for randomly missing measurements. A nonlinear dynamic system with the state delay is given as

$$x(k+1) = f(x(k), x(k-1)) + Gw(k) \tag{4.1}$$

The measurement model for each sensor with missing data is given as

$$z_m(k) = \beta_m H_m x(k) + v_m(k) \tag{4.2}$$

In eqn. (4.2), m = 1,2 is the number of sensors. The process noise $w(.)$, and measurement noise $v(.)$ are assumed to be WG stochastic processes with the respective covariance matrices as Q and R.

The quantity β_m is a Bernoulli sequence, which takes values 0 and 1 randomly; with $E\{\beta_m(k)=1\} = a_m(k)$ and $E\{\beta_m(k)=0\} = 1-a_m(k)$, with '$a$' as the percentage of measurements that arrive to the sensor fusion node; this also signifies that a few measurements are randomly missing. The constant 'a' is assumed to be known or pre-specified.

4.2.1 Modified Gain Fusion Algorithm (GFA)

A gain fusion algorithm that is inherently a combination of the filtering and data-fusion processing capability [4] is modified here to incorporate the state delay and randomly missing measurements. It considers the local filters and one global fusion filter; in each scan, the local filters receive the measurements and after processing these, communicate the estimated states to the global filter. The time propagation of the global estimates is obtained as follows:

The state estimate is obtained in the usual way as

$$\hat{x}(k+1) = F_0\hat{x}^f(k) + F_1\hat{x}^f(k-1) \tag{4.3}$$

In eqn. (4.3), the Jacobians are given as

$$F_0 = \frac{\partial f(.,.)}{\partial x}, at\, x = \hat{x}^f(k);\; F_1 = \frac{\partial f(.,.)}{\partial x},\; at\, x = \hat{x}^f(k-1)$$

The state-error covariance estimate is given as

$$\tilde{P}^f(k+1) = F_0\hat{P}^f(k)F_0^T + F_1\hat{P}^f(k-1)F_1^T + GQG^T \tag{4.4}$$

Then, the local sensor level (i = 1, 2,) filters are reset as

$$\tilde{x}_m(k+1) = \hat{x}^f(k+1) \tag{4.5}$$

$$\tilde{P}_m(k+1) = \hat{P}^f(k+1) \tag{4.6}$$

The measurement updates of the local gains and the states is obtained as

$$K_m = \left(\frac{1}{\gamma_m}\right)\beta_m \tilde{P}^f(k+1)H_m^T(\beta_m{}^2 H_m \tilde{P}^f(k+1)H_m^T + \frac{1}{\gamma_m}R_m)^{-1} \tag{4.7}$$

$$\hat{x}(k+1) = \hat{x}^f(k+1) + K_m(z_m - \beta_m H_m \hat{x}^f(k-1)) \tag{4.8}$$

Then, the global fusion of 'm' local estimates is obtained as

$$\hat{x}(k+1) = \tilde{x}^f(\sum_i^m (\hat{x}^i k+1) - (m-1)\tilde{x}^f(k-1)) \tag{4.9}$$

$$\hat{P}^f(k+1) = [I - \sum_i^m \beta_m K^i H]P(k+1)[I - \sum_i^m \beta_m K^i H]^T + \sum_i^m K^i R^i K^{iT} \tag{4.10}$$

The 'gamma' in eqn. (4.7) is the theoretical upper bound of the state-error energy gain from the 'input' noise energies to the output energy.

4.2.2 Modified H-Infinity *A Posteriori* Filtering Algorithm (HIPF)

This HI filter is based on minimization of the HI norm, and it is supposed to be a robust algorithm [5]. The basic filter is modified to suit eqns. (4.1) and (4.2). The (covariance) state-error Gramian (SEG) propagation is obtained for each sensor as

$$\begin{aligned} P_m(k+1) &= F_0 P_m(k) F_0^T + F_1 P_m(k-1) F_1^T + GQG^T \\ &- F_0 P_m(k)[\beta_m H_m^T \;\; L_m^T] R_m^{-1} \begin{bmatrix} \beta_m H_m \\ L_m \end{bmatrix} P_m(k) F_0^T \\ &- F_1 P_m(k-1)[\beta_m H_m^T \;\; L_m^T] R_m^{-1} \begin{bmatrix} \beta_m H_m \\ L_m \end{bmatrix} P_m(k-1) F_1^T \end{aligned} \tag{4.11}$$

In eqn. (4.11) the composite measurement covariance matrix is obtained as

$$R_m = \begin{bmatrix} I & 0 \\ 0 & -\gamma^2 I \end{bmatrix} + \begin{bmatrix} \beta_m H_m \\ L_m \end{bmatrix} P_m(k)[\beta_m H_m^T \;\; L_m^T] \tag{4.12}$$

In eqn. (4.12), γ is the factor that specifies the bound on the energy (variance) gain from the input energies due to the disturbances, $w(.)$, and $v(.)$, and the input error in the state initial condition to the output state-error energy. The HIPF filter gain is obtained as

$$K_m = \beta_m P_m(k+1)H_m^T(\beta_m^2 H_m P(k+1)H_m^T + I)^{-1} \tag{4.13}$$

Then, the measurement update of the state is given as

$$\hat{x}_m(k+1) = \tilde{x}_m(k) + K_m(z_m(k+1) - \beta_m H_m \tilde{x}(k)) \tag{4.14}$$

In eqn. (4.14), the previous state estimate is obtained as

$$\tilde{x}_m(k) = F_0 \hat{x}_m(k) + F_1 \hat{x}_m(k-1) \tag{4.15}$$

The HIPF as such obtains the individual state and covariance estimates for each sensor, hence it requires to obtain the fused state and fused covariance estimates, and this is obtained by the state-vector fusion formulae. The fused estimate of the states is given as

$$\hat{x}^f(k+1) = \hat{x}_1(k+1) + \hat{P}_1(k+1)[\hat{P}_1(k+1) + \hat{P}_2(k+1)]^{-1}\{\hat{x}_2(k+1) - \hat{x}_1(k+1)\} \tag{4.16}$$

The fused estimate of the SEG matrix is given as

$$\hat{P}^f(k+1) = \hat{P}_1(k+1) - \hat{P}_1(k+1)[\hat{P}_1(k+1) + \hat{P}_2(k+1)]^{-1}\hat{P}_1^T(k+1) \tag{4.17}$$

As can be seen from eqns. (4.16), and (4.17), that the fusion formulae utilize the individual estimates from the two sensors' estimates obtained in eqns. (4.11) and (4.14).

4.2.3 Modified H-Infinity Global Filtering-cum-Fusion Algorithm (HIGF)

This is a decentralized algorithm and is also based on the HI concept [6], and has the central fusion equations. The modified form suitable to eqns. (4.1) and (4.2) is given next. The local filters are given for each senor as

$$\tilde{x}_m(k+1) = F_0\hat{x}_m(k) + F_1\hat{x}_m(k-1) \tag{4.18}$$

$$\tilde{P}_m(k+1) = F_0\hat{P}_m(k)F_0^T + F_1\hat{P}_mF_1^T + GQG^T \tag{4.19}$$

The SEG update is given as

$$\hat{P}_m^{-1}(k+1) = \tilde{P}_m^{-1}(k+1) + [\beta_m H_m^T \;\; L_m^T]\begin{bmatrix} I & 0 \\ 0 & -\gamma^2 I \end{bmatrix}\begin{bmatrix} \beta_m H_m \\ L_m \end{bmatrix} \tag{4.20}$$

The local filter gains are given as

$$A_m = I + \frac{1}{\gamma^2}\hat{P}_m(k+1)L_m^T L_m; \quad K_m = \beta_m A_m^{-1}\hat{P}_m(k+1)H_m^T \tag{4.21}$$

$$\hat{x}_m(k+1) = \tilde{x}_m(k) + K_m(z_m(k+1) - \beta_m H_m\tilde{x}(k+1)) \tag{4.22}$$

The time propagation of the fused states and the SEG matrix is given as

$$\tilde{x}^f(k+1) = F_0\hat{x}^f(k) + F_1\hat{x}^f(k-1) \tag{4.23}$$

$$\tilde{P}^f(k+1) = F_0\hat{P}^f(k)F_0^T + F_1\hat{P}^f(k-1)F_1^T + GQG^T \tag{4.24}$$

The measurements update of the fusion states and covariance is given as

$$(\hat{P}^f(k+1))^{-1} = (\tilde{P}^f(k+1))^{-1} + \sum_{i=1}^{m}\{\hat{P}_i^{-1}(k+1) - \tilde{P}_i^{-1}(k+1)\} + \frac{m-1}{\gamma^2}L^T L \tag{4.25}$$

The global gain is obtained as $A^f = I + \frac{1}{\gamma^2}\hat{P}^f(k+1)L^T L$

The global-measurement update fused state is given as

$$\begin{aligned}\hat{x}^f(k+1) &= [I-(A^f)^{-1}\hat{P}^f(k+1)\beta_m^2 H^{fT} H_f]\tilde{x}^f(k+1) \\ &+(A^f)^{-1}\hat{P}^f(k+1)\sum_{i=1}^{m}\{\hat{P}_i^{-1}(k+1)A_i\hat{x}_i(k+1) \\ &-[\hat{P}_i^{-1}(k+1)A_i+\beta_m^2 H_i^T H]\breve{x}_i(k)\}\end{aligned} \tag{4.26}$$

In eqn. (4.26) the last term is given as

$$\breve{x}_i(k) = F_0\hat{x}_i(k) + F_1\hat{x}_i(k-1) \tag{4.27}$$

For this HIGF algorithm, the SVF formulae (as needed in the case of HIPF) are not required. Also, such explicit state-vector fusion formulae are not required in the case of the GFA.

4.2.4 Modified H-Infinity *A Posteriori* Risk-Sensitive Filtering Algorithm (HIRF)

This risk-sensitive filter is also based on the HI concept [5]; this feature is dependent on a real parameter, which determines whether more or less weightage should be given to higher or smaller errors in the estimation. The basic filter is modified to suit the eqns. (4.1) and (4.2). The SEG matrix propagation is obtained for each sensor as

$$\begin{aligned}P_m(k+1) &= F_0P_m(k)F_0^T + F_1P_m(k-1)F_1^T + GQG^T - F_0P_m(k)[\beta_m H_m^T\ L_m^T]R_m^{-1}\begin{bmatrix}\beta_m H_m \\ L_m\end{bmatrix} \\ &P_m(k)F_0^T - F_1P_m(k-1)[\beta_m H_m^T\ L_m^T]R_m^{-1}\begin{bmatrix}\beta_m H_m \\ L_m\end{bmatrix}P_m(k-1)F_1^T\end{aligned} \tag{4.28}$$

In eqn. (4.28) the composite-measurement covariance matrix is obtained as

$$R_m = \begin{bmatrix}R_i & 0 \\ 0 & \alpha^{-1}I\end{bmatrix} + \begin{bmatrix}\beta_m H_m \\ L_m\end{bmatrix}P_m(k)[\beta_m H_m^T\ L_m^T] \tag{4.29}$$

In eqn. (4.29), α is a risk-sensitive parameter [5]. The HI filter gain is obtained as

$$K_m = \beta_m P_m(k+1)H_m^T(\beta_m^2 H_m P(k+1)H_m^T + I)^{-1} \tag{4.30}$$

Then, the measurement update of the state is given as

$$\hat{x}_m(k+1) = \tilde{x}_m(k) + K_m(z_m(k+1) - \beta_m H_m\tilde{x}(k)) \tag{4.31}$$

In eqn. (4.31), the previous state estimate is obtained as

$$\tilde{x}_m(k) = F_0\hat{x}_m(k) + F_1\hat{x}_m(k-1) \tag{4.32}$$

The HIRF also obtains the individual state and covariance estimates for each sensor, hence it is required to obtain the fused state and fused covariance estimates, and this is given by the SVF formulae (4.16) and (4.17).

4.3 NONLINEAR OBSERVER BASED ON A CONTINUOUS-TIME DATA-FUSION FILTER

Next, a continuous-time data-fusion filter [7] is studied and a nonlinear observer is proposed that incorporates randomly missing measurements. Let the nonlinear state model be given as (considered for say one local senor only):

$$\begin{aligned} \dot{x}(t) &= f(x(t),u(t)) \\ y(t) &= \beta H x(t) \end{aligned} \tag{4.33}$$

In eqn. (4.33), often control input is ignored. The scalar quantity β is a Bernoulli sequence; in which case it could have been written as $y\,(k) = \beta\,(k)\,H\,(k) \times (k)$, without loss of any generality. A nonlinear observer for the system of eqn. (4.33) can be given as

$$\begin{aligned} \dot{\hat{x}}(t) &= f(\hat{x}(t),u(t)) + L(t)(y(t) - \hat{y}(t)) \\ \hat{y}(t) &= \beta H \hat{x}(t) \end{aligned} \tag{4.34}$$

In eqn. (4.34), $L\,(t)$ is an observer gain matrix of appropriate dimension, and is given as

$$L(t) = P(t) H^T R^{-1} \tag{4.35}$$

The matrix $P(t)$ is obtained as the solution of the observer Riccati-type differential (ORD) equation, which is based on the continuous-time data-fusion filter [7]:

$$\frac{dP^{-1}(t)}{dt} = -P^{-1}(t)A(t) - A^T(t)P^{-1}(t) + \beta^2 H^T R^{-1} H - P^{-1}(y) Q P^{-1}(t) \tag{3.36}$$

One can use $Y(t) = P^{-1}(t)$ to obtain the equivalent equation, which is called the matrix Riccati differential equation:

$$\dot{Y}(t) = -Y(t)A(t) - A^T(t)Y(t) + \beta^2 H^T R^{-1} H - Y(t) Q Y(t) \tag{3.37}$$

It is seen that the observer gain is obtained from the continuous-time gain-fusion algorithm [7]. The required Jacobian for eqn. (4.37) is obtained as

$$A(t) = \frac{\partial f(.)}{\partial \hat{x}(t)} \tag{4.38}$$

By subtracting eqn. (4.34) from eqn. (4.33), the following observer error dynamics are obtained

$$\begin{aligned} \dot{e}(t) &= A(t)e(t) - L(t)H\beta(x(t) - \hat{x}(t)) + \phi(.) \\ &= A(t)e(t) - \beta L(t)He(t) + \phi(.) \end{aligned} \tag{4.39}$$

In eqn. (4.39), new nonlinear function is

$$\phi(.) = -A(t)e(t) + f(x,t) - f(\hat{x},t) \tag{4.40}$$

The form of eqn. (4.40) is obtained by adding and subtracting the terms related to the Jacobian in the original equations of the error dynamics. In eqn. (4.40), the short forms for f and ϕ (.) can be used for simplicity

$$\phi(.) = \phi(x(t), \hat{x}(t), t); f(x,t) = f(x(t),t) \tag{3.41}$$

It is noted that the state errors are given as

$$e(t) = x(t) - \hat{x}(t) \tag{4.42}$$

4.3.1 Asymptotic Stability Analysis

It is necessary to consider the following conditions [8,9] for the study of the local asymptotic behaviour of the observer error dynamics of eqn. (4.39):

1. The solution of the matrix RTD eqn. (4.36) should be bounded

$$p_l I \leq P(t) \leq p_u I \tag{4.43}$$

In eqn. (4.43), $p_l, p_u > 0$; are constants (since $P(t)$ is theoretically, positive definite and symmetrical matrix), and are the lower and upper bounds, respectively.

2. The nonlinearity eqn. (4.41) of the error dynamics is bounded

$$\|\phi(.)\| \leq \rho_1 \|x(t) - \hat{x}(t)\|^2 \tag{4.44}$$

In eqn. (4.44), the bounding constant is taken greater than zero. Then, the nonlinear observer error dynamics eqn. (4.39) are locally asymptotically stable, if basically conditions 1 and 2 are satisfied. First, the normalized LE functional is considered to establish the asymptotic stability of the error dynamics eqn. (4.39) as

$$V(t) = e^T(t)Y(t)e(t) \tag{4.45}$$

In eqn. (4.45), $Y(t)$ is the normalizing matrix and is recognized as an information matrix, and is given as $Y(t) = P^{-1}(t)$. The matrix $P(.)$ is called Gramian matrix, and $Y(.)$ the 'information matrix', or the information Gramian. In such cases, because the deterministic observers are considered, the variables $x(.)$, and $y(.)$ are called the generalized 'random' variables [5]. It is seen that the LE functional is positive definite because it is governed by condition 1, the inequality of eqn. (4.43):

$$\frac{1}{p_u}\|e(t)\|^2 \leq e^T(t)Y(t)e(t) \leq \frac{1}{p_l}\|e(t)\|^2 \tag{4.46}$$

The time derivative of the LE functional, eqn. (4.45), under the constraints governed by error dynamics eqns. (4.39), (4.35) and (4.37), should be negative definite. This time derivative is obtained, after substituting these constraints and simplifying (without any approximations, and keeping track of matrix/vector algebra consistency) as

$$\begin{aligned}\dot{V}(t) = &-e^T(t)Y(t)QY(t)e(t) + \beta^2 e^T(t)H^T R^{-1} He(t) \\ &- \beta e^T(t)H^T L^T Ye(t) - \beta e^T(t)Y(t)LHe(t) + 2e^T(t)Y(t)\phi(.)\end{aligned} \tag{4.47}$$

The last term is so written, because the matrix $Y(.)$ is symmetrical. Now, for eqn. (4.47), it is assumed that $\|R^{-1}\| \leq 1/r; \|H^T H\| \leq h^2$, ($r$ and h being positive constants) and $\|e(t)\|^2 \leq \varepsilon^2$; and since, these are known and pre-specified quantities, or should be finite; one obtains, using the inequality from eqn. (4.44), and the observer gain from eqn. (4.35), in terms of respective norms:

$$\dot{V}(t) \le -\left\{ \frac{q_l}{p_u^2} - \frac{2\rho_1}{p_u} \|e(t)\| \right\} \|e(t)\|^2 - \frac{(2-\beta)\beta h^2}{r} \|e(t)\|^2 \tag{4.48}$$

In eqn. (4.48), q_l is the smallest (positive) eigenvalue of the matrix Q, that is positive definite and symmetric. For $\|e(t)\| \le \varepsilon = d$, the following condition from eqn. (4.48) results

$$\dot{V}(t) \le -\left\{ \frac{q_l}{p_u^2} - \frac{2\rho_1}{p_u} d \right\} \|e(t)\|^2 - \left[\frac{(2-\beta)\beta h^2}{r} \right] \|e(t)\|^2 \tag{4.49}$$

$$\dot{V}(t) \le -\left\{ d < \frac{q_l}{2\rho_1 p_u}; \beta < 2 \right\} \|e(t)\|^2 \tag{4.50}$$

Since, various constants and bounds (defined earlier) appearing in the {.;.} of eqn. (4.49) are positive, then for the specified conditions on 'd' and β in eqn. (4.50), it is seen that the time derivative of the Lyapunov energy functional is locally negative definite as in eqn. (4.50). Hence, the error dynamics of the nonlinear observer for the system with randomly missing measurements are locally asymptotically stable. This result establishes that the continuous-time global data-fusion filtering algorithm [7] that receives the estimates from the local filters, would also be asymptotically stable; because the asymptotic convergence result is based on the observer error dynamics that use the gain and the covariance matrix from the local continuous-time data-fusion filter. This result also signifies the fact that, if the local filters are asymptotically stable, then the global filter that is an optimal combination of these local filters would also be asymptotically stable. This kind of cross-utilization of the observer theoretic analysis in predicting the stability of the global data-fusion algorithm is a novel interpretation. Such studies have utilization in applications of the observers and the data-fusion algorithms in communications systems, wireless sensor networks, mechanical/aerospace engineering (target tracking), and robotics.

4.4 PERFORMANCE EVALUATION OF THE FOUR MODIFIED HI-BASED FUSION ALGORITHMS

These algorithms are implemented in MATLAB and validated using simulated data. The target tracking dynamic system model considered is given as

$$x(k+1) = F(x) + Gw(k) \tag{4.51}$$

The system matrices are

$$F = \begin{bmatrix} 1 & T \\ 0 & 1 \end{bmatrix}; \; G = \begin{bmatrix} T^2/2 \\ T \end{bmatrix} \tag{4.52}$$

The measurement model for each sensor is given as

$$z_m(k) = H_m x(k) + v_m(k) \tag{4.53}$$

For simplicity, the state delay is not considered in the MATLAB implementation; however, it can be easily incorporated. The sampling interval is 0.5 sec; and additive random noise processes are considered in the state and measurements. The two-sensor situation is considered with dissimilar measurement noise variances. The initial conditions for the states are $x(0)$=[200 0.5]. The state $x1(.)$ is a position and the state $x2(.)$ is a velocity; and position state is taken as observations. The measurement fit error and state-vector error (%) performance metrics are computed as PFE(fe)= 100×cov(measurements' error)/cov(true or actual measurements). If the covariance measure is found to be ill-conditioned, then 'norm' can be used instead. Also, the H-infinity norm is evaluated as

$$HI(norm) = \frac{\sum_{k=0}^{N} (\hat{x}^f(k) - x(k)^T(\hat{x}^f(k) - x(k))}{(\hat{x}_0^f - x)^T P_0^f(\hat{x}_0^f - x) + \sum_{k=0}^{N} w^T(k)w(k) + \sum_{i}^{m}\sum_{k=0}^{N} v_i^T(k)v_i(k)} \tag{4.54}$$

The HI norm is the ratio of the output error (state-estimate error) energy to the total input energies of the disturbances; this includes the error in the state initial condition and all the noise variances. The performance evaluation of the four filters is carried out for no measurement-data loss, and for randomly missing measurements (at the level of 0.95), for each sensor in turn. The sensors are differentiated here in terms of their different measurement variances; R1, and R2. The performance metrics are given in Table 4.1; from where it can be seen that the GFA algorithm performs better (the metrics' values are marked in bold) than the three HI-based filters; in both the situations, without and with measurement loss. This might be due to the fact that there is information feedback in GFA from the global to the local filters. The performance of the HIPF, HIRF (appropriate choice of the risk factor is required) and HIGF filtering/fusion algorithms is nearly similar in most cases, as these are based on the same theory of H-infinity.

As a typical case, Figures 4.1–4.4 show the filter performance plots for the case of HIGF when there is random data loss in sensor 1, which shows satisfactory tracking. Also, it can be seen that the fusion performance of all four algorithms, is as per the accepted inferences (from the literature on data fusion), and is very satisfactory; as can also be seen in Figure 4.3.

It has been found, the gain fusion GFA filtering-cum-fusion algorithm performs better than the other three HI-based algorithms; however, the latter three also exhibit consistent and good performance within themselves in the two cases of the data loss studied.

4.5 NONLINEAR OBSERVERS WITH STATE DELAY AND MISSING DATA

In control and estimation-cum-filtering theory, the objective of an observer is to reconstruct the state of a dynamic system using the knowledge of the system IO data/signals. For linear systems, the system states are generally estimated using the Luenberger observer or KF. For a nonlinear system one uses an extended Luenberger observer or EKF. However, there are not many results for the state estimation of nonlinear systems with delayed states and randomly missing measurements. There are real-life dynamic systems like transportation, chemical reactors, biological systems, computer networks, and communication systems, including WSNs, wherein state delays and/or missing measurements could occur; e.g. sometimes, in a data-communication channel a few or many measurements might be missing (from one or more sensors). In such cases, it becomes important to study and evaluate the performance of the data-processing algorithms in the presence (and despite) of missing measurements. That is, in the absence of certain signals/data during certain time intervals, only the random (measurement) noises are made available for these channels.

This joint/combined/synergy aspect of treating state delay and randomly missing measurements has not gained much attention in the context of nonlinear systems, though some work has been done

TABLE 4.1
Performance Metrics for the Four Modified HI-Based Fusion Filters

Cond.	GFA			HIPF			HIRF			HIGF		
No loss of data	Hnorm	Pfe S1	Pfe S2	Hnorm	Pfe S1	Pfe S2	Hnorm	Pfe S1	Pfe S2	Hnorm	Pfe S1	Pfe S2
	0.004	**0.0801**	**0.1123**	0.0051	0.0815	0.1151	0.0058	0.0821	0.1151	0.0146	0.081	0.1148
	SVE in	Pos	Vel	SVE in	Pos	Vel	SVE in	Pos	Vel	SVE in	Pos	Vel
	S1	**0.0569**	**3.746**	S1	0.078	6.241	S1	0.0798	7.600	S1	0.0757	4.176
	S2	**0.0579**	**3.666**	S2	0.0916	6.463	S2	0.0909	6.2594	S2	0.0838	4.467
	F	**0.0565**	**3.65**	F	0.0649	5.942	F	0.0687	6.5977	F	0.071	6.658
Data loss in Sensor 1:S1; F-fused	Hnorm	Pfe S1	Pfe S2	Hnorm	Pfe S1	Pfe S2	Hnorm	Pfe S1	Pfe S2	Hnorm	Pfe S1	Pfe S2
	0.0041	**0.0264**	**0.1123**	0.0051	0.0271	0.1151	0.0058	0.0273	0.1151	0.0713	0.027	0.1148
	SVE in	Pos	Vel	SVE in	Pos	Vel	SVE in	Pos	Vel	SVE in	Pos	Vel
	S1	**0.0577**	**3.7197**	S1	0.0814	5.96	S1	0.0817	7.299	S1	0.0801	4.2782
	S2	**0.0588**	**3.6391**	S2	0.0916	6.4625	S2	0.0909	6.2594	S2	0.0838	4.487
	F	**0.0573**	**3.6235**	F	0.0654	5.7935	F	0.0690	5.4209	F	0.1568	6.6116
Data loss in Sensor 2:S2; F-fused	Hnorm	Pfe S1	Pfe S2	Hnorm	Pfe S1	Pfe S2	Hnorm	Pfe S1	Pfe S2	Hnorm	Pfe S1	Pfe S2
	0.0041	**0.08**	**0.037**	0.0051	0.0815	0.0378	0.0058	0.0821	0.0378	0.0709	0.0810	0.0378
	SVE in	Pos	Vel	SVE in	Pos	Vel	SVE in	Pos	Vel	SVE in	Pos	Vel
	S1	**0.0576**	**3.724**	S1	0.0779	6.241	S1	0.0798	7.6	S1	0.0757	4.176
	S2	**0.0588**	**3.6553**	S2	0.096	6.2277	S2	0.0958	6.1444	S2	0.0901	4.637
	F	**0.0572**	**3.6342**	F	0.0648	5.7434	F	0.0690	6.4876	F	0.1563	6.5616

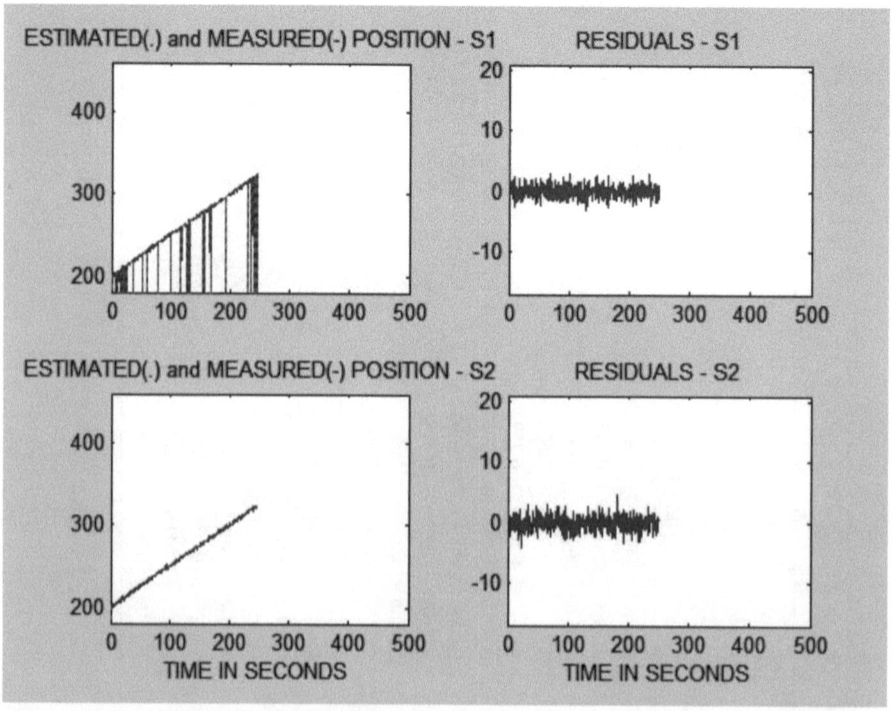

FIGURE 4.1 Time histories of states for HIGF.

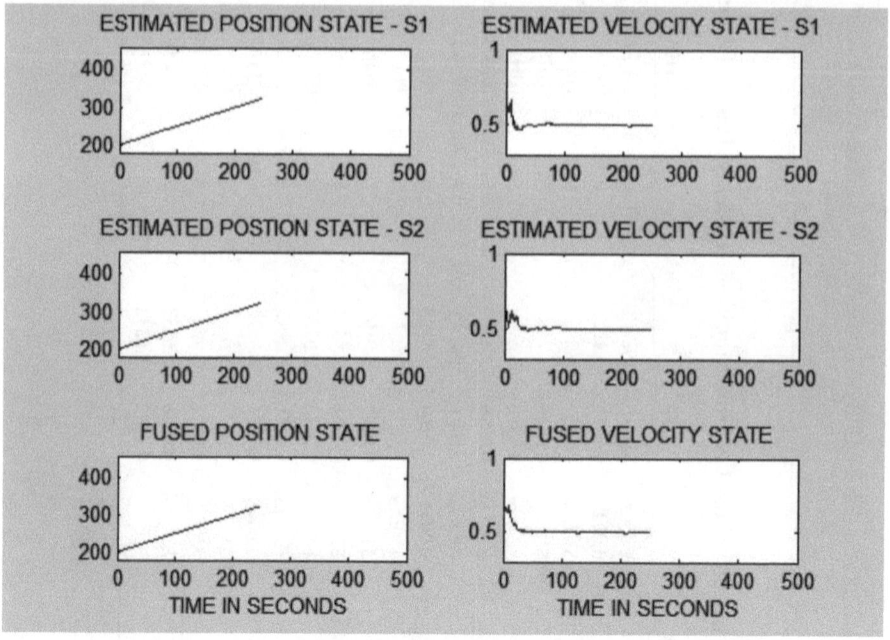

FIGURE 4.2 Measurements and residuals for HIGF.

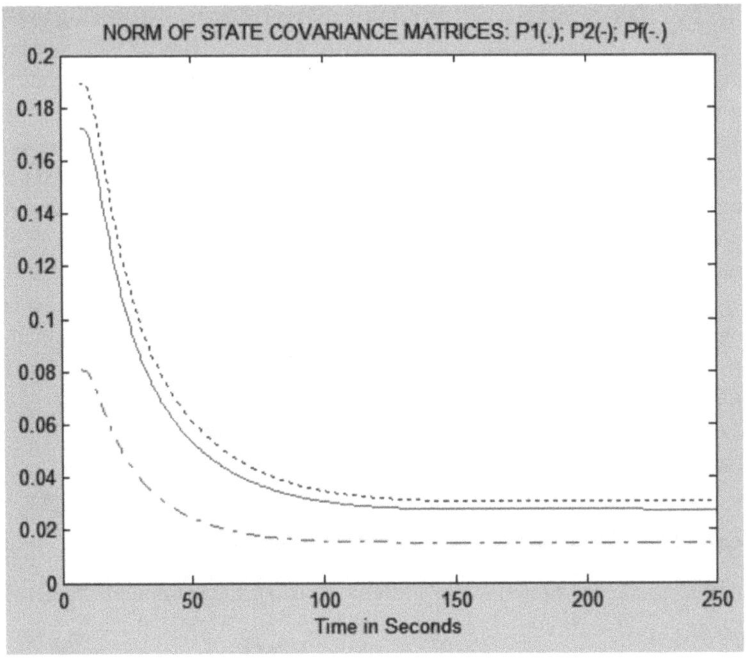

FIGURE 4.3 Norms of $P(.)$, the Gramians: individual and fused.

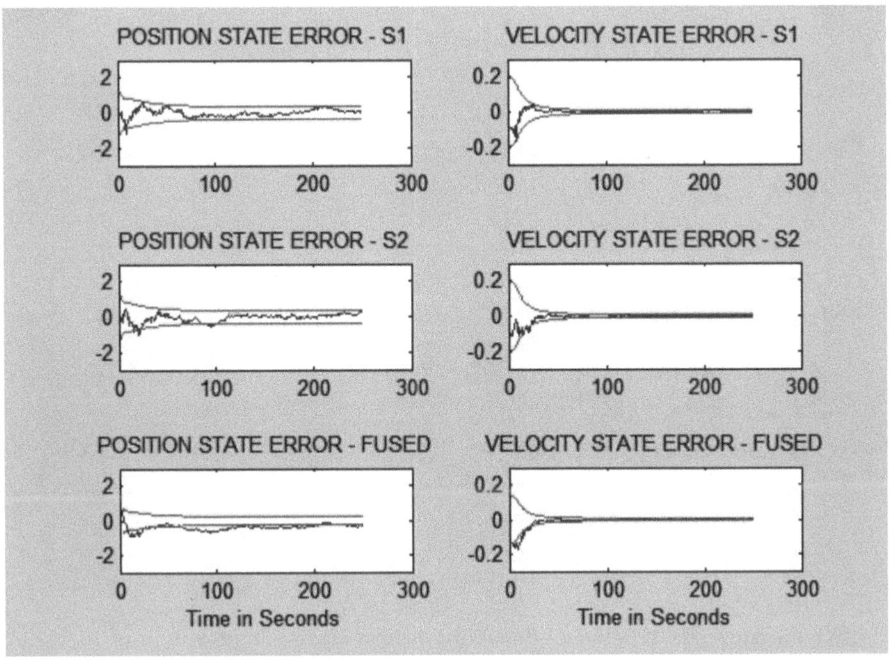

FIGURE 4.4 The state errors with their theoretical bounds.

in certain special cases: (a) availability of intermittent observations, (b) missing data in online condition monitoring and (c) some packet dropouts. Here, the state delay as well as randomly missing measurements/data in a nonlinear observer in a synergistic manner are considered. Some data may be missing due to the following: (a) a failure of a senor, and/or (b) there might be a problem in one or more communication channel/s; in such cases, the received data might be only the noise and the real signal is missing. So, it is very important to handle the situation of missing data in a nonlinear observer in some optimal way. Also, the time delay could be encountered in some real-time systems, due to latency time of certain channels. Thus, the time delay is a key factor that influences the overall system stability and performance; and if these aspects of system's state delay and missing data are not handled appropriately in a tracking algorithm, then the track would diverge. Hence, a nonlinear observer that would handle state delay as well as randomly missing measurements is presented here; also the asymptotic conditions for the observer error dynamics is derived. The performance is illustrated with implementation in MATAB.

Also, nonlinear observers for nonlinear continuous-time system with state delay and randomly missing measurements in the MLF and SVF are proposed.

4.5.1 Nonlinear System and Observer Error Dynamics

Let the nonlinear delayed state model with randomly missing measurements be given as

$$\begin{aligned} \dot{x}(t) &= f(x(t), x(t-\tau), u(t)) \\ y(t) &= \beta H x(t) \end{aligned} \tag{4.55}$$

In eqn. (4.55), often the control input is ignored. It is assumed that the nonlinear function f is continuously differentiable. Then, a nonlinear observer for the system of eqn. (4.55) is proposed as

$$\begin{aligned} \dot{\hat{x}}(t) &= f(\hat{x}(t), \hat{x}(t-\tau), u(t)) + L(t)(y(t) - \hat{y}(t)) \\ \hat{y}(t) &= \beta H \hat{x}(t) \end{aligned} \tag{4.56}$$

In eqn. (4.56), $L(t)$ is an observer gain matrix of appropriate dimension, and is given as

$$L(t) = P(t) H^T R^{-1} \tag{4.57}$$

In eqn. (4.57), R is some positive definite matrix, and $P(t)$ is obtained as the solution of the observer Riccati differential (ORD) equation

$$\dot{P}(t) = P(t) A_0^T(t) + A_0(t) P(t) - a^2 P(t) H^T R^{-1} H P(t) + S + A_1(t) A_1^T(t) \tag{4.58}$$

The observer gain is obtained from a standard EKF. The required Jacobians for eqn. (4.58) are obtained as

$$A_0(t) = \frac{\partial f(.)}{\partial \hat{x}(t)}; \quad A_1(t) = \frac{\partial f(.)}{\partial \hat{x}(t-\tau)} \tag{4.59}$$

In eqn. (4.58), the additional term arises due to the state delay. By subtracting eqn. (4.56) from eqn. (4.55) the following error dynamics are obtained:

$$\begin{aligned} \dot{e}(t) &= A_0(t) e(t) + A_1(t) e(t-\tau) - L(t) H a (x(t) - \hat{x}(t)) + \phi(.) \\ &= A_0(t) e(t) + A_1(t) e(t-\tau) - a L(t) H e(t) + \phi(.) \end{aligned} \tag{4.60}$$

In eqn. (4.60), there is a new nonlinear function as

$$\phi(.) = -A_0(t)e(t) - A_1(t)e(t-\tau) + f(x,t,t-\tau) - f(\hat{x},t,t-\tau) \tag{4.61}$$

The form of eqn. (4.61) is obtained by adding and subtracting the terms related to the Jacobians in the original equations of the error dynamics. In eqn. (4.61) the full forms for f and ϕ (.) are given for clarity

$$\phi(.) = \phi\left(x(t), x(t-\tau) - \hat{x}(t), \hat{x}(t-\tau)\right) \tag{4.62}$$

$$f(x,t,t-\tau) = f(x(t), x(t-\tau))$$
$$f(\hat{x},t,t-\tau) = f(\hat{x}(t), \hat{x}(t-\tau))$$

However, for simplicity compact form of f is used by avoiding the arguments since, these would be implied any way from its defining form eqn. (4.55), and eqn. (4.56), and $u(t)$ is also omitted. Now the state errors are given as

$$e(t) = x(t) - \hat{x}(t);\ e(t-\tau) = x(t-\tau) - \hat{x}(t-\tau) \tag{4.63}$$

4.5.2 Asymptotic Stability Analysis

The following conditions for the local asymptotic behaviour of the observer error dynamics of eqn. (4.60) are considered:

1. The solution of the ORD eqn (4.58) is bounded

$$p_l I \le P(t) \le p_u I \tag{4.64}$$

   with $p_l, p_u > 0$ as positive constants (since $P(t)$ is also theoretically, positive definite and symmetrical matrix), and are the lower and upper bounds respectively; and I is the identity matrix.

2. The nonlinearity eqn. (4.62) of the error dynamics is bounded

$$\|\phi(.)\| \le \rho_1 \left\|x(t) - \hat{x}(t)\right\|^2 + \rho_2 \left\|x(t-\tau) - \hat{x}(t-\tau)\right\|^2 \tag{4.65}$$

   with the bounding constants equal to or greater than zero. Then, the nonlinear observer error dynamics eqn. (4.60) are locally asymptotically stable, if the conditions 1 and 2 are satisfied. The following normalized LE functional to establish the asymptotic stability of the error dynamics eqn. (4.60) is considered

$$V(t) = e^T(t)Y(t)e(t) \tag{4.66}$$

In eqn. (4.66), $Y(t)$ is the normalizing/weighting matrix and can be recognized as an information matrix, it is given as $Y(t) = P^{-1}(t)$. In the case of KF, $P(t)$ is considered as the covariance matrix of the state-error vector. Here, also it can be considered so, however, since stochastic noise processes are not dealt, this matrix can be called as the information Gramian matrix, and still retain the name 'information matrix' for the inverse of matrix $P(t)$. In such a case the variables $x(.)$, and $y(.)$ can be considered as the generalized 'random' variables. It is seen that LE functional is positive definite because it is governed by condition 1, the inequality of eqn. (4.64) as follows:

$$\frac{1}{p_u}\|e(t)\|^2 \le e^T(t)Y(t)e(t) \le \frac{1}{p_l}\|e(t)\|^2 \tag{4.67}$$

The idea in obtaining the asymptotic result is that the time derivative of the LE functional, eqn. (4.66), under the constraints governed by error dynamics eqn. (4.60), and eqns. (4.57) and (4.58), should be negative definite. The time derivative is obtained as

$$\begin{aligned}\dot{V}(t) = {} & e^T(t)\dot{Y}(t)e(t) + e^T(t)\{A_0^T(t)Y(t) + Y(t)A_0(t)\\ & - bH^T L^T(t)Y(t) - bY(t)L(t)H\}e(t) + e^T(t-\tau)A_1^T(t)Y(t)e(t)\\ & + e^T(t)Y(t)A_1(t)e(t-\tau)\\ & + 2e^T(t)Y(t)\phi(.)\end{aligned} \tag{4.68}$$

Next, substitute $\dot{Y}(t) = -Y(t)\dot{P}(t)Y(t)$, and eqns. (4.57) and (4.58) in eqn. (4.68) to obtain

$$\begin{aligned}\dot{V}(t) = {} & -e^T(t)Y(t)A_1(t)A_1^T(t)Y(t)e(t) + e^T(t-\tau)A_1^T(t)Y(t)e(t) + e^T(t)Y(t)A_1(t)e(t-\tau)\\ & - e^T(t-\tau)e(t-\tau) + e^T(t-\tau)e(t-\tau) - e^T(t)Y(t)SY(t)e(t)\\ & - b^2 e^T(t)H^T R^{-1}He(t) + 2e^T(t)Y(t)\phi(.)\end{aligned} \tag{4.69}$$

In obtaining eqn. (4.69), because of the substitution of eqn. (4.57), and eqn. (4.58), several common terms cancel out. In eqn. (4.69), the term $e^T(t-\tau)e(t-\tau)$, is added and subtracted and due to the structure of the first four terms of eqn. (4.69), it can be combined in the compact form $-\|A_1^T(t)Y(t)e(t) - e(t-\tau)\|^2$, and by using this in eqn. (4.69) the following equation is obtained:

$$\begin{aligned}\dot{V}(t) = {} & -\|A_1^T(t)Y(t)e(t) - e(t-\tau)\|^2 - e^T(t)Y(t)SY(t)e(t)\\ & - b^2 e^T(t)H^T R^{-1}He(t) + e^T(t-\tau)e(t-\tau) + 2e^T(t)Y(t)\phi(.)\end{aligned} \tag{4.70}$$

The following equivalent inequality is obtained by dropping the first compact term (this does not affect the inequality) from eqn. (4.70):

$$\begin{aligned}\dot{V}(t) \le {} & -e^T(t)Y(t)SY(t)e(t) - b^2 e^T(t)H^T R^{-1}He(t)\\ & + e^T(t-\tau)e(t-\tau) + 2e^T(t)Y(t)\phi(.)\end{aligned} \tag{4.71}$$

In eqn. (4.71), it assumed that $\|R^{-1}\| \le 1/r; \|H^T H\| \le h^2$, ($r$ and h being positive constants) and $\|e(t)\|^2 \le \varepsilon^2$ and since, these are known and pre-specified quantities, or should be finite, the following equation is obtained:

$$\dot{V}(t) \le -e^T(t)Y(t)SY(t)e(t) - \frac{h^2\varepsilon^2 b^2}{r} + \|e(t-\tau)\|^2 + 2e^T(t)Y(t)\phi(.) \tag{4.72}$$

Then, using the inequality from eqn. (4.65), the following is obtained:

$$\begin{aligned}\dot{V}(t) \le {} & -e^T(t)Y(t)SY(t)e(t) + 2e^T(t)Y(t)\{\rho_1\|e(t)\|^2 + \rho_2\|e(t-\tau\|^2\}\\ & -\frac{h^2 b^2}{r}\varepsilon^2 + \|e(t-\tau\|^2\end{aligned} \tag{4.73}$$

$$\dot{V}(t) \le -e^T(t)Y(t)SY(t)e(t) + 2\left\{\frac{\rho_1}{p_u}\|e(t)\|\|e(t)\|^2 + \frac{\rho_2}{p_u}\|e(t)\|\|e(t-\tau\|^2\right\} - \frac{h^2b^2}{r}\varepsilon^2 + \|e(t-\tau\|^2 \tag{4.74}$$

$$\dot{V}(t) \le -\frac{s_l}{p_u^2}\|e(t)\|^2 + 2\left\{\frac{\rho_1}{p_u}\|e(t)\|\|e(t)\|^2 + \frac{\rho_2}{p_u}\|e(t)\|\|e(t-\tau\|^2\right\} - \frac{h^2b^2}{r}\varepsilon^2 + \|e(t-\tau\|^2 \tag{4.75}$$

$$\dot{V}(t) \le -\left\{\frac{s_l}{p_u^2} - \frac{2\rho_1}{p_u}\|e(t)\|\right\}\|e(t)\|^2 - \left\{\frac{h^2b^2}{r} - 1 - \frac{2\rho_2}{p_u}\|e(t)\|\right\}\|e(t-\tau\|^2 \tag{4.76}$$

In eqn. (4.76), s_l is the smallest (positive) eigenvalue of the matrix S, that is positive definite. For $\|e(t)\| \le \varepsilon$, the following condition is obtained from eqn. (4.76)

$$\dot{V}(t) \le -\left\{\frac{s_l}{p_u^2} - \frac{2\rho_1}{p_u}\varepsilon\right\}\|e(t)\|^2 - \left\{\frac{h^2b^2}{r} - 1 - \frac{2\rho_2}{p_u}\varepsilon\right\}\|e(t-\tau\|^2 \tag{4.77}$$

$$\dot{V}(t) \le -\left\{\frac{s_l}{2\rho_1 p_u}; \frac{(h^2b^2 - r)p_u}{2\rho_2 r}\right\}\|e(t)\|^2 \tag{4.78}$$

Since, all the constants and bounds appearing in the {.;.} in eqn. (4.78) are positive, and the bounding constants in eqn. (4.65) is positive (in the case that these are really equal to zero, the positive values are assigned, with no loss of any generality, since these constants are arbitrary, and it is always sure that $h^2b^2 > r$), the time derivative of the Lyapunov energy functional is bounded from above as in eqn. (4.78). Since, the LEF is positive definite, and its time derivative is locally negative definite as in eqn. (4.78), the observer error dynamics of the newly proposed nonlinear observer for system with state delay, and randomly missing measurements/data is locally asymptotically stable.

4.6 NONLINEAR OBSERVERS FOR SENSOR DATA FUSION

Multisensory data fusion (MSDF) is an evolving technology at software/algorithms and hardware levels and is defined as follows: an act, that could be additive, multiplicative, operative, and/or logical by which (a) the quantitative information in the sense of Fisher's information matrix is enhanced by fusing/combining data from more than one sensor or source, and/or (b) the prediction accuracy is enhanced, compared to the usage of a single sensor data or source. In this section we consider nonlinear observers for MSDF.

4.6.1 Observer for Measurement Data-Level Fusion

Here we consider the delayed state model, and missing measurements/data in the MLF. The state model and the measurement model for the two-sensor scheme is given as

$$\begin{aligned} \dot{x}(t) &= f(x(t), x(t-\tau), u(t)) \\ y^i(t) &= \beta^i H^i x(t) \end{aligned} \tag{4.79}$$

In eqn. (4.79), $y^i(t)$, $i = 1,2$ are the output from the first ($i = 1$), and the second ($i = 2$) sensors. A nonlinear observer for the system of eqn. (4.79) is specified as

$$\begin{aligned}\dot{\hat{x}}_f(t) &= f(\hat{x}_f(t), \hat{x}_f(t-\tau), u(t)) + L(t)(y(t) - \hat{y}(t)) \\ \hat{y}(t) &= \beta H \hat{x}_f(t)\end{aligned} \tag{4.80}$$

The 'f' denotes the fused state obtained as a result of the MLF of the two measurements from the two sensors, combined at the data level, and eqn. (4.80) is rewritten as

$$\begin{aligned}\dot{\hat{x}}_f(t) &= f(\hat{x}_f(t), \hat{x}_f(t-\tau), u(t)) + L_c(t)(y_{cm}(t) - aH_c\hat{x}_f(t)) \\ \hat{y}_c(t) &= aH_c\hat{x}_f(t)\end{aligned} \tag{4.81}$$

In eq. (4.81), y_{cm} and H_c are appropriate composite vectors/matrices to account for the direct data-level fusion of the measurements/data coming from the two sensors, and a^i can be appropriately accounted for. In eqn. (4.81), $L_c(t)$ is the observer gain matrix of appropriate dimension:

$$L_c(t) = P(t)H_c^{\,T}R_c^{-1} \tag{4.82}$$

In the case of the observer gain, eqn. (4.82), R_c is some positive definite matrix reflecting the combination of the measurements/data, and $P(t)$ is obtained as the solution of the ORD equation

$$\dot{P}(t) = P(t)A_0^T(t) + A_0(t)P(t) - b^2P(t)H_c^{\,T}R^{-1}H_cP(t) + S + A_1(t)A_1^T(t) \tag{4.83}$$

By subtracting eqns. (4.81) from (4.80), the following error dynamics are obtained:

$$\begin{aligned}\dot{e}_f(t) &= A_0(t)e_f(t) + A_1(t)e_f(t-\tau) - L_c(t)H_cb(x_f(t) - \hat{x}_f(t)) + \phi_f(.) \\ &= A_0(t)e_f(t) + A_1(t)e_f(t-\tau) - bL_c(t)H_ce_f(t) + \phi_f(.)\end{aligned} \tag{4.84}$$

In eqn. (4.84), 'f' denotes the fused condition, and it does not have any effect on the dimension of the state vector and hence state error, it only reflects the fact that the observer state has the combined effect of two measurements. In eqn. (4.84), the new nonlinear function is as

$$\phi_f(.) = -A_0(t)e_f(t) - A_1(t)e_f(t-\tau) + f(x_f, t, t-\tau) - f(\hat{x}_f, t, t-\tau) \tag{4.85}$$

In eqn. (4.85), the full forms for nonlinear functions and $\phi_f(.)$ are given for clarity as

$$\begin{aligned}\phi_f(.) &= \phi(x_f(t), x_f(t-\tau), \hat{x}_f(t), \hat{x}_f(t-\tau)) \\ f(x_f, t, t-\tau) &= f(x_f(t), x_f(t-\tau)) \\ f(\hat{x}_f, t, t-\tau) &= f(\hat{x}_f(t), \hat{x}_f(t-\tau))\end{aligned} \tag{4.86}$$

The state errors are given as

$$e_f(t) = x_f(t) - \hat{x}_f(t);\ \ e_f(t-\tau) = x_f(t-\tau) - \hat{x}_f(t-\tau) \tag{4.87}$$

The theoretical development of Section 4.5.2 is equally applicable to the observer error dynamics of eqn. (4.84), and hence, it is ascertained by induction that observer error dynamics of the newly proposed nonlinear observer eqn. (4.25) for systems with state delay, and randomly missing measurements/data in the data-level fusion is locally asymptotically stable.

4.6.2 Observer for State-Vector-Level Fusion

In the SVF, the measurements coming from two sensors are individually processed at each local sensor node, and then the estimated state vector is obtained from the individual estimate by the SVF formula. The nonlinear dynamics model is

$$\begin{aligned} \dot{x}^i(t) &= f^i(x^i(t), x^i(t-\tau)) \\ y^i(t) &= \beta^i H^i x^i(t) \end{aligned} \tag{4.88}$$

In eqn. (4.88), i =1, 2 are the two sensors, the measurement data from which are processed by an individual observer at each sensor node:

$$\begin{aligned} \dot{\hat{x}}^i(t) &= f^i(\hat{x}^i(t), \hat{x}^i(t-\tau)) + L^i(t)(y^i(t) - \hat{y}^i(t)) \\ \hat{y}^i(t) &= \beta^i H^i \hat{x}^i(t) \end{aligned} \tag{4.89}$$

Once, the state estimates are obtained at each sensor node (of course concurrently by two processors), these estimates can be fused by using the following formula like in the KF:

$$\hat{x}_f = \hat{x}^1 - \hat{P}^1\left(\hat{P}^1 + \hat{P}^2\right)^{-1}\left(\hat{x}^2 - \hat{x}^1\right) \tag{4.90}$$

$$\hat{P}_f = \hat{P}^1 - \hat{P}^1\left(\hat{P}^1 + \hat{P}^2\right)^{-1}\hat{P}^{1T} \tag{4.91}$$

In eqn. (4.90), the individual state estimate are obtained from the corresponding observer), that has processed the measurements from the corresponding sensor (i =1,2), and in eqn. (4.91), the Gramians (P for i = 1,2), are obtained by solving the corresponding matrix ORD equation. So, $P(.)$, i =1,2; as the weighting matrices are used in the fusion rule eqn. (4.90). Although, in the case of KF/EKF, $P(.)$ are considered as the covariance matrices; so far as the SVF is concerned, these are the appropriate weighting factors obtained from the covariance matrices, and incidentally happened to be the covariance matrices themselves. In the case of the observers, these weighting factors obtained from the Gramians (representing some uncertainty or dispersion of the estimate from the true value) are considered, and just happen to be the Gramians themselves. Again, it is ascertained that the theoretical development of Section 4.5.2 is equally applicable to the observers of eqn. (4.89), since each is an individual observer, and hence, by induction observer error dynamics of the systems with state delay, and randomly missing measurements/data for the SVF are locally asymptotically stable.

4.6.3 Performance Evaluation

The performance of the presented observer is validated using numerical simulations carried out in MATLAB. The simulations are done for a period of 4 sec with a sampling interval of 0.01 sec. The following nonlinear dynamic system are considered

$$\begin{aligned} \dot{x}_1(t) &= -(x_1(t) + 3.3)(x_1(t) + x_2(t)) \\ \dot{x}_2(t) &= -10x_2(t-\tau) + 10x_2(t) - (3x_2(t-\tau) - 10)x_1(t-\tau)) \\ y(t) &= x_1(t) \end{aligned} \tag{4.92}$$

The model in eqn. (4.92) is the prey, $x1(t)$ – predator, $x2(t)$ population dynamics model [4]. For data simulation as well as observer states, the dynamic equations are solved by using Euler integration method, and hence, these equations and the Jacobians are properly represented in the discrete form as

$$\begin{aligned} x_1(k) &= (-x_1(k-1)-3.3)(x_1(k-1)+x_2(k-1)) \\ x_2(k) &= -10x_2(k-2)+10x_2(k-1)-3x_2(k-2)x_1(k-2)+10x_1(k-2) \end{aligned} \tag{4.93}$$

$$A_0(k) = \begin{bmatrix} -2x_1(k-1)-3.3-x_2(k-1) & -x_1(k-1)-3.3 \\ 0 & 10 \end{bmatrix} \tag{4.94}$$

$$A_1(k) = \begin{bmatrix} 0 & 0 \\ -3x_2(k-2)+10 & -10-3x_1(k-2) \end{bmatrix} \tag{4.95}$$

The value of R (or 'r') is used as a tuning parameter. The state initial conditions used for the simulation and the nonlinear observer are chosen appropriately. In order to implement the observer algorithm, we need to solve the matrix ORD equation, and for this the following transformation are to be used:

$$a = P(t)d \tag{4.96}$$

to obtain the following differential equations:

$$\dot{d} = -A_0^T d + b^2 H^T R^{-1} H\, a \tag{4.97}$$

$$\dot{a} = (I + A_1 A_1^T)d + A_0 a \tag{4.98}$$

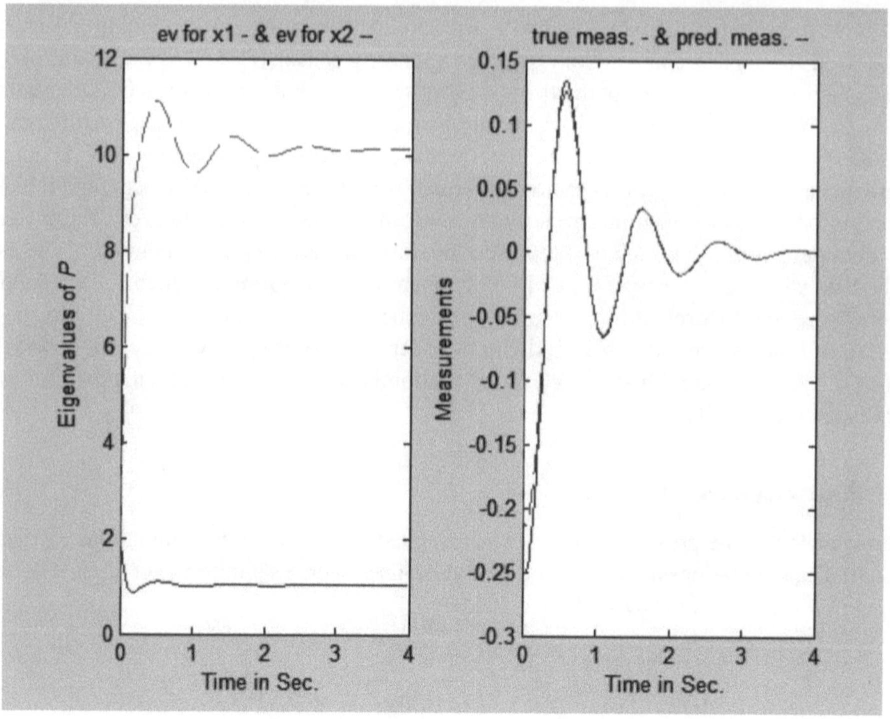

FIGURE 4.5 Time history match of the true (-) and observer states (-.): no missing data.

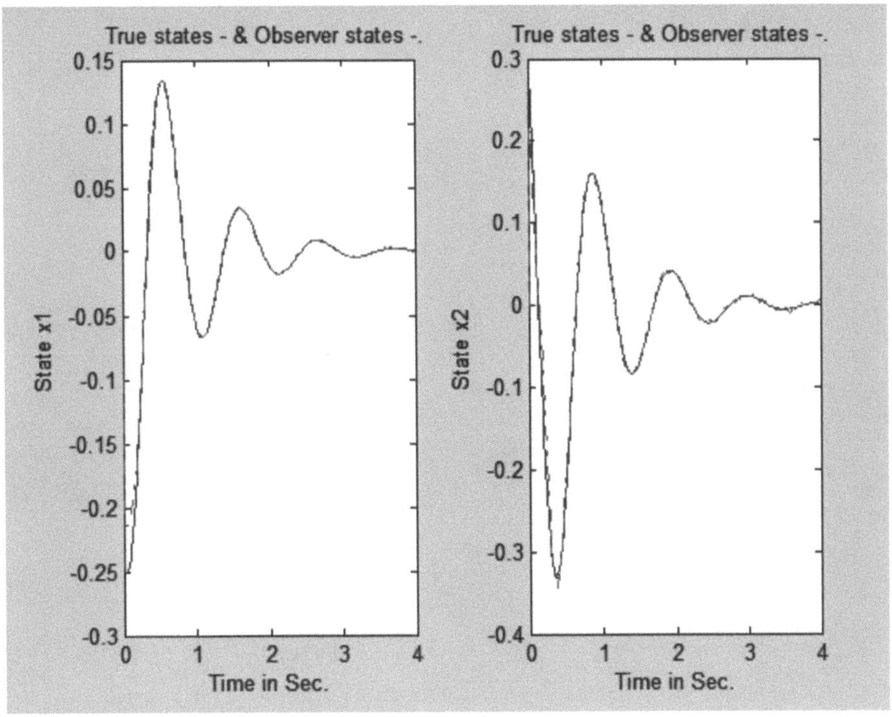

FIGURE 4.6 Eigenvalues of the P showing convergence related to state x_1 (-), and state x_2 (--); and the true (-), and predicted measurements (-.): no missing data.

Eqns. (4.97) and (4.98) are solved by using the transition matrix method, then using eqn. (4.96), $P(t)$ is obtained. The performance of the observer with no missing measurements/data is illustrated in Figure 4.5.

Figure 4.6 depicts the convergence of the eigenvalues of the matrix P (left graph). Next, the measurement data are missed at some level randomly. The performance of the observer with missing measurements/data is illustrated in Figure 4.7. Figure 4.8 depicts the convergence of the eigenvalues of the matrix P (left graph), and the true and predicted measurements when the measurement data are missing at some level.

From these results, it is clear that the proposed nonlinear observer is asymptotically stable and satisfies the condition eqn. (4.78), when $b = 1$ (no missing data), and when b = some value (some data are missing).

4.7 NONLINEAR OBSERVERS FOR DATA FUSION BASED ON THE HI NORM

Here, we consider the state delay as well as randomly missing measurements in a nonlinear observer in the HI framework, and derive asymptotic results for the convergence of the observer error dynamics.

4.7.1 Nonlinear System and Observer Error Dynamics

Let the nonlinear delayed-state model with randomly missing measurements be given as

$$\begin{aligned} \dot{x}(t) &= f(x(t), x(t-\tau), u(t)) + g(x)w(t) \\ z = h_1(x); \quad y(t) &= \beta h_2(x) + q(x)w(t) \end{aligned} \tag{4.99}$$

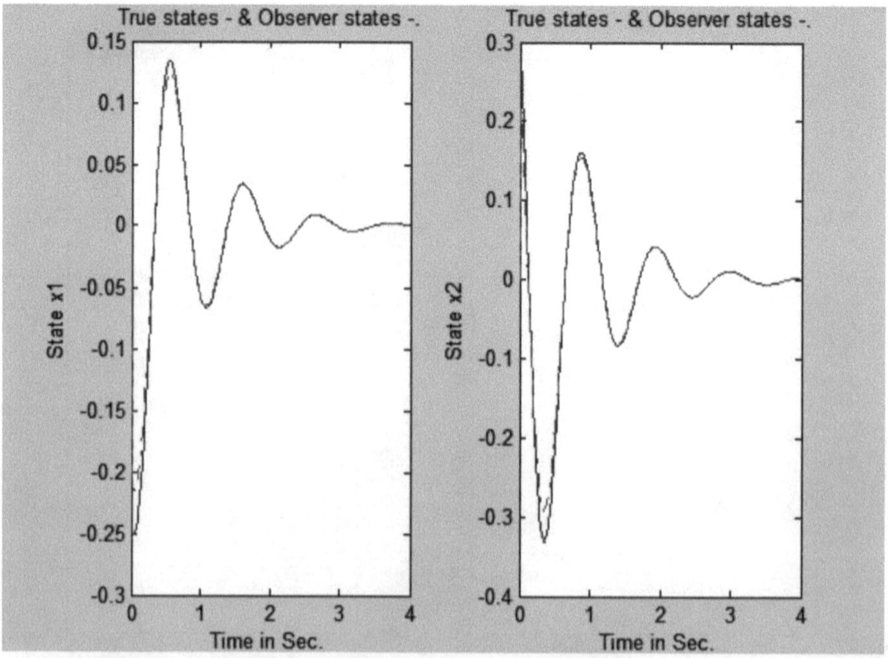

FIGURE 4.7 Time history match of the true (-) and observer (-.) states: some missing data.

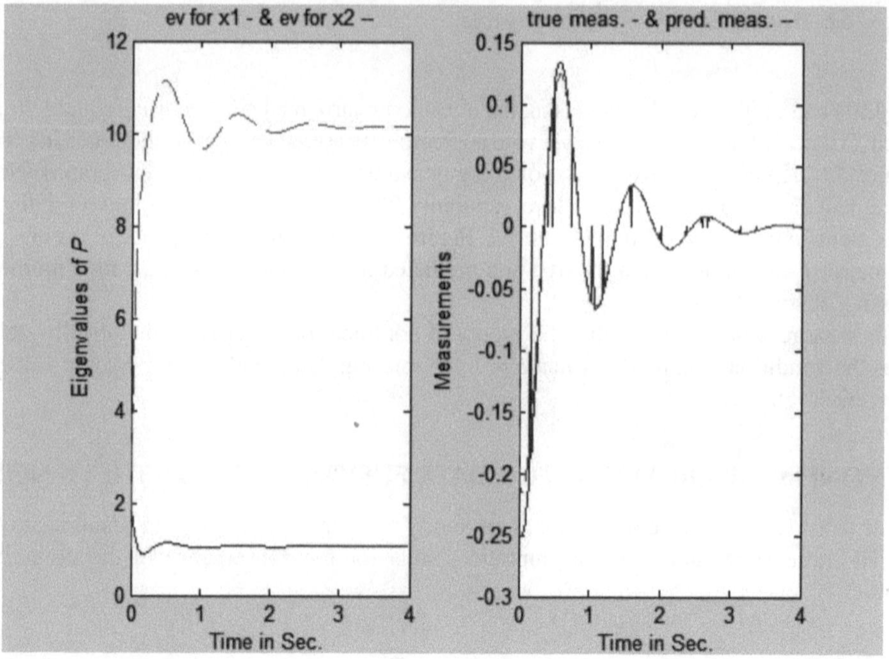

FIGURE 4.8 Eigenvalues of P showing convergence related to state x_1 (-), and state x_2 (--); and the true (-), and predicted measurements (--): some missing data.

In eqn. (4.99), the scalar quantity β is a Bernoulli sequence (in which case one could have written $y(k) = \beta(k)H(k)x(k)$, after linearization of h_2). The variables in eqn. (4.99), have usual appropriate dimensions, and presently we consider that these belong to real 2D space (say H_2 vector spaces), however, in the context of HI theory these variables are considered as generalized random variables. In eqn. (4.99), $w(.)$ is considered as an unknown disturbance, $y(.)$ is a measurement vector, and $z(.)$ is the quantity to be estimated, and is often called the estimate of the state vector x. It is also assumed that the nonlinear function f, g, h_1, h_2, and q are continuously differentiable. A nonlinear observer for the system of eqn. (4.99) is proposed as

$$\begin{aligned}\dot{\hat{x}}(t) &= f(\hat{x}(t), \hat{x}(t-\tau), u(t)) + L(t)(y(t) - \hat{y}(t)) \\ \hat{z}(t) &= h_1(\hat{x}); \quad \hat{y}(t) = \beta h_2(\hat{x})\end{aligned} \tag{4.100}$$

In eqn. (4.100), $L(t)$ is an observer gain matrix of appropriate dimension, and is determined by using the cost function based on the HI norm concept. The idea is to seek that the L_2 gain from the input disturbance energy to the output estimation-error energy obeys the following inequality:

$$\int_{t_0}^{T} \|\tilde{z}(\tau)\|^2 \, d\tau \leq \int_{t_0}^{T} \|w(\tau)\|^2 \, d\tau, \; T > t_0 \tag{4.101}$$

In eqn. (4.101), the estimation error is defined as

$$\tilde{z}(t) = z(t) - \hat{z}(t) = h_1(x) - h_1(\hat{x}) \tag{4.102}$$

The cost function for eqn. (4.101) can be formulated as

$$J(w, L) \triangleq \frac{1}{2}\int_{t_0}^{T} (\|\tilde{z}(t)\|^2 - \gamma^2 \|w(t)\|^2) dt \tag{4.103}$$

In eqn. (4.105), γ (>0) specifies the error energy gain from the input to the output and can be considered as a tuning parameter. This means that the actual error energy gain is bounded from above by the square of this factor, so the observer is not optimal but it is sub-optimal, and yet robust. Finally, the observer gain that includes the missing data factor can be chosen as

$$L(t) = aP(t)H^T(t) \tag{4.104}$$

In eqn. (4.104), $P(t)$ is obtained as the solution of the observer Riccati differential (ORD) equation

$$\begin{aligned}\dot{P}(t) &= P(t)A_0(t) + A_0^T(t)P(t) + P(t)\left[\frac{1}{\gamma^2}C^T(t)C(t) - L^T(t)L(t)\right] \\ &\quad P(t) + A_1(t)A_1^T(t) + B(t)B^T(t); \\ &\quad \text{with } P(0) = P_0; P(T) = 0\end{aligned} \tag{4.105}$$

Also, in eqn. (4.104) matrix R that is found in the KF gain formula ($P(t)H^T(t)R^{-1}$) is not present; however it can be easily incorporated for the sake of tuning the observer, and can be regarded as a weighting matrix/factor. In that case the observer gain would be $L(t) = bP(t)H^T(t)R^{-1}$, and this would not pose any problem in the proposed derivation of the asymptotical result for this nonlinear observer. The alternative form of the ORD eqn (4.104) is

$$\dot{P}(t) = P(t)A_0^T(t) + A_0(t)P(t) + P(t)\left[\frac{1}{\gamma^2}C^T(t)C(t) - b^2H^T(t)H(t)\right]$$
$$P(t) + A_1(t)A_1^T(t) + B(t)B^T(t) \tag{4.106}$$

Now, the observer gain eqn. (4.104) is based eqn. (4.105) and hence, in turn on the H-infinity norm-cum-filtering theory. Various Jacobians needed in eqn. (4.106) are obtained as

$$A_0(t) = \frac{\partial f(.)}{\partial \hat{x}(t)};\quad A_1(t) = \frac{\partial f(.)}{\partial \hat{x}(t-\tau)};$$

$$B(t) = \frac{\partial g(.)}{\partial \hat{x}(t)};\quad C(t) = \frac{\partial h_1(.)}{\partial \hat{x}(t)};\quad H(t) = \frac{\partial h_2(.)}{\hat{x}(t)};\quad D(t) = \frac{\partial q(.)}{\partial \hat{x}(t)} \tag{4.107}$$

In eqn. (4.106), an additional term, $A_1(.)$ arises due to the state delay. One can see that although various quantities are a function of ($\hat{x}$, t), the dependence only on t for simplicity is mentioned. By subtracting eqn. (4.100) from eqn. (4.99) one obtains the following error dynamics:

$$\begin{aligned}\dot{e}(t) &= A_0(t)e(t) + A_1(t)e(t-\tau) - L(t)H(t)b(x(t) - \hat{x}(t)) + \phi(.) + (B(t)x - L(t)D(t)x)w \\ &= A_0(t)e(t) + A_1(t)e(t-\tau) - bL(t)H(t)e(t) + \phi(.) + (B(t) - L(t)D(t))xw\end{aligned} \tag{4.108}$$

Then drop the last term from eqn. (4.108), to obtain the error dynamics for the observer without the disturbance term; also the nonlinear function is

$$\phi(.) = -A_0(t)e(t) - A_1(t)e(t-\tau) + f(x,t,t-\tau) - f(\hat{x},t,t-\tau) \tag{4.109}$$

The full forms of f and ϕ (.) are given for clarity

$$\phi(.) = \phi\big(x(t), x(t-\tau) - \hat{x}(t), \hat{x}(t-\tau)\big) \tag{4.110}$$

$$\begin{aligned}f(x,t,t-\tau) &= f(x(t), x(t-\tau)) \\ f(\hat{x},t,t-\tau) &= f(\hat{x}(t), \hat{x}(t-\tau))\end{aligned} \tag{4.111}$$

However, for simplicity one can use a very compact form of f by avoiding the arguments.

The state errors are given as

$$e(t) = x(t) - \hat{x}(t);\quad e(t-\tau) = x(t-\tau) - \hat{x}(t-\tau) \tag{4.112}$$

4.7.2 Asymptotic Results

Let us consider the following conditions for the local asymptotic behaviour of the observer error dynamics of eqn. (4.108). The solution of the ORD eqn. (4.106) is considered bounded

$$p_l I \le P(t) \le p_u I \tag{4.113}$$

with $p_l, p_u > 0$ as positive constants (since $P(t)$ is also theoretically, positive definite and symmetric matrix), and p_l, p_u are the lower and upper bounds, respectively; and I is the identity matrix. The nonlinearity in the error dynamics is bounded

$$\|\phi(.)\| \le \rho_1 \left\|x(t) - \hat{x}(t)\right\|^2 + \rho_2 \left\|x(t-\tau) - \hat{x}(t-\tau)\right\|^2 \tag{4.114}$$

with the bounding constants equal to or greater than zero. Then, the nonlinear observer error dynamics are locally asymptotically stable, if conditions 1 and 2 are satisfied. First, consider the following normalized LE functional to establish the asymptotic stability of the error dynamics:

$$V(t) = e^T(t)Y(t)e(t) \tag{4.115}$$

$Y(t)$ is the normalizing/weighting matrix and can be recognized as an information matrix/Gramian, and is given as $Y(t)=P^{-1}(t)$. From condition 1, the inequality follows

$$\frac{1}{p_u}\|e(t)\|^2 \le e^T(t)Y(t)e(t) \le \frac{1}{p_l}\|e(t)\|^2 \tag{4.116}$$

For simplicity, write the error dynamics as

$$\dot{e}(t) = A_0 e(t) + A_1 e(t-\tau) - bLHe(t) + \phi(.) + (B - LD)xw \tag{4.117}$$

Then the time derivative of eqn. (4.115) is given as

$$\dot{V}(t) = e^T(t)\dot{Y}(t)e(t) + e^T(t)Y(t)\dot{e}(t) + \dot{e}^T(t)Y(t)e(t)$$

$$\begin{aligned}\dot{V}(t) = {} & e^T(t)\dot{Y}(t)e(t) + e^T(t)\{A_0^T Y(t) + Y(t)A_0 \\ & - aH^T L^T Y(t) - aY(t)LH\}e(t) + e^T(t-\tau)A_1^T Y(t)e(t) \\ & + e^T(t)Y(t)A_1 e(t-\tau) \\ & + 2e^T(t)Y(t)\phi(.) + 2e^T(t)Y(t)(B - LD)xw\end{aligned} \tag{4.118}$$

Following the procedure of the previous sections, we obtain

$$\begin{aligned}\dot{V}(t) = {} & -e^T(t)Y(t)A_1(t)A_1^T(t)Y(t)e(t) + e^T(t-\tau)A_1^T(t)Y(t)e(t) + e^T(t)Y(t) \\ & A_1(t)e(t-\tau) - e^T(t-\tau)e(t-\tau) + e^T(t-\tau)e(t-\tau) - e^T(t) \\ & Y(t)BB^T Y(t)e(t) + \frac{1}{\gamma^2}e(t)C^T Ce(t) - a^2 e^T(t)H^T He(t) + 2e^T(t)Y(t)\phi(.) \\ & + 2e^T(t)Y(t)(B - LD)xw\end{aligned} \tag{4.119}$$

With further simplification without any approximation one obtains

$$\begin{aligned}\dot{V}(t) = {} & -\left\|A_1^T(t)Y(t)e(t) - e(t-\tau)\right\|^2 - e^T(t)Y(t)SY(t)e(t) \\ & + \frac{1}{\gamma^2}e^T(t)C^T Ce(t) - a^2 e^T(t)H^T He(t) + e^T(t-\tau)e(t-\tau) \\ & + 2e^T(t)Y(t)\phi(.) + 2e^T(t)(Y(t)B - aH^T D)xw\end{aligned} \tag{4.120}$$

In eqn. (4.120), $S = BB^T$, then, get the following equivalent inequality by dropping the first compact term (this does not affect the inequality), we get

$$\dot{V}(t) \le -e^T(t)Y(t)SY(t)e(t) + \frac{1}{\gamma^2} e(t)C^T Ce(t) - a^2 e^T(t)H^T He(t) \\ + e^T(t-\tau)e(t-\tau) + 2e^T(t)Y(t)\phi(.) + 2e^T(t)\{Y(t)B - bH^T D)xw \tag{4.121}$$

Now, in eqn. (4.121), assume that $\|C^T C\| \le c^2$; $\|H^T H\| \le h^2$; (c and h being positive constants), $\|e(t)\|^2 \le \varepsilon^2$, and since, these are known and pre-specified quantities, or should be finite, one obtains

$$\dot{V}(t) \le -e^T(t)Y(t)SY(t)e(t) + \frac{1}{\gamma^2} c^2\varepsilon^2 - b^2h^2\varepsilon^2 + \|e(t-\tau)\|^2 \\ + 2e^T(t)Y(t)\phi(.) + 2e^T(t)\{Y(t)B - bH^T D\}xw \tag{4.122}$$

$$\dot{V}(t) \le -e^T(t)Y(t)SY(t)e(t) + 2\left\{\frac{\rho_1}{p_u}\|e(t)\|\|e(t)\|^2 + \frac{\rho_2}{p_u}\|e(t)\|\|e(t-\tau\|^2\right\} \\ + \frac{1}{\gamma^2}c^2\varepsilon^2 - b^2h^2\varepsilon^2 + \|e(t-\tau\|^2 \\ + 2e^T(t)\{Y(t)B - bH^T D\}xw$$

$$\dot{V}(t) \le -\frac{s_l}{p_u^2}\|e(t)\|^2 + 2\{\frac{\rho_1}{p_u}\|e(t)\|\|e(t)\|^2 + \frac{\rho_2}{p_u}\|e(t)\|\|e(t-\tau\|^2\} + \frac{1}{\gamma^2}c^2\varepsilon^2 - a^2h^2\varepsilon^2 \\ + \|e(t-\tau\|^2 + 2e^T(t)\{Y(t)B - aH^T D\}xw \tag{4.123}$$

$$\dot{V}(t) \le -\left\{\frac{s_l}{p_u^2} - \frac{2\rho_1}{p_u}\|e(t)\|\right\}\|e(t)\|^2 - \left\{a^2h^2 - 1 - \frac{1}{\gamma^2}c^2 - \frac{2\rho_2}{p_u}\|e(t)\|\right\} \\ \|e(t-\tau\|^2 + 2e^T(t)\{Y(t)B - aH^T D\}xw \tag{4.124}$$

In eqn. (4.123), s_l is the smallest (positive) eigenvalue of the matrix S, that is positive definite. For $\|e(t)\| \le \varepsilon$, the following condition from eqn. (4.124) is obtained

$$\dot{V}(t) \le -\left\{\frac{s_l}{p_u^2} - \frac{2\rho_1}{p_u}\varepsilon\right\}\|e(t)\|^2 - \left\{a^2h^2 - 1 - \frac{1}{\gamma^2}c^2 - \frac{2\rho_2}{p_u}\varepsilon\right\}\|e(t-\tau\|^2 \\ +2\varepsilon\left(\frac{B}{p_u} - aH^T D\right)xw \tag{4.125}$$

$$\dot{V}(t) \le -\{\frac{s_l}{2\rho_1 p_u}; \frac{(\gamma^2(a^2h^2-1) - c^2)p_u}{2\gamma^2\rho_2}\}\|e(t)\|^2 \tag{4.126}$$

In eqn. (4.125), it is assured that the last term in the parenthesis is zero, by having $B = bH^T Dp_u$ (or alternatively one can heuristically assume that the estimation error, the state and the unknown disturbance are uncorrelated, and drop the last term) and since, all the constants and bounds appearing in the {.;.} in eqn. (4.126) are positive, and also, regarding the other bounding constants as positive (in the case that these are really equal to zero, one can assign them slightly positive values, without loss of any generality; and since these constants are arbitrary, and one can always assure that

$\gamma^2 h^2 a^2 > \gamma^2 + c^2$), it is seen that the time derivative of the Lyapunov energy functional is bounded from above as in eqn. (4.126). Since, the Lyapunov energy functional is positive definite, and its time derivative is locally negative definite as in eqn. (4.126), the observer error dynamics of the newly proposed HI-based nonlinear observer for the system with state delay, and randomly missing measurement data is locally asymptotically stable.

4.8 NONLINEAR OBSERVERS FOR DATA FUSION BASED ON HI SCHEMES

We present the nonlinear observers based on the HI scheme for two levels of MSDF.

4.8.1 Measurement Data-Level Fusion

Consider the model with state delay, and missing data in the measurement data-level fusion. The state-space model and the measurement model for the two-sensor scheme is made simpler and given as

$$\begin{aligned} \dot{x}(t) &= f(x(t), x(t-\tau), u(t)) + g(x)w(t) \\ z &= h_1(x); \quad y^i(t) = \beta^i h_2^i(x) \end{aligned} \tag{4.127}$$

A nonlinear observer for the system of eqn. (4.127) can be specified as

$$\begin{aligned} \dot{\hat{x}}_f(t) &= f(\hat{x}_f(t), \hat{x}_f(t-\tau), u(t)) + L_c(t)(y_{cm}(t) - \hat{y}_c(t)) \\ \hat{y}_c(t) &= \beta_c H_c \hat{x}_f(t) \end{aligned} \tag{4.128}$$

In eqn. (4.128) the subscript '*f*' denotes the fused state obtained as a result of the MLF of the two measurements from the two sensors, combined at the data level, and eqn. (4.128) is rewritten as

$$\begin{aligned} \dot{\hat{x}}_f(t) &= f(\hat{x}_f(t), \hat{x}_f(t-\tau), u(t)) + L_c(t)(y_{cm}(t) - aH_c\hat{x}_f(t)) \\ \hat{y}_c(t) &= aH_c\hat{x}_f(t) \end{aligned} \tag{4.129}$$

In eqn. (4.128), y_{cm} and H_c are appropriate composite vectors/matrices, which account for the direct data-level fusion (MLF) of the measurement data coming from the two sensors, and a^i can be appropriately accounted for; $L_c(t)$ is the observer gain matrix of appropriate dimension, and is given as per the HI theory as seen in the previous section

$$L_c(t) = P_f(t) H_c^T \tag{4.130}$$

In eqn. (4.130), $P_f(t)$ is obtained as the solution of the ORD equation

$$\begin{aligned} \dot{P}_f(t) = P_f(t)A_0^T(t) + A_0(t)P_f(t) + P_f(t)\left[\frac{1}{\gamma^2}C^T(t)C(t) - a^2 H_c^T(t)H_c(t)\right] \\ P_f(t) + A_1(t)A_1^T(t) + S \end{aligned} \tag{4.131}$$

One obtains the following error dynamics

$$\begin{aligned} \dot{e}_f(t) &= A_0(t)e_f(t) + A_1(t)e_f(t-\tau) - L_c(t)H_c(t)b(x_f(t) - \hat{x}_f(t)) + \phi_f(.) + B(t)x_f w \\ &= A_0(t)e_f(t) + A_1(t)e_f(t-\tau) - bL_c(t)H_c(t)e_f(t) + \phi_f(.) + B(t)x_f w \end{aligned} \tag{4.132}$$

One has the nonlinear function as

$$\phi_f(.) = -A_0(t)e_f(t) - A_1(t)e_f(t-\tau) + f(x_f,t,t-\tau) - f(\hat{x}_f,t,t-\tau) \tag{4.133}$$

$$\begin{aligned} \phi_f(.) &= \phi(x_f(t), x_f(t-\tau), \hat{x}_f(t), \hat{x}_f(t-\tau)) \\ f(x_f,t,t-\tau) &= f(x_f(t), x_f(t-\tau)) \\ f(\hat{x}_f,t,t-\tau) &= f(\hat{x}_f(t), \hat{x}_f(t-\tau)) \end{aligned} \tag{4.134}$$

Note that the state errors are given as

$$e_f(t) = x_f(t) - \hat{x}_f(t); \; e_f(t-\tau) = x_f(t-\tau) - \hat{x}_f(t-\tau) \tag{4.135}$$

One can observe that the theoretical development of previous sections is equally applicable to the observer error dynamics of eqn. (4.132), and hence, one infers and ascertains by induction that observer error dynamics of the newly proposed nonlinear observer based on H-infinity filter theory for systems with state delay, and randomly missing measurement data in the data-level fusion is locally asymptotically stable.

4.8.2 State-Vector-Level Fusion

Consider in the SVF, that the measurements coming from two sensors are separately processed at each local sensor node, and then the estimated state vector is obtained from the individual estimates by the SVF formula. Consider the nonlinear dynamics model

$$\begin{aligned} \dot{x}^i(t) &= f^i(x^i(t), x^i(t-\tau)) + g(x^i)w(t) \\ z^i &= h_1^i(x^i); \quad y^i(t) = \beta^i h_2^i(x^i) \end{aligned} \tag{4.136}$$

The observer is given as

$$\begin{aligned} \dot{\hat{x}}^i(t) &= f^i(\hat{x}^i(t), \hat{x}^i(t-\tau)) + L^i(t)(y^i(t) - \hat{y}^i(t)) \\ \hat{z}^i &= h_1^i(\hat{x}^i); \; \hat{y}^i(t) = \beta^i h_2{}^i \hat{x}^i(t) \end{aligned} \tag{4.137}$$

After the state estimates are obtained by each sensor node (of course concurrently by two processors), one can fuse these state estimates by using the formula as is done in the case of KF SVF

$$\hat{x}_f = \hat{x}^1 + \hat{P}^1\left(\hat{P}^1 + \hat{P}^2\right)^{-1}\left(\hat{x}^2 - \hat{x}^1\right) \tag{4.138}$$

$$\hat{P}_f = \hat{P}^1 - \hat{P}^1\left(\hat{P}^1 + \hat{P}^2\right)^{-1}\hat{P}^{1T} \tag{4.139}$$

In eqn. (4.138) one has the individual state estimates obtained from the corresponding observer that has processed the measurements from the corresponding sensor ($i = 1,2$), and in eqn. (4.139), one has the Gramians (P for $i = 1,2$), obtained by solving the corresponding matrix ORD equation. Hence, here, look upon $P(.)$, $i = 1,2$; as the weighting matrices used in the fusion rule eqn. (4.138). In the case of KF/EKF, $P(.)$ are considered as the covariance matrices; however, for the SVF these are the appropriate weighting factors/matrices obtained from the covariance matrices, and happen to be the covariance matrices themselves. In the case of the nonlinear observers, one can consider these weighting factors/matrices as obtained from the Gramians (representing some uncertainty or dispersion of the estimate from the true value), and just happen to be the Gramians themselves, from the H-infinity filtering theory. Here, also, it is ascertained that the theoretical development of Section 4.3 is equally applicable to the observers of eqn. (4.137), since each is an individual observer and hence, by straightforward induction, the observer error dynamics of the systems with state delay, and randomly missing measurement data for the state-vector fusion are locally asymptotically stable.

4.8.3 Performance Evaluation

The presented HI-based nonlinear observer is validated using numerical simulations carried out in MATLAB. The simulations are done for a period of 4 sec with a sampling interval of 0.01 sec. Consider the following nonlinear dynamic system for simulation purpose:

$$\begin{aligned} \dot{x}_1(t) &= -(x_1(t)+3.3)(x_1(t)+x_2(t)) \\ \dot{x}_2(t) &= -10x_2(t-\tau)+10x_2(t)-(3x_2(t-\tau)-10)x_1(t-\tau)) \\ y(t) &= x_1(t) \end{aligned} \tag{4.140}$$

For simulation as well as observer states, the dynamic equations are solved by using the Euler integration method, and hence, these equations and the Jacobians are appropriately represented in the discrete form as

$$\begin{aligned} x_1(k) &= (-x_1(k-1)-3.3)(x_1(k-1)+x_2(k-1)) \\ x_2(k) &= -10x_2(k-2)+10x_2(k-1)-3x_2(k-2)x_1(k-2)+10x_1(k-2) \end{aligned} \tag{4.141}$$

$$A_0(k) = \begin{bmatrix} -2x_1(k-1)-3.3-x_2(k-1) & -x_1(k-1)-3.3 \\ 0 & 10 \end{bmatrix} \tag{4.142}$$

$$A_1(k) = \begin{bmatrix} 0 & 0 \\ -3x_2(k-2)+10 & -10-3x_1(k-2) \end{bmatrix} \tag{4.143}$$

The state initial conditions for the simulation and the observer are chosen appropriately. To implement the observer algorithm, the matrix ORD equation is solved using the transition matrix method. The performance of the observer with missing measurements at some level is illustrated in Figure 4.9a. Figure 4.9b depicts the convergence of the eigenvalues of the matrix P (left graph), and the predicted measurements (--); with missing data (level 20%, no gamma factor used), true and predicted measurements. The performance of the observer with missing measurement data at some level and with the gamma factor used is illustrated in Figure 4.9c. Figure 4.9d depicts the convergence of the eigenvalues of the matrix P (left graph), and the true and predicted measurements when the measurement data are missing at some level, and with gamma factor used. Table 4.1 shows the percentage fit errors for the states x_1, and x_2 and the measurement data fit error computed by using the formula: PFE = 100 × norm(error)/norm (true signal).

From Figures 4.9a–4.9d, it is clear that the proposed nonlinear observer is asymptotically stable when some measurement data are missing at certain level. From Table 4.2 it can be seen that the PFEs of the HI-based nonlinear observer are nearly similar or slightly more compared to the nonlinear observer when no gamma factor is used; in some cases it is even lesser, as shown in bold numbers. The case of the no gamma factor corresponds to the nonlinear observer that is based on the usage of observer gain from the continuous time KF theory. Obviously, the measurements' PFEs show higher values when these data are missing. The studies presented here pave the way for bringing synergy in the theory of observers, EKF and H_∞ filtering.

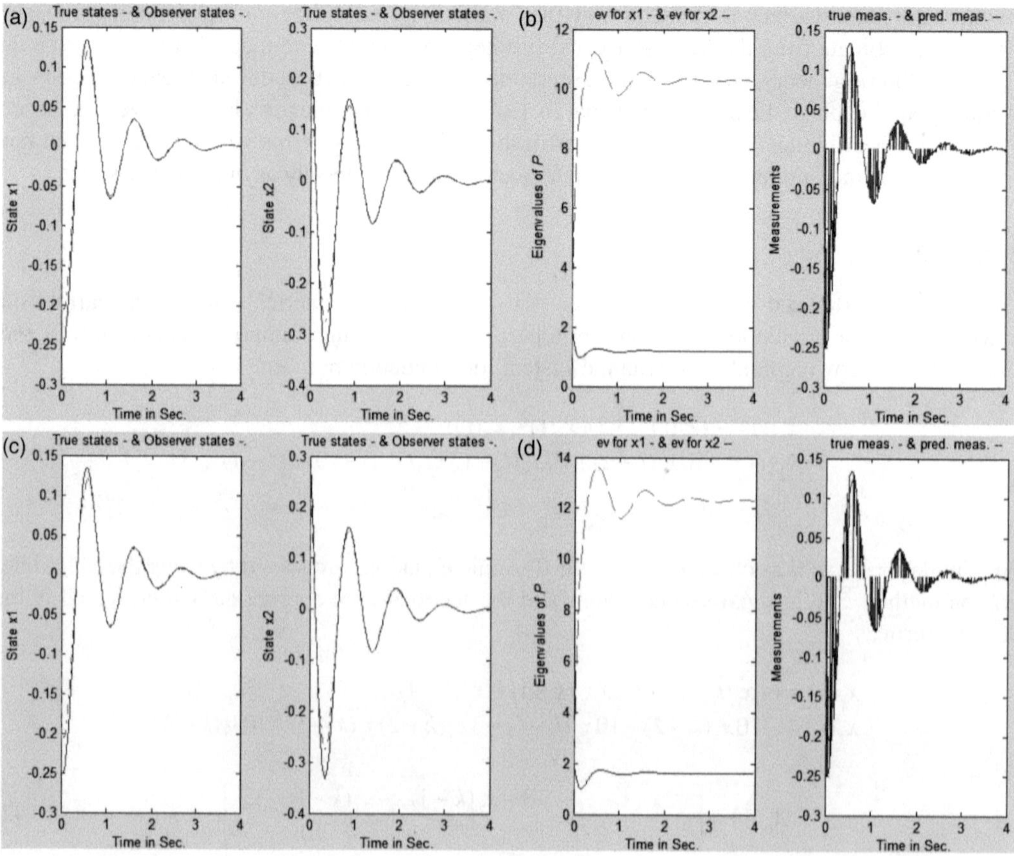

FIGURE 4.9a Time history match of the true (-) and observer states (-.): with missing data (level 20%, no gamma factor used).

FIGURE 4.9b Eigenvalues of the P showing convergence of the condition of asymptotic stability, related to state x_1 (-), and state x_2 (--), left graph; the right graph: the true (-) and predicted measurements (--).

FIGURE 4.9c Time history match of the true (-) and observer (-.) states: with missing data (level 20%, with gamma factor used).

FIGURE 4.9d Eigenvalues of the P showing convergence of the condition of asymptotic stability, related to state x_1 (-), and state x_2 (--), the left graph; the right graph: true (-) and predicted measurements (--); with missing data (level 20%, with gamma factor used).

TABLE 4.2
PFE Metrics for the Nonlinear Observers

	No Gamma Factor Used			Gamma Factor Used $\gamma > 0$ (HI-Based Observer)		
Data Missing	**State x_1**	**State x_2**	**Meas. y**	**State x_1**	**State x_2**	**Meas. y**
No	4.32	4.95	6.2	4.96	**4.6**	6.6
Yes, 5 %	4.2	4.76	23.4	4.94	**4.3**	23.61
Yes, 20 %	5.86	7.32	49.7	**5.5**	**5.6**	50.35

REFERENCES

1. Raol, J.R., *Data Fusion Mathematics – Theory and Practice*, CRC Press, Florida, July 2015.
2. Nelwamondo, F.V., and Marwala, T., Techniques for handling missing data: applications to online condition monitoring. *International Journal of Innovative Computing Information and Control*, Vol. 4, No. 6, pp. 1507–1526, 2008.
3. Shivashankarappa, N., Synergy of Out of Sequence Measurements and Missing Data in Wireless Sensor Network. Ph.D. Thesis, Dept. of Electronics Engineering, Jain University, Bangalore, India, Jan. 2018.
4. Paik, B.S., and Oh, J.H. Gain fusion algorithm for decentralized parallel Kalman filters, *IEE Proceedings – Control Theory and Applications*, Vol. 147, No. 1, pp. 97–103, Jan. 2000.
5. Hassibi, B., Sayad, A.H., and Kailath, T., Linear estimation in Krein spaces-Part II: Applications, *IEEE Transactions on Automatic Control*, Vol. 41, No. 1, pp. 34–49, Jan. 1996.
6. Jin, S.H., Park, J.B., Kim, K.K., and Yoon, T.S., Kreing space approach to decentralized H-Infintity estimation, *IEE Proceedings – Control Theory and Applications*, Vol. 148, No. 6, pp. 502–508, Nov. 2001.
7. Vershinin, Y.A., A data fusion algorithm for multisensor systems. In *Proc. of the 5th International Conference on Information Fusion, Annapolis, MD, USA, 2002*, pp. 341–345, July 2002.
8. Raff, T., and Allgower, F., An EKF-based observer for nonlinear time-delay systems. In *Proc. of the American Control Conference, Minneapolis, MN, USA, 2006*, Vol. 4, pp. 3130–3133, July 2006, http://bibbase.org/network/publication/raff-allgwer-anekfbasedobserverfornonlineartimedelay-systems-2006, accessed January 2016.
9. Reif, K., Sonnemann, F., and Unbehauen, R., An EKF-based nonlinear observer with a prescribed degree of stability, Automatica, Vol. 34, pp. 1119–1123, Sept. 1998, www.researchgate.net/publication/ 220157426, accessed January 2017.

5 Gaussian-Sum Extended Kalman Filter with Lyapunov Stability Analysis*

5.1 INTRODUCTION

In many aviation scenarios, a reasonable understanding of control and aircraft dynamics to improve the overall performance of modern aircraft systems is needed. Aircraft/aerodynamic parameter estimation (APE) is a challenging research area and is an integral part of aircraft-system identification (SID). It is the process of finding the optimal partial derivatives of the aerodynamic coefficients in the flight mechanical model of an airplane. The information of these coefficients helps in flight testing-based data analysis, and verification of an overall aircraft-system performance [1,2]. Also, the estimation of these parameters is of great importance in the design, flight test-based clearance and initial operational certification (IOC) processes of an aircraft, especially for fighter aircraft. Improved measurement techniques and faster data-processing capabilities would aid further to the success of flight vehicle SID; other factors that would contribute to the SID are advanced developments in the fields of state/parameter estimation, filtering, and control theory [1,3,4]. For a linear state-space model and Gaussian noise assumption, the optimal observer is the KF; however, the nonlinear aerodynamic models along with the biased and noisy sensor measurements render APE a nonlinear filtering problem [4].

The use of a Gaussian-sum filter (GSF) as a state/parameter estimation tool for aerodynamic modelling is relatively unexplored in the literature. In this chapter, GSF has been employed for aircraft parameter estimation. The stability analysis of GSF has also been discussed using the normalized Lyapunov energy functional and the conditions for the convergence of the GSF are derived by using the proposed nonlinear observer, which is a novel approach. An approach to prune the weights of GSF based on the Pearson coefficient is considered. The performances of the three recursive parameter-estimation algorithms: EKF, the simplified version of the GSF, and the improved version of the GSF are evaluated for aircraft parameter estimation using realistically simulated flight data in the MATLAB implementations. The numerical results show that pruning the weights reduces the computational load by nearly 75%, and still provides good performance compared to the conventional GSF.

An EKF is one of the most widely adopted approach for solving the nonlinear state/parameter-estimation problem, and it is based on the first-order approximation, with respect to the states of the nonlinear system and measurement dynamics. This approximated model is then propagated forward using KF equations along with the observations and is used to estimate the mean and covariance for the state of the system, due to the underlying Gaussian pdf. The EKF algorithm provides good estimates if the errors due to linearization of the associated nonlinear functions are reasonably small. This will be the case when good initial state estimates are available, i.e. when the initial covariance is small, and also when the process noise is not very large. However, in some applications these

* This chapter is contributed by Sara M. George, S. S. Selvi and Jitendra R. Raol.

DOI: 10.1201/9781003620211-5

conditions might not be satisfied. The unscented Kalman filter (UKF) overcomes largely these deficiencies by using a set of carefully selected sample points (known as Sigma points) to approximate the pdf of the random variables.

A variant to nonlinear filtering, GSF, can be used for APE, where the prior and posterior densities are approximated as weighted sum of several Gaussians [8]. A GSF can incorporate any Kalman-based nonlinear filter so that the prior and posterior pdfs can be represented using Gaussian pdfs. This approach has shown that as the number of Gaussian components increases the Gaussian-sum approximation converges uniformly to any probability density function [4]. However, the large number of (filter) components increase the complexity and computational time for estimation; an effective solution to this is also considered in this chapter. The GSF and an improved GSF make use of the sum of Gaussians to approximate the posterior pdf and use three steps to achieve the global update of the posterior density:

(iv) A conventional state estimator like EKF is used for each Gaussian to update the prior density of each Gaussian.
(v) Innovation statistics of each Gaussian is calculated and this is used to update the weights of the corresponding Gaussian.
(vi) The posterior moments of each Gaussian and the updated weights together provide the global point estimate.

The GSFs: Gaussian-sum extended Kalman filter (GSEKF), Gaussian-sum unscented Kalman filter (GS-UKF), and Gaussian-sum particle filter (GS-PF) [5], belong to the class of nonlinear approximations filters. We study two types of GSFs and derive the convergence results for one of these filters using the normalized Lyapunov energy functional (LEF), and the performances of these filters are evaluated using MATLAB implementations with realistic simulated data for aircraft parameter estimation.

5.2 GAUSSIAN-SUM FILTER (GSF)

The main objective of the GSF is to approximate an arbitrary *a posteriori* pdf and its moments [6,7], and the concept of approximating this pdf with a weighted sum of Gaussian functions was introduced in [4]. As the number of Gaussian components increases, the sum tends to a valid density function. Consider a nonlinear signal discrete-time model corrupted by noise in both the states and the measurements:

$$x_{k+1} = f(x_k) + g_k \tag{5.1}$$

$$y_{k+1} = h(x_{k+1}) + v_{k+1} \tag{5.2}$$

In eqns. (5.1) and (5.2), $f(.)$ and $h(.)$ are the nonlinear vector-valued functions; and g_k is the process noise and v_{k+1} is the measurement noise (temporally uncorrelated with each other), zero-mean random sequences with covariance matrices Q_k and R_{k+1}, respectively:

$$Q_k = E[g_k g_k^T]; R_{k+1} = E[v_{k+1} v_{k+1}^T] \tag{5.3}$$

In the subsequent analysis, the covariance matrices will be represented by only Q and R for brevity. For a stochastic dynamic system, the *a posteriori* (after incorporating the observations into the KF/EKF) normal/Gaussian pdf is expressed as

$$p(\hat{x}_k \mid z_k) = \sum_{i=1}^{N} w_k^i \mathrm{N}\left(\hat{x}_k; x_k^i, \hat{P}_k^i\right) \tag{5.4}$$

Here, N is the total number of Gaussian densities used; and w_k^i, x_k^i and $\hat{P}_k^i$ represent, respectively, the components' conditional weight/s, mean/s and covariance/s of the ith Gaussian with respect to the first k measurements. The usual condition on the weight is

$$\sum_{i=1}^{N} w_k^i = 1; w_k^i \geq 0 \tag{5.5}$$

Since, all the components of the mixture pdf in eqn. (5.4) are Gaussian, only estimates of their means and covariance/s need to be considered between k and k+1 [6]. The prior (before the observations are incorporated into the EKF) mean $\tilde{x}_{k+1}$ and the prior covariance $\tilde{P}_{k+1}$ of the ith Gaussian component are calculated using the time update equations of the EKF.

5.2.1 Time Propagation/Evolution

$$\tilde{x}_{k+1}^i = f_k(\hat{x}_k^i) \tag{5.6}$$

$$\tilde{P}_{k+1}^i = F_k \hat{P}_k^i F_k^T + Q_k \tag{5.7}$$

The recursion in eqn. (5.6) is performed by choosing proper initial condition of $x(.,.)$.

And the Jacobian is given as

$$F_k = \frac{\partial f_k\left(\hat{x}_k^i\right)}{\partial \hat{x}_k^i} \tag{5.8}$$

The Runge–Kutta algorithm can be used for integration of eqn. (5.6), and the state-transition matrix F_k can be approximated using Taylor series expansion. In time update, the weights are not updated:

$$\tilde{w}_{k+1}^i = \hat{w}_k^i \tag{5.9}$$

5.2.2 Measurement/Data Update

$$\hat{x}_{k+1}^i = \tilde{x}_{k+1}^i + K_{k+1}^i [y_{k+1} - h_{k+1}(\tilde{x}_{k+1}^i)] \tag{5.10}$$

$$S_{k+1}^i = H_{k+1} \tilde{P}_{k+1}^i H_{k+1}^T + R_{k+1} \tag{5.11}$$

$$K_{k+1}^i = \tilde{P}_{k+1}^i H^T{}_{k+1} (S_{k+1}^i)^{-1} \tag{5.12}$$

$$\hat{P}_{k+1}^i = (I - K_{k+1}^i H_{k+1}) \tilde{P}_{k+1}^i \tag{5.13}$$

$$H_{k+1} = \frac{\partial h_{k+1}(\tilde{x}^i_{k+1})}{\partial \tilde{x}^i_{k+1}} \tag{5.14}$$

In the subsequent analysis, the measurement vector/matrix H, the process noise covariance matrix Q, and the measurement noise covariance matrix R will be represented without any time index/subscript for the sake of simplicity; the variation of H, since it is time dependent will be taken care of in the program code for the implementation of the GSF.

5.2.3 Gaussian-Sum Estimate

The new weights are calculated as

$$\hat{w}^i_{k+1} = \frac{\tilde{w}^i_{k+1}\mathbb{N}(y_{k+1} - h_{k+1}(\tilde{x}^i_{k+1}), S^i_{k+1})}{\sum_{i=1}^{N} \tilde{w}^i_{k+1}\mathbb{N}(y_{k+1} - h_{k+1}(\tilde{x}^i_{k+1}), S^i_{k+1})} \tag{5.15}$$

The final state estimate at each time step is the weighted average from eqn. (5.10) and eqn. (5.15):

$$\hat{x}_{k+1} = \sum_{i=1}^{N} \hat{w}^i_{k+1}\hat{x}^i_{k+1} \tag{5.16}$$

5.3 GSF WITH PRUNING

The required number of components in the GSF can be set depending on the application. However, as this number increases, the computational complexity and time also increase, hence pruning of the number of filters is needed, and this can help in reducing the computational complexity, and possibly maintaining the estimation accuracy. In order to differentiate between less important weights and the important ones, an optimal threshold value needs to be evaluated; as in typical pruning methods, all the weights with the magnitudes below a specified threshold need not be considered [8] for the update of the final state estimate. However, pruning of the component filters based on the amplitude of the weights only may prune some weights, which might be important for contribution to the final state estimate.

In this chapter, an algorithm called GSF with pruning (GSFP) wherein the weights and weight changes are considered together for pruning is studied. Weight changes are measured by how much the weight has changed from one iteration to the next after each update. Pearson's correlation coefficient is used to set the threshold; i.e. between (−1 and +1). If the weight amplitude and the weight-change relationship of a weight w^i_k is unimportant, then the corresponding correlation coefficient would be closer to 0, and the corresponding filter components are considered not so valuable in the estimation process. The correlation coefficient to be calculated for each Gaussian component is given as

$$\text{coeff}_w(w^i, \Delta w^i) = \frac{m(\sum w^i \Delta w^i) - \sum w^i \sum \Delta w^i}{\sqrt{(m\sum (w^i)^2 - (\sum w^i)^2)(m\sum \Delta(w^i)^2 - (\sum \Delta w^i)^2)}} \tag{5.17}$$

In eqn. (5.17) m is the number of timesteps between the consecutive pruning of filters, and Δw is the weight-change magnitude:

$$\Delta w(k)_i = | w(k)_i - w(k-1)_i | \tag{5.18}$$

The value of m and the threshold are decided by experimentation, i.e. by trial and error/sensitivity analysis. The comparison of the correlation value with the threshold will decide which component filters will be pruned, and this is performed once in m timesteps. In cases where the initial state is approximately known, the number of filters required in the GSF is less and pruning of the less important weights contributes to reducing the computational cost. However, when the initial state is not known and is taken very far away from the true value, the number of Gaussian components required is more to provide better performance. In such cases, overfitting might happen and pruning will remove such weights, which in turn would have increased/maintained the accuracy along with decreasing the computational cost. Therefore, pruning helps in improving the performance of the filter especially when the initial guess of the state estimate is not close to the true state.

5.3.1 Algorithm for GSFP

(a) Run N Gaussian-sum EKF for m timesteps.
(b) Compute the Pearson coefficient, eqn. (5.17), $\text{coeff}_W(w^i, \Delta w^i)$ related to magnitude of the weight and change of the weight.
(c) Truncate filters, which have $|\text{coeff}_W(w^i, \Delta w^i)| <$ threshold.
(d) Normalize the weights of the remaining filters $N = N - N_T$ where N_T is the number of truncated filters.
(e) Go back to step (a).

As the Gaussian components with the $\text{coeff}_W(w_i, \Delta w_i)$ close to zero do not contribute to improve the accuracy of estimation, the threshold can be set to a value very close to zero. The appropriate value of the threshold can be decided without loss of accuracy via some trial-sensitivity analysis. The decision on the number of iterations is also done using such trials.

5.4 ASYMPTOTIC STABILITY ANALYSIS OF GSF VIA AN OBSERVER

First, an appropriate observer is proposed, and then the conditions for the convergence of the GSF is established via the convergence of this observer. An observer for the nonlinear system of eqns. (5.1) and (5.2) is considered:

$$\hat{x}^i_{k+1} = f(\hat{x}^i_k) + K^i_{k+1}(y_{k+1} - \tilde{y}^i_{k+1}) = f(\hat{x}^i_k) + K^i_{k+1} r^i_{k+1} \tag{5.19}$$

$$\hat{y}^i_{k+1} = h(\hat{x}^i_{k+1}) \cong H\hat{x}^i_{k+1} \tag{5.20}$$

The observer gain is taken as the gain of the GSF-KF with associated covariance matrices:

$$K^i_{k+1} = \tilde{P}^i_{k+1} H^T (S^i_{k+1})^{-1} \tag{5.21}$$

$$S^i_{k+1} = H\, \tilde{P}^i_{k+1} H^T + R \tag{5.22}$$

The state errors are

$$\hat{e}^i_{k+1} = x_{k+1} - \hat{x}^i_{k+1} \tag{5.23}$$

Since, there are N filters the weights of each filter have to be considered to show the combined final effect. The combined weights, *a posterior* error, *a priori* error, residual error, *a posterior* state covariance and *a priori* state covariance are defined for the sake of clarity and completion for the time index 'k':

$$w_k = [w_k^1 \, w_k^2 \ldots w_k^N] \tag{5.24}$$

$$\hat{e}_k = \sum_{i=1}^{N} w_k^i \hat{e}_k^i \tag{5.25}$$

$$\tilde{e}_k = \sum_{i=1}^{N} w_k^i \tilde{e}_k^i \tag{5.27}$$

$$\hat{P}_k = \sum_{i=1}^{N} w_k^i \hat{P}_k^i \tag{5.28}$$

$$\tilde{P}_k = \sum_{i=1}^{N} w_k^i \tilde{P}_k^i \tag{5.29}$$

The observer error is obtained by subtracting eqn. (5.19) from eqn. (5.1):

$$\hat{e}_{k+1}^i = x_{k+1} - \hat{x}_{k+1}^i = f(x_k) - f(\hat{x}_k^i) - K_{k+1}^i(y_{k+1} - \tilde{y}_{k+1}^i) \tag{5.30}$$

In fact, since the total estimate from the observer has to be considered for comparison with the actual states, we use a short-hand symbol for sigma operation, see eqn. (5.25): $E\{.\} = \sum_{i=1}^{N} w_k^i(.)$, and with this notation we rewrite eqn. (5.30) as

$$\hat{e}_{k+1} = E\{\hat{e}_{k+1}^i\} = f(x_k) - E\{f(\hat{x}_k^i)\} - E\{K_{k+1}^i(y_{k+1} - \tilde{y}_{k+1}^i)\} \tag{5.31}$$

$$\begin{aligned} &\hat{e}_{k+1} = F_x x_k + \phi_k - E\{K_{k+1}^i(y_{k+1} - \tilde{y}_{k+1}^i)\}; \\ &\text{wherein } \ \phi_k = -F_x x_k + f(x_k) - E\{f(\hat{x}_k^i)\} \end{aligned} \tag{5.32}$$

The nonlinear function as 'phi' in eqn. (5.32) is included for facilitating the convergence analysis, otherwise both eqns. (5.31), and (5.32) are exactly identical. For the sake of simplicity in the analytical development, the process noise term is avoided, but can be easily incorporated. In eqn. (5.32) one can substitute for $K(.)$ in eqn. (5.33) from eqn. (5.21) and eqn. (5.22) to obtain, after the simplification of the measurement terms, i.e. the residuals:

$$\hat{e}_{k+1} = F_x x_k + \phi_k - E\{\tilde{P}_{k+1}^i H^T (H \, \tilde{P}_{k+1}^i H^T + R)^{-1} H \tilde{e}_{k+1}^i)\} \tag{5.33}$$

We would utilize Lyapunov–Krasovskii's approach to derive the conditions for the convergence of the observer error, eqn. (5.33), for which we use the following development initially:

Let the error state in general be described as

$$e(k+1) = \mathrm{G}e(k) \tag{5.34}$$

Let us define a feasible normalized Lyapunov energy function (LEF) as

$$V(e(k)) = e^T(k) Y(k) e(k) \tag{5.35}$$

In eqn. (5.35), $Y(.)$ is the information matrix (as the inverse of the conventional covariance matrix P) and it is a positive definite matrix, and we use the following expression as the (equivalent or analogous) derivative of the LEF:

$$\Delta\{V(e(k))\} = V(e(k+1)) - V(e(k)) \tag{5.36}$$

In the continuous-time case for the similar convergence analysis, one uses the time derivative of the LEF (defined in the continuous-time domain) and seeks the conditions such that this time derivative (differential) is negative definite, thereby establishing the convergence of the observer error dynamics like in eqn. (5.33). In the discrete-time case we use a differencing approach; substituting for $V(.)$s' in eqn. (5.36) we obtain

$$\begin{aligned}\Delta\{V(e(k))\} &= V(e(k+1)) - V(e(k)) \\ &= e^T(k+1)Ye(k+1) - e^T(k)Ye(k) \\ &= e^T(k)G^TYGe(k) - e^T(k)Ye(k) \\ &= e^T(k)[G^TYG - Y]e(k)\end{aligned} \tag{5.37}$$

Since the 'derivative' term of the error should be negative definite for the convergence of the observer error dynamics we have

$$\Delta\{V(e(k))\} = -e^T(k)[-(G^TYG - Y)]e(k) \tag{5.38}$$

Then for the convergence of the observers' error dynamics we have the following condition:

$$-(G^TYG - Y) = Y - G^TYG \tag{5.39}$$

should be positive definite (PD). This means that in terms of the magnitudes of the involved terms, the following should hold true:

$$| Y - G^TYG | > 0 \tag{5.40}$$

Since the magnitude of any matrix is measured by its norm, one can arrive at the final condition as

$$\|Y - G^TYG\| > 0 \Rightarrow \|G\|^2 < 1 \tag{5.41}$$

Next, the error dynamics eqn. (5.33) should be brought in the form of eqn. (5.41), due to eqn. (5.34), so that one can easily derive the conditions for the convergence, and in order to do this it is necessary to define certain bounds:

1. The steady state solution, $P(\infty)$ of eqn. (5.7) must be in the range:

$$p_l I \le P(.) \le p_u I \tag{5.42}$$

   Here, $p_u, p_l > 0$ are +ve numbers, $P(.)$ is inherently symmetric/positive definite (PD) matrix; the constants specify the bounds.
2. The term $F_k x_k$ should also be bounded as

$$\|F_k x_k\| = \rho_1 \|e_k\|^2 \tag{5.43}$$

The variable 'ρ_l' is the upper bound of the associated nonlinear function as a scalar factor of the state-error norm.

3. The nonlinear function 'phi' in eqn. (5.32) is bounded, because it is composed of the bounded functions of the dynamic system:

$$\|\phi(.)\| \le \rho_2 \left\|x_k - \hat{x}_k\right\|^2 = \rho_2 \left\|e_k\right\|^2 \tag{5.44}$$

The variable 'ρ_2' is the upper bound of the associated nonlinear function as a scalar factor of the state-error norm. Also, the following bounds on the norms of matrices H and R are needed to simplify the error dynamics:

$$\|H^T H\| \le h^2; \quad \|R\| = r; \quad \left\|e_k\right\|^2 \le \varepsilon^2 \tag{5.45}$$

$$\left\|e_k\right\| \le \varepsilon = \text{some constant}$$

Next, the observer's error dynamics of eqn. (5.33) should be brought in the form of eqn. (5.41):

$$\begin{aligned} \hat{e}_{k+1} &= F_x x_k + \phi_k - E\{\tilde{P}^i_{k+1} H^T (H\,\tilde{P}^i_{k+1} H^T + R)^{-1} H \tilde{e}^i_{k+1})\} = G\hat{e}_k \\ \left\|\hat{e}_{k+1}\right\| &= \|G\| \left\|\hat{e}_k\right\| \end{aligned} \tag{5.46}$$

It is better to recast eqn. (5.46) in terms of the norms as follows:

$$\left\|\hat{e}_{k+1}\right\| \equiv \left\|F_x x_k\right\| + \left\|\phi_k\right\| - \left\|E\{\tilde{P}^i_{k+1} H^T (H\,\tilde{P}^i_{k+1} H^T + R)^{-1} H\tilde{e}^i_{k+1})\}\right\| \tag{5.47}$$

Substituting various bounds defined earlier into eqn. (5.47) one obtains

$$\left\|\hat{e}_{k+1}\right\| \equiv \rho_1 \left\|\hat{e}_k\right\|^2 + \rho_2 \left\|\hat{e}_k\right\|^2 - \left\|E\{\tilde{P}^i_{k+1} H^T (H\,\tilde{P}^i_{k+1} H^T + R)^{-1} H\tilde{e}^i_{k+1})\}\right\| \tag{5.48}$$

After substituting other bounds from eqn. (5.45) and simplifying the resultant expression, and utilizing the definition of error norm, one obtains

$$\left\|\hat{e}_{k+1}\right\| \equiv \rho_1 \left\|e_k\right\|^2 + \rho_2 \left\|e_k\right\|^2 - \left(\frac{p_l h^2}{(p_l h^2 + r)} \varepsilon\right) \tag{5.49}$$

$$\left\|\hat{e}_{k+1}\right\| \equiv \left\{\rho_1 \varepsilon + \rho_2 \varepsilon - \frac{p_l h^2}{(p_l h^2 + r)}\right\} \varepsilon = \|G\| \left\|e_k\right\| \tag{5.50}$$

Eqn. (5.50) is now in the required form of the second equation of (5.46). From eqn. (5.50) one obtains the following expression:

$$\|G\| \equiv \left\{\rho_1 \varepsilon + \rho_2 \varepsilon - \frac{p_l h^2}{(p_l h^2 + r)}\right\} \tag{5.51}$$

Now applying the condition of eqn. (5.41) and considering the positive square root of 1, one obtains

$$\|G\|^2 \equiv \left\{\rho_1\varepsilon+\rho_2\varepsilon-\frac{p_{_I}h^2}{(p_{_I}h^2+r)}\right\}^2 < 1$$
$$\left\{\rho_1\varepsilon+\rho_2\varepsilon-\frac{p_{_I}h^2}{(p_{_I}h^2+r)}\right\} sqrt(1) = \pm 1 \tag{5.52}$$
$$\left\{\rho_1\varepsilon+\rho_2\varepsilon-\frac{p_{_I}h^2}{(p_{_I}h^2+r)}\right\} < 1$$

Thus, the convergence condition for the observer's error dynamics is obtained as

$$(\rho_1+\rho_2)\varepsilon < 1+\frac{p_{_I}h^2}{(p_{_I}h^2+r)} \tag{5.53}$$

All the coefficients appearing in eqn. (5.53) are well defined as these are the positive bounding constants in eqns. (5.42)–(5.45). The condition of eqn. (5.53) means that if it is satisfied, then the time derivative, analogously to the differential in eqn. (5.38) of the Lyapunov energy functional would be negative definite and the observer error dynamics would converge and the error would (asymptotically) tend to zero. This in turn, also establishes that the GSF would also converge, since its gain and covariance matrix were used in the observer error dynamics. This is a novel way of establishing the asymptotic stability of the discrete-time GSF via the use of the deterministic observer. Next, the results of the numerical evaluations of some GSFs are presented.

5.5 AIRCRAFT PARAMETER ESTIMATION

The longitudinal motion of aircraft is represented using the following postulated aircraft model [1–2]:

$$\begin{aligned}
\dot{u} &= \frac{\bar{q}S}{m}C_X - qw - g\sin\theta \\
\dot{w} &= \frac{\bar{q}S}{mV}C_Z + qu + g\cos\theta \\
\dot{\theta} &= q \\
\dot{q} &= \frac{\bar{q}S\bar{c}}{I_{yy}}C_m
\end{aligned} \tag{5.54}$$

In eqn. (5.54), C_x, C_z and C_m are the nondimensional aerodynamic coefficients expressed as

$$\begin{aligned}
C_Z &= C_{Z_0} + C_{Z_\alpha}\alpha + + C_{Z_q}\frac{q\bar{c}}{2V} + C_{Z_{\delta_e}}\delta_e \\
C_X &= C_{X_0} + C_{X_\alpha}\alpha + C_{X_{\alpha^2}}\alpha^2 \\
C_m &= C_{m_0} + C_{m_\alpha}\alpha + + C_{m_{\alpha^2}}\alpha^2 + C_{m_q}\frac{q\,\bar{c}}{2V} + C_{m_{\delta_e}}\delta_e
\end{aligned} \tag{5.55}$$

Here, the state variables are true air velocity V, angle of attack α, pitch angle θ, and pitch rate q. The elevator deflection δ_e is the control input to aircraft dynamics. Other relevant parameters are as follows: mass of the aircraft is 'm' and its geometrical parameters consist of wing area S, wing chord $\bar{c}$ and moment of inertia I_{yy}. The dynamic pressure $\bar{q}$ is given by $\frac{1}{2}\rho V^2$, where ρ is the density of air and V is the true air speed; and all the variables have the SI-metric units, and have appropriate dimensions. The measurement vector is given by

$$\begin{aligned} u &= u \\ w &= w \\ q &= q \\ \theta &= \theta \\ a_x &= \frac{\bar{q}S}{m} C_X + \frac{F_T}{m} \cos\sigma_T \equiv a_{x_m} \\ a_z &= \frac{\bar{q}S}{m} C_Z - \frac{F_T}{m} \sin\sigma_T \equiv a_{z_m} \\ \dot{q} &= \dot{q} \end{aligned} \tag{5.56}$$

In eqn. (5.56) σ_T is the tilt angle of the engine.

However, for the sake of simplicity, the engine effects are not simulated in the present case. It is presumed that each component of the measurement vector has some bias and/or random noise, but it is not shown in the model of eqn. (5.56). The unknown parameter vector β that is to be estimated consists of the aerodynamic derivatives:

$$\beta = [C_{X_0}\ C_{X_\alpha}\ C_{X_{\alpha^2}}\ C_{Z_0}\ C_{Z_\alpha}\ C_{Z_{\delta_e}}\ C_{m_0}\ C_{m_\alpha}\ C_{m_{\alpha^2}}\ C_{m_q}\ C_{m_{\delta_e}}] \tag{5.57}$$

5.6 RESULTS AND DISCUSSIONS

To evaluate the efficacy of GSF and GSFP, short period data of a LTA are simulated and the nondimensional longitudinal parameters are estimated. The 4DOF longitudinal axis model is used for estimation; using eqns. (5.54) to (5.56), the data are generated with a sampling interval of 0.03 sec for a duration of nearly 7 sec by giving a doublet input to the elevator control surface. The initial states for $[u,\ w,\ q,\ \theta]$ are chosen as in [2]. Random Gaussian noise with standard deviation 0.001 (as the process noise) is added to the states u, w, q and θ. Random noise with SNR = 10 dB is added to these measurements, eqn. (5.56), to generate noisy data. A fourth-order RK integration method is used to obtain the longitudinal flight variables u, w, q and θ from eqn. (5.54). Two examples are considered to demonstrate the effectiveness of GSF over EKF and the performance improvement using GSFP; the metric PEEN (parameter estimation error norm) is evaluated for this and is defined as

$$\text{PEEN} = 100 \times \frac{\text{norm}\,(\beta_{\text{true}} - \beta_{\text{est}})}{\text{norm}\,(\beta_{\text{true}})} \tag{5.58}$$

where β_{true} is the vector of true parameters and β_{est} is the vector of estimated parameters.

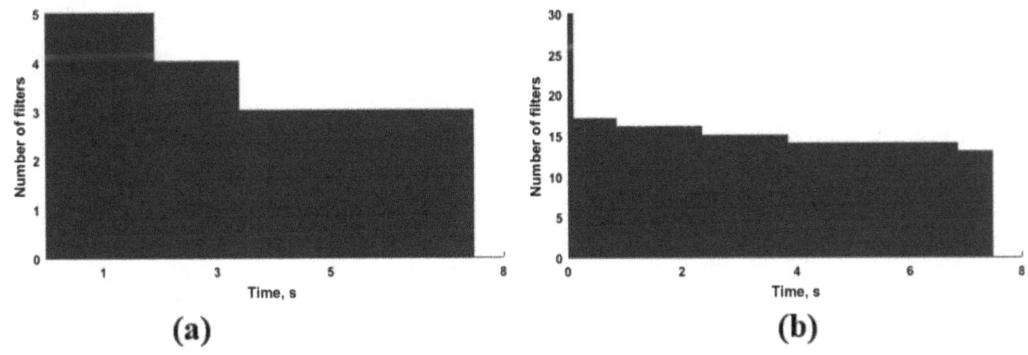

FIGURE 5.1a and 5.1b Filter pruning for GSFP: (a) Example 5.1, (b) Example 5.2.

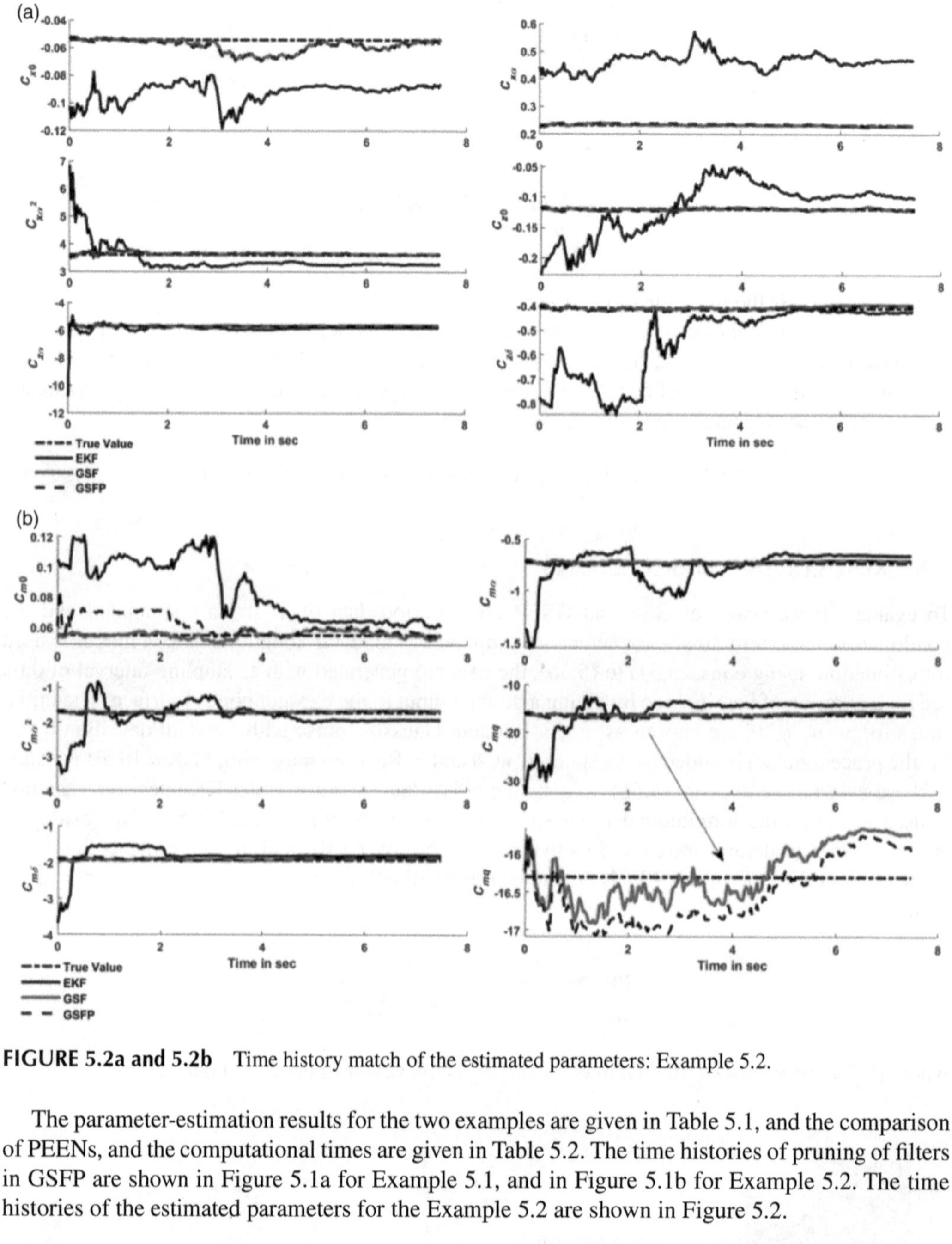

FIGURE 5.2a and 5.2b Time history match of the estimated parameters: Example 5.2.

The parameter-estimation results for the two examples are given in Table 5.1, and the comparison of PEENs, and the computational times are given in Table 5.2. The time histories of pruning of filters in GSFP are shown in Figure 5.1a for Example 5.1, and in Figure 5.1b for Example 5.2. The time histories of the estimated parameters for the Example 5.2 are shown in Figure 5.2.

Example 5.1 (benign initial conditions): in this example, the initial states/parameters for the UDF-EKF are taken 10% away from their true values (used for generating the simulated data); and the initial covariance matrix $P(0)$ is chosen to reflect these uncertainties in the initial values. The estimated values of the parameters are compared with the true values as in the second and third columns of Table 5.1. It is seen that the estimates are fairly close to the true values even when there is some noise (SNR = 10) in the measurement data, and the PEEN related to this is 5.63 (% is implied), which is reasonably small. In the case of the GSF, the number of Gaussian components required for effective filtering is less, as the initial state is close to the true value used for simulation; only

TABLE 5.1
Estimated Parameters for GSF/GSFP: Example 5.1, and Example 5.2

Parameter	True Values	EUD (~EKF) SNR = 10	EUD (~EKF) SNR = 10 (&)	GSF-E-1	GSFP-E-1	GSF-E-2	GSFP-E-2
C_{x0}(*)	-0.0540	*-0.0592 (#)*	-0.0882	-0.054	-0.054	-0.053	-0.054
$C_{x}\alpha$	0.2330	*0.2543*	0.4937	0.234	0.234	0.231	0.233
$C_{x}\alpha^2$	3.6089	*3.7058*	3.2183	3.653	3.653	3.633	3.647
C_{z0}	-0.1200	*-0.1249*	-0.1012	-0.122	-0.122	-0.123	-0.123
$C_{z}\alpha$	-5.6800	*-5.7247*	-5.7807	-5.667	-5.667	-5.644	-5.654
$C_{z}\delta_{e}$	-0.4070	*-0.5049*	-0.4138	-0.402	-0.402	-0.395	-0.399
C_{m0}	0.0550	*0.0576*	0.0606	0.054	0.054	0.054	-0.056
$C_{m}\alpha$	-0.7290	*-0.7092*	-0.654	-0.715	-0.715	-0.707	-0.708
$C_{m}\alpha^2$	-1.7150	*-1.7843*	-1.9417	-1.687	-1.687	-1.667	-1.671
C_{mq}	-16.300	*-15.3075*	-13.992	-16.069	-16.069	-15.76	-15.95
$C_{m}\delta_{e}$	-1.9400	*-1.8873*	-1.8097	-1.9037	-1.9037	-1.882	-1.867
PEEN(%)		*5.6329*	13.3006	1.345	1.345	3.097	2.065

(*) The aerodynamic derivatives are expressed in the form of the text for convenience/
(#) Estimates in *italics*, column 3, are from [2]. (&) The initial values are far away from true values.

TABLE 5.2
Comparison: PEENs and Computational Times

	Number of Filters	PEEN in %		Comp. Times	
		GSF	GSFP	GSF	GSFP
Example 5.1	3	3.228	3.228	1.05	1.17
	5	1.345	1.345	1.52	1.45
	6	1.345	1.345	1.56	1.22
Example 5.2	25	3.539	3.246	4.88	2.12
	30	3.097	2.068	5.62	2.81
	35	3.097	2.065	6.033	2.86

five filters have been used for GSF and GSFP. The estimated parameters in Table 5.1 are those of the average of the last ten data points. The GSF offers somewhat better estimation than EKF. When three Gaussian components were used, PEEN of GSF and GSFP was 3.228, though the value is higher than that of with five components, the estimation accuracy is still better than EKF. When the number of components was increased to six, the PEEN of GSF and GSFP remained the same as that obtained using five components. This shows that, even if the number of Gaussian components is increased beyond five, the performance is not improving.

Example 5.2 (far away initial conditions): in this example, the initial state is chosen far (~100 %) from the true values. A greater number of Gaussian components are required by GSF/GSFP, since the initial states are not accurately known and are presumed to be far from the true values. In this example, 50 filters have been used for GSF and GSFP. A comparison is done by increasing and decreasing the number of filters and the results are presented in Tables 5.1 and 5.2. Even in this case, the results are fairly satisfactory, and definitely better than those of the EKF. The time histories of the estimated parameters of EKF, GSF, and GSFP are shown in Figure 5.2.

5.6.1 Effects of Pruning

Calculating $\text{coeff}_W(w_i, \Delta w_i)$ will help to locate the filters that do not contribute to the estimation process, and the pruning helps to negate overfitting, and still maintain the accuracy. In some cases, the GSFP may give a better performance than GSF; as in the inset of Figure 5.2, the estimates of C_{mq} by GSF and GSFP are shown to illustrate this. The computational load of GSF is more than that of GSFP. Figure 5.1b shows the number of filters that get pruned as the time elapses in GSFP; the pruning of filters that do not contribute significantly to the estimation process decreases the computational load of GSFP compared to GSF. To better visualize the effect of change in the number of filters, further study was performed with 25 and 35 Gaussian components. It was observed that when 25 components were used, the performance accuracy decreased as compared to 30 Gaussian components. There was no performance improvement when the number of filters was increased from 30 to 35. This implies that 30 Gaussian components is sufficient for this case. The comparison is presented in Table 5.2. The comparative CPU time requirements for the Examples 5.1 and 5.2 are shown in Table 5.2. The execution time for EKF is the least. The computational time of GSF and GSFP depends upon the number of filters used. In the case of three Gaussian components, the execution time by GSFP is marginally more than the time taken by GSF as the number of components that get pruned is much less. As the number of Gaussian components increases, the computational load would significantly reduce due to pruning.

Thus, in this chapter the asymptotic stability/convergence condition for the basic GSF has been derived using Lyapunov–Krasovskii's approach via proposition of a nonlinear observer that used the gain and the covariance matrices from the basic GSF/EKF, and the normalized Lyapunov energy functional. Then, the performances of a Gaussian-sum filter (GSF) and an improved Gaussian-sum filter (GSFP) for estimation of aircraft parameters from the realistically simulated flight data have been validated. The improved GSF gives better performance than GSF when the number of filters required for estimation is more, especially when the initial states are not known very accurately *a priori*. These estimation results confirm that the GSFP approach has a great potential for estimation of the aerodynamic derivatives from flight data using the nonlinear model. The results obtained are very encouraging and point towards the future research aiming the implementation of advanced approximations methods and modelling techniques for more complex aerodynamic and flight mechanics problems, where aircraft parameter estimation (APE) is a special case, for which the analytical and numerical convergence results have been established in the present chapter.

EXERCISES

Exercise 5.1

Explain the significance of aircraft parameter estimation (APE) in the context of aircraft system identification (SID).

Exercise 5.2

Provide an overview of the Gaussian-sum filter (GSF) and its application in APE.

Exercise 5.3

Discuss the conditions for the convergence of the GSF as derived through the proposed nonlinear observer.

Exercise 5.4

Discuss the implications of weight pruning on computational load and the performance of GSF for APE.

Exercise 5.5

Describe the approach to prune the weights of GSF based on Pearson's coefficient.

REFERENCES

1. Sara, M.G., Approximations in Nonlinear Filtering: Some Novel Approaches. Ph.D. Thesis, Electrical and Electronics Engineering Sciences, Visvesvaraya Technological University (VTU), Jnana Sangama, VTU Main Road, Machhe, Belagavi, Karnataka, 2023.
2. Raol, J.R., Girija, G., and Singh, J., *Modelling and Parameter Estimation of Dynamic Systems*, The IEE/IET Control Series, Vol. 65, London, UK, 2004.
3. Majeed, M., and Narayan Kar, I., Aerodynamic parameter estimation using adaptive unscented Kalman filter, *Aircraft Engineering and Aerospace Technology*, Vol. 85, No. 4, pp. 267–279, 2013.
4. Alspach, D.L., and Sorenson, H.W., Nonlinear Bayesian estimation using Gaussian sum approximations, *IEEE Transactions on Automatic Control*, Vol. 17, No. 4, pp. 439–448, Aug. 1972.
5. Kotecha, J. H., and Djuric, P. M., Gaussian particle filtering, *IEEE Transactions on Signal Processing*, Vol. 51, No. 10, pp. 2592–2601, Oct. 2003, doi: 10.1109/TSP.2003.816758.
6. Terajanu, G., Singla, P., Singh, T., and Scott, P.D., Adaptive Gaussian sum filter for nonlinear Baysian estimation, *IEEE Transactions on Automatic Control, USA*, Vol. 56, No. 9, pp. 2151–2156, 2011.
7. Bhaumik, S., and Date, P., *Nonlinear Estimation: Methods and Applications with Deterministic Sample Points*, 1st Edition, Chapman and Hall, CRC, Boca Raton, July 17, 2019.
8. Clark, D.E., and Bell, J., Convergence results for the particle PHD filter, *IEEE Transactions on Signal Processing, USA*, Vol. 54, No. 7, pp. 2652–2661, 2006.

6 Gaussian-Sum Information Filter with Lyapunov Stability Analysis

6.1 INTRODUCTION

Flight data measured from onboard sensors can contain a considerable amount of noise. Further, there might be biases and unobserved states in the system model, which must be estimated. Hence, filtering techniques are generally employed to estimate the required aerodynamic derivatives. When a linear state-space model with WGN is considered, KF is the best choice as an optimal observer/estimator. Information filtering (IF) is an alternative method of implementing KF, in its equivalent information domain. Initialization of IF is less tedious compared to KF and the update stage is computationally economic. The linear IF is algebraically equivalent to the linear KF, however IF is faster than KF especially when the state-vector dimension is greater than the measurement vector dimension, due to the difference in computational burden [1]. The nonlinear aerodynamic models along with the biased and noisy sensor measurements render aircraft parameter estimation a nonlinear filtering problem [2,3]. A frequently used algorithm to tackle nonlinear filtering is EKF, and often aircraft aerodynamic parameter estimation is carried out using EKF [4]. EKF is based on a first-order approximation of the system dynamics, which may lead to linearization error.

The extended information filter (EIF) equations are derived from the EKF equations. Since the filter equations are algebraically equivalent, the EIF and EKF theoretically calculate the same estimates as well as the same estimation-error covariances. The EIF uses the propagation of the inverse of the covariance matrix called the information matrix, instead of the covariance matrix used in the EKF. The state prediction/estimation in terms of information reduces the computational load particularly for nonlinear systems [5], and this aspect is often exploited in the applications of the information concept in multisensory data fusion. However, the prediction part of EIF is a bit more complex in this form and it relies on a propagation matrix, which is dependent of observation made [5,6].

In order to overcome the complexities in local approximation-based solutions, approaches based on global approximation can be employed in estimation [7]. A Gaussian-sum filter (GSF) can incorporate any Kalman-based nonlinear filter so that the prior and posterior pdfs can be represented using Gaussian pdfs. GSF makes use of the sum of Gaussians to approximate the posterior density, motivated by the result that a sum of Gaussians can approximate any arbitrary density to an arbitrary degree of accuracy [7]. This approach has shown that as the number of Gaussian components increases the Gaussian-sum approximation converges uniformly to any probability density function [7,8]. This was treated in Chapter 5.

In this chapter, a Gaussian-sum extended information filter (GS-EIF) is studied, the convergence results are derived using the normalized Lyapunov energy functional (LEF), and the performance of this filter is evaluated and compared with conventional EIF using MATLAB implementations with realistically simulated data for aircraft parameter estimation.

DOI: 10.1201/9781003620211-6

6.2 GS EXTENDED INFORMATION FILTER (GDEIF)

The main objective of the GSF is to approximate an arbitrary *a posteriori* pdf and its moments [8,9]. The concept of approximating this pdf with a weighted sum of Gaussian functions was introduced in [7]. Here GSF has been incorporated with extended information filter; the prior and posterior pdfs can be represented using Gaussian pdfs. Let the nonlinear system be described as

$$x_{k+1} = f\left(x_k\right) + g_k \tag{6.1}$$

$$z_{k+1} = h\left(x_{k+1}\right) + v_{k+1} \tag{6.2}$$

In eqns. (6.1) and (6.2), $f(.)$ and $h(.)$ are the nonlinear vector valued functions; and g_k, the process noise and v_{k+1}, the measurement noise (temporally/spatially uncorrelated with each other) are zero-mean white Gaussian random sequences with covariance matrices Q_k and R_k, respectively:

$$Q_k = E\left[g_k g_k^T\right]; R_{k+1} = E\left[v_{k+1} v_{k+1}^T\right] \tag{6.3}$$

For a stochastic dynamic system, the *a posteriori* (after incorporating the observations into the EIF) normal/Gaussian pdf is expressed as

$$p(\hat{x}_k | z_k) = \sum_{j=1}^{N} w_k^j \mathrm{N}\left(\hat{x}_k; x_k^j, \hat{P}_k^j\right) \tag{6.4}$$

Here, N is the total number of Gaussian densities used; and w_k^j, x_k^j and $\hat{P}_k^j$ represent, respectively, the conditional weight/s, mean/s and covariance/s of the *j*thGaussian component with respect to the first k measurements. The usual condition on the weight is

$$\sum_{j=1}^{N} w_k^j = 1; w_k^j \geq 0 \tag{6.5}$$

Since, all the components of the mixture pdf in eqn. (6.4) are Gaussian, only estimates of their means and covariance/s need to be considered between k and $k+1$ [7].

The extended information filter has two phases: prediction phase and correction phase. The prediction phase uses the *posteriori* estimate at the previous timestep to produce *a priori* estimate at the current timestep. In the correction phase, measurement at the current timestep is taken to correct the *a priori* estimate so that it will be more accurate, and it will result in a new *posteriori* estimate.

6.2.1 Time Propagation/Evolution

The predicted information state $y(.)$, and information matrix $Y(.)$ are defined as

$$\tilde{y}_{k+1}^j = \hat{Y}_k^j f_k(\hat{x}_k^j) \tag{6.6}$$

$$\tilde{Y}_{k+1}^j = \left(F_k\left(\hat{Y}_k^j\right)^{-1} F_k^T + Q_k\right)^{-1} \tag{6.7}$$

where the Jacobian is given as

$$F_k = \frac{\partial f_k\left(\hat{x}_k^j\right)}{\partial \hat{x}_k^j}, \text{ from eqn. (6.1)} \tag{6.8}$$

In eqn. (6.6), $Y(.) = P^{-1}(.)$, (P is the KF covariance matrix), $Y(.)$ can be calledinformation Gramian (IG). The Runge–Kutta algorithm can be used for integration of eqn. (6.6), and the state-transition matrix φ_k can be approximated using Taylor series expansion. In time update, the weights are not updated:

$$\tilde{w}_{k+1}^j = \hat{w}_k^j \tag{6.9}$$

6.2.2 Measurement/Data Update

The information-pair (information state y, or i, and the information matrix Y, or I; as intermediate computations are termed for the sake of convenience) associated with an observation is given by

$$i_k^j = H_k^T R_k^{-1}\left(e_k^j + H_k \tilde{x}_k^j\right) \tag{6.10}$$

$$I_k = H_k^T R_k^{-1} H_k \tag{6.11}$$

Innovation sequence (in the covariance domain) is

$$e_k^j = z_k - h(\tilde{x}_k^j) \tag{6.12}$$

Here, measurement matrix is obtained as the Jacobian

$$H_{k+1} = \frac{\partial h_{k+1}(\tilde{x}_{k+1}^j)}{\partial \tilde{x}_{k+1}^j} \tag{6.13}$$

The final estimated information state (y) and information matrix (Y) are given as

$$\hat{y}_k^j = \tilde{y}_k^j + i_k^j \tag{6.14}$$

$$\hat{Y}_k^j = \tilde{Y}_k^j + I_k^j \tag{6.15}$$

The estimated covariance state x, can be calculated using the estimated information state and the information matrix as follows:

$$\hat{x}_k^j = \left(\hat{Y}_k^j\right)^{-1} \hat{y}_k^j \tag{6.16}$$

This state is the one that is normally appreciated in our analysis of the results and even further used in the control and forecasting applications in/for real systems and situations; however, the use of the information concept is highly motivating and useful in decentralized and distributed sensor data-fusion algorithms and applications.

6.2.3 Gaussian-Sum Estimates

The new weights for eqn. (6.4), and eqn. (6.5) are calculated as

$$\hat{w}_{k+1}^{j} = \frac{\tilde{w}_{k+1}^{j}\mathbb{N}\left(z_{k+1} - h_{k+1}\left(f_k\left(\hat{x}_k^{j}\right)\right), S_{k+1}^{j}\right)}{\sum_{j=1}^{N}\tilde{w}_{k+1}^{j}\mathbb{N}\left(z_{k+1} - h_{k+1}\left(f_k\left(\hat{x}_k^{j}\right)\right), S_{k+1}^{j}\right)} \tag{6.17}$$

$$\text{where,}\quad S_{k+1}^{j} = H_k(\tilde{Y}_{k+1}^{j})^{-1}H_k^T + R_k \tag{6.18}$$

The final state estimate at each timestep is the weighted average from eqns. (6.16) and (6.17):

$$\hat{x}_{k+1} = \sum_{j=1}^{N}\hat{w}_{k+1}^{j}\hat{x}_{k+1}^{j} \tag{6.19}$$

In the subsequent analysis, the measurement vector/matrix H, the process noise covariance matrix Q, and the measurement noise covariance matrix R are represented without any time index/subscript for the sake of simplicity; also the variation of H, since it is time dependent will be taken care of in the program code for the implementation of the GSEIF.

6.3 ASYMPTOTIC STABILITY ANALYSIS OF GSEIF VIA AN OBSERVER

An appropriate observer is proposed and the conditions for the convergence of the GSEIF are established via the convergence of this observer, which for the nonlinear system of eqns. (6.1) and (6.2) is given as

$$\hat{x}_{k+1}^{j} = f(\hat{x}_k^{j}) + K_{k+1}^{j}(z_{k+1} - \hat{z}_{k+1}^{j}) = f(\hat{x}_k^{j}) + K_{k+1}^{j}r^{j}{}_{k+1} \tag{6.20}$$

$$\hat{z}_{k+1}^{j} = h(\hat{x}_{k+1}^{j}) \cong H\hat{x}_{k+1}^{j} \tag{6.21}$$

Here, the observer is proposed as if it is for a typical EKF, with its gain as the EKF-gain for the discrete KF; this does not matter since the observer is used as a vehicle for establishing the convergence of EIF, and especially the GSEIF, as long as the gain and other IG matrices are from the main IF; thus this joint scheme can be utilized for obtaining the stability conditions for the inherent IF. The observer gain is taken as

$$K_{k+1}^{j} = (\tilde{Y}_{k+1}^{j})^{-1}(H_{k+1}^{j})^T(S_{k+1}^{j})^{-1} \text{ with } S_{k+1}^{i} = H_{k+1}^{j}(\tilde{Y}_{k+1}^{j})^{-1}(H_{k+1}^{j})^T + R$$

Since there are N filters, the weights of each filter have to be considered to show the combined final effect. The combined weights, *a posterior* error, *a priori* error, residual error, *a posterior* information Gramian (related to the *posterior* information state error) and *a priori* information Gramian (related to the *priori* information state error) are defined/specified here for the sake of clarity and completion for the time index 'k':

$$w_k = \left[w_k^1\, w_k^2 \,\ldots w_k^N\right] \tag{6.22}$$

$$\hat{e}_k = \sum_{j=1}^{N} w_k^j \hat{e}_k^j \tag{6.23}$$

$$\tilde{e}_k = \sum_{j=1}^{N} w_k^j \tilde{e}_k^j ; r_k = \sum_{j=1}^{N} w_k^j r_k^j \tag{6.24}$$

$$\hat{Y}_k = \sum_{j=1}^{N} w_k^j \hat{Y}_k^j \tag{6.25}$$

$$\tilde{Y}_k = \sum_{j=1}^{N} w_k^j \tilde{Y}_k^j \tag{6.26}$$

The observer error is obtained by subtracting eqn. (6.20) from eqn. (6.1)

$$\hat{e}_{k+1}^j = x_{k+1} - \hat{x}_{k+1}^j = f(x_k) - f(\hat{x}_k^j) - K_{k+1}^j (z_{k+1} - \tilde{z}_{k+1}^j) \tag{6.27}$$

Since, the total estimate from the observer has to be considered for comparison with the actual states, a short-hand notation/symbol for sigma operation is used as in eqn. (6.23), and others: $E\{.\} = \sum_{j=1}^{N} w_k^j (.)$, and with this notation eqn. (6.27) is rewritten as

$$\hat{e}_{k+1} = E\{\hat{e}_{k+1}^j\} = f(x_k) - E\{f(\hat{x}_k^j)\} - E\{(z_{k+1} - \tilde{z}_{k+1}^j)\} \tag{6.28}$$

$$\hat{e}_{k+1} = F_x x_k + \phi_k - E\{(z_{k+1} - \tilde{z}_{k+1}^j)\} \tag{6.29}$$

$$\text{where} \quad \phi_k = -F_x x_k + f(x_k) - E\{f(\hat{x}_k^j)\} \tag{6.30}$$

The nonlinear function as 'phi' in eqn. (6.30) is included for facilitating the convergence analysis, otherwise both eqn. (6.28) and eqn. (6.29) are exactly identical. For the sake of simplicity in the analytical development, the process noise term is avoided, but can be easily incorporated. In eqn. (6.29) one can substitute for *K*(.) to obtain after the simplification of the measurement terms, i.e. the residuals

$$\hat{e}_{k+1} = F_x x_k + \phi_k - E\{\tilde{P}_{k+1}^j H^T (H \tilde{P}_{k+1}^j H^T + R)^{-1} H \tilde{e}_{k+1}^j)\}$$

Substituting for *P* in terms of *Y* one obtains

$$\hat{e}_{k+1} = F_x x_k + \phi_k - E\{(\tilde{Y}_{k+1}^j)^{-1} H^T (H(\tilde{Y}_{k+1}^j)^{-1} H^T + R)^{-1} H \tilde{e}_{k+1}^j)\}$$

Then substitute for the information matrix from eqn. (7), to obtain

$$\hat{e}_{k+1} = F_x x_k + \phi_k - E\{(F_k (\hat{Y}_k^j)^{-1} F_k^T + Q) H^T [H(F_k (\hat{Y}_k^j)^{-1} F_k^T + Q) H^T + R]^{-1} H \tilde{e}_{k+1}^j)\} \tag{6.31}$$

Since, Lyapunov–Krasovskii's approach is utilized to derive the conditions for the convergence of the observer error, eqn. (6.31), the following development is in order.

Let the error state in general be described as

$$e(k+1) = Ge(k) \tag{6.32}$$

Let us define a feasible normalized Lyapunov energy function (LEF) as

$$V\left(e(k)\right) = e^T(k)Y(k)e(k) \tag{6.33}$$

In eqn. (6.33) $Y(.)$ is the information matrix and it is a positive definite matrix, and we use the following expression as the (equivalent or analogous) derivative of the LEF

$$\Delta\left\{V\left(e(k)\right)\right\} = V\left(e(k+1)\right) - V\left(e(k)\right) \tag{6.34}$$

In the continuous-time case for the similar convergence analysis one uses the time derivative of the LEF (defined in the continuous-time domain) and seeks the conditions such that this time derivative is negative definite, thereby establishing the convergence of the observer error dynamics like in eqn. (6.31). For the discrete case we use the differencing operation. Substituting for $V(.)$s' in eqn. (6.34) we obtain

$$\begin{aligned}\Delta\left\{V\left(e(k)\right)\right\} &= V\left(e(k+1)\right) - V\left(e(k)\right)\\ &= e^T(k+1)Ye(k+1) - e^T(k)Ye(k)\\ &= e^T(k)G^TYGe(k) - e^T(k)Ye(k)\\ &= e^T(k)\left[G^TYG - Y\right]e(k)\end{aligned} \tag{6.35}$$

Since the 'derivative' term of the error should be negative definite for the convergence of the observer error dynamics we have

$$\Delta\left\{V\left(e(k)\right)\right\} = -e^T(k)\left[-\left(G^TYG - Y\right)\right]e(k) \tag{6.36}$$

Then for the convergence of the observers' error dynamics we have the following condition:

$$-\left(G^TYG - Y\right) = Y - G^TYG \tag{6.37}$$

should be positive definite (PD). This means that in terms of the magnitudes of the involved terms, the following should hold true:

$$\left|Y - G^TYG\right| > 0 \tag{6.38}$$

Since the magnitude of any matrix is measured by its norm, one can arrive at the final condition as

$$\left\|Y - G^TYG\right\| > 0 \Rightarrow \left\|G^2 < 1\right\| \tag{6.39}$$

Next, the error dynamics eqn. (6.31) should be brought in the form of eqn. (6.38) and/or eqn. (6.32), so that one can easily derive the conditions for the convergence, and in order to do this it is necessary to define certain bounds:

1. The information matrix of eqn. (6.7) must be in the range:

$$\rho_l I \le Y(.) \le \rho_u I \tag{6.40}$$

Here, $\rho_u, \rho_l > 0$ are +ve numbers, $Y(.)$ is inherently symmetric/positive definite (PD) matrix; and these constants specify the bounds.

2. The term $F_k x_k$ should also be bounded as

$$\left\|F_k x_k\right\| = \rho_1 \left\|e_k^2\right\| \tag{6.41}$$

The variable ρ_1 is the upper bound of the associated nonlinear function as a scalar factor of the state-error norm.

3. The nonlinear function 'phi' in eqn. (30) is bounded, because it is composed of the bounded functions of the dynamic system:

$$\left\|\phi(.)\right\| \leq \rho_2 \left\|x_k - \hat{x}_k^2\right\| = \rho_2 \left\|e_k^2\right\| \tag{6.42}$$

The variable ρ_2 is the upper bound of the associated nonlinear function as a scalar factor of the state error norm. Also, the following bounds on the norms of matrices H and R are needed to simplify the error dynamics

$$\left\|H^T H\right\| \leq h^2 \tag{6.43}$$

$$\left\|R\right\| = r; \left\|e_k^2\right\| \leq \varepsilon^2 \tag{6.44}$$

$$\left\|e_k\right\| \leq \varepsilon = \text{some constant}$$

Next, the observer's error dynamics of eqn. (6.31) should be brought in the form of eqn. (6.39)

$$\begin{aligned} \hat{e}_{k+1} &= F_x x_k + \phi_k - E\{(F_k(\hat{Y}_k^j)^{-1} F_k^T + Q)H^T[H(F_k(\hat{Y}_k^j)^{-1} F_k^T + Q)H^T + R]^{-1} H\tilde{e}_{k+1}^j)\} \\ &= G\hat{e}_k \end{aligned} \tag{6.45}$$

$$\left\|\hat{e}_{k+1}\right\| = \left\|G\right\| \hat{e}_k \| \tag{6.46}$$

Expressing eqn. (6.45) in terms of norms

$$\begin{aligned} \left\|\hat{e}_{k+1}\right\| = \left\|F_x x_k\right\| + \left\|\phi_k\right\| - \Big\|E\{(F_k(\hat{Y}_k^j)^{-1} F_k^T + Q)H^T[H(F_k(\hat{Y}_k^j)^{-1} F_k^T \\ + Q)H^T + R]^{-1} H\tilde{e}_{k+1}^j)\}\Big\| \end{aligned} \tag{6.47}$$

Substituting various bounds defined earlier into eqn. (6.47) one obtains

$$\begin{aligned} \left\|\hat{e}_{k+1}\right\| = \rho_1 \left\|\hat{e}_k\right\|^2 + \rho_2 \left\|\hat{e}_k\right\|^2 - \Big\|E\{(F_k(\hat{Y}_k^j)^{-1} F_k^T + Q)H^T[H(F_k(\hat{Y}_k^j)^{-1} F_k^T + Q) \\ H^T + R]^{-1} H\tilde{e}_{k+1}^j)\}\Big\| \end{aligned} \tag{6.48}$$

After substituting other bounds from eqns. (6.43) and (6.44) and simplifying the resultant expression, and utilizing the definition of error norm, one obtains

$$\| \hat{e}_{k+1} \| = \rho_1 \left\| \hat{e}_k \right\|^2 + \rho_2 \left\| \hat{e}_k \right\|^2 - N\left[\left\{ \left(\frac{a^2}{\rho_l} + q \right) h^2 \right\} \left\{ \left(\frac{a^2}{\rho_l} + q \right) h^2 + r \right\}^{-1} \right] \varepsilon = \left\| G \right\| \left\| e_k \right\| \tag{6.49}$$

$$= \left\| G \right\| \left\| e_k \right\| \tag{6.50}$$

In eqn.(6.49), $|F| = a$.

Eqn. (6.50) is now in the required form of eqn. (6.46). From eqn. (6.50) one obtains the following expression:

$$\| G \| = \rho_1 \varepsilon + \rho_2 \varepsilon - N\left[\left\{ \left(\frac{a^2}{\rho_l} + q \right) h^2 \right\} \left\{ \left(\frac{a^2}{\rho_l} + q \right) h^2 + r \right\}^{-1} \right] \tag{6.51}$$

Applying the condition of eqn. (6.39) to eqn. (6.51) and considering the positive square root of 1, one obtains

$$\| G \|^2 = \left\{ \rho_1 \varepsilon + \rho_2 \varepsilon - N\left[\left\{ \left(\frac{a^2}{\rho_l} + q \right) h^2 \right\} \left\{ \left(\frac{a^2}{\rho_l} + q \right) h^2 + r \right\}^{-1} \right] \right\}^2 < 1 \tag{6.52}$$

$$\| G \| = \left\{ \rho_1 \varepsilon + \rho_2 \varepsilon - N\left[\left\{ \left(\frac{a^2}{\rho_l} + q \right) h^2 \right\} \left\{ \left(\frac{a^2}{\rho_l} + q \right) h^2 + r \right\}^{-1} \right] \right\} < \mathrm{sqrt}(1) = +1 \tag{6.53}$$

$$\| G \| = \{ \rho_1 \varepsilon + \rho_2 \varepsilon - N\left[\left\{ \left(\frac{a^2}{\rho_l} + q \right) h^2 \right\} \left\{ \left(\frac{a^2}{\rho_l} + q \right) h^2 + r \right\}^{-1} \right] < 1 \tag{6.54}$$

Thus, the convergence condition for the observer's error dynamics is obtained as

$$\| G \| = \{ \rho_1 \varepsilon + \rho_2 \varepsilon < 1 + N\left[\left\{ \left(\frac{a^2}{\rho_l} + q \right) h^2 \right\} \left\{ \left(\frac{a^2}{\rho_l} + q \right) h^2 + r \right\}^{-1} \right] \tag{6.55}$$

All the coefficients appearing in eqn. (6.55) are well defined as these are the positive bounding constants in eqns. (6.40)–(6.44). The condition of eqn. (6.55) means that if it is satisfied then, the time derivative, analogously the differential in eqn. (6.36) of the Lyapunov energy functional would be negative definite and the observer error dynamics would converge and the error would (asymptotically) tend to zero. This in turn, also establishes that the GSEIF would also converge, since its information matrix was used in the observer error dynamics. This is a novel way of establishing the asymptotic stability of the discrete-time GSEIF via the use of the deterministic observer.

6.4 AIRCRAFT PARAMETER ESTIMATION

The longitudinal motion of aircraft is represented using the following postulated aircraft model [1,2]:

$$\dot{u} = \frac{\bar{q}S}{m} C_X - qw - g\sin\theta$$

$$\dot{w} = \frac{\bar{q}S}{mV}C_Z + qu + g\cos\theta$$

$$\dot{\theta} = q$$

$$\dot{q} = \frac{\bar{q}S\bar{c}}{I_{yy}}C_m \qquad (6.56)$$

In eqn. (6.56), C_x, C_z and C_m are the nondimensional aerodynamic coefficients expressed as

$$C_Z = C_{Z_0} + C_{Z_\alpha}\alpha + +C_{Z_q}\frac{q\bar{c}}{2V} + C_{Z_{\delta_e}}\delta_e$$

$$C_X = C_{X_0} + C_{X_\alpha}\alpha + C_{X_{\alpha^2}}\alpha^2$$

$$C_m = C_{m_0} + C_{m_\alpha}\alpha + +C_{m_{\alpha^2}}\alpha^2 + C_{m_q}\frac{q\bar{c}}{2V} + C_{m_{\delta_e}}\delta_e \qquad (6.57)$$

Here, the state variables are true air velocity V, angle of attack α, pitch angle θ and pitch rate q. The elevator deflection δ_e is the control input to aircraft dynamics. Other relevant parameters are as follows: the mass of the aircraft is m and its geometrical parameters consist of wing area S, wing chord $\bar{c}$ and moment of inertia I_{yy}. The dynamic pressure $\bar{q}$ is given by $\frac{1}{2}\rho V^2$, where ρ is the density of air and V is the true air speed; and all the variables have the SI-metric units, and have appropriate dimensions. The measurement vector is given by

$$u = u$$

$$w = w$$

$$q = q$$

$$\theta = \theta$$

$$a_x = \frac{\bar{q}S}{m}C_X$$

$$a_y = \frac{\bar{q}S}{m}C_Z$$

$$\dot{q} = \dot{q} \qquad (6.58)$$

It is presumed that each component of the measurementvector has some bias and/or random noise, but it is not shown in the model of eqn. (6.58).The unknown parameter vector β that is to be estimated consists of the aerodynamic derivatives:

$$\beta = [C_{X_0}\ C_{X_\alpha}\ C_{X_{\alpha^2}}\ C_{Z_0}\ C_{Z_\alpha}\ C_{Z_{\delta_e}}\ C_{m_0}\ C_{m_\alpha}\ C_{m_{\alpha^2}}\ C_{m_q}\ C_{m_{\delta_e}}] \qquad (6.59)$$

6.5 RESULTS AND DISCUSSIONS

To evaluate the efficacy of EIF and GSEIF, the short period data of a light transport aircraft (LTA) are simulated and the nondimensional longitudinal parameters are estimated. The 4DOF longitudinal axis model is used for estimation; using eqns. (6.56) to (6.58), the data are generated with a sampling interval of 0.03 sec for a duration of nearly 7 sec by giving a doublet input to the elevator control surface. The initial states for $[u, w, q, \theta]$ are chosen as in [2]. Random Gaussian noise with standard deviation 0.001 (as the process noise) is added to the states u, w, q and θ. Random noise with SNR = 10 dB is added to these measurements, eqn. (6.58) to generate noisy data. A fourth-order RK integration method is used to obtain the longitudinal flight variables u, w, q and θ from eqn. (6.57). Two examples are considered to demonstrate the effectiveness of GSEIF over EIF; the metric PEEN (parameter estimation error norm) is evaluated for this and is defined

$$\text{PEEN} = 100 \times \frac{\text{norm}(\beta_{\text{true}} - \beta_{\text{est}})}{\text{norm}(\beta_{\text{true}})} \tag{6.60}$$

Where β_{true} is the vector of true parameters and β_{est} is the vector of estimated parameters. The parameter-estimation results for the examples taken are shown in Table 6.1, and the comparison of PEENs, and the computational times is depicted in Table 6.2. The time histories of the estimated parameters for Example 6.2 are shown in Figure 6.1.

Example 6.1 (benign initial conditions): In this example, the initial states/the parameters for the EIF are taken 10% away from their true values (used for generating the simulated data); and the initial covariance matrix $P(0)$ is chosen to reflect these uncertainties in the initial values. The estimated values of the parameters are compared with the true values as in second column of Table 6.1. It is seen that the estimates are fairly close to the true values even when there is some noise (SNR = 10) in the measurement data, and the PEEN related to this is 4.12%, which is reasonably small. In the case of the GSEIF, the number of Gaussian components required for effective filtering is less, as the initial state is close to the true value used for simulation; only five filters have been used for

TABLE 6.1
Estimated Parameters for EIF/GSEIF: Example 6.1 and Example 6.2

Parameter	True Values	EIF-Ex1	GSEIF-Ex1	EIF-Ex2	GSEIF-Ex2
C_{x0}(*)	-0.0540	-0.0617	-0.0556	-0.0733	-0.0547
$C_{x}\alpha$	0.2330	0.2557	0.2329	0.3696	0.2300
$C_{x}\alpha^2$	3.6089	3.7092	3.6474	3.4706	3.6288
C_{z0}	-0.1200	-0.1239	-0.1209	-0.1404	-0.1216
$C_{z}\alpha$	-5.6800	-5.6988	-5.6594	-5.6538	-5.6389
$C_{z}\delta_e$	-0.4070	-0.4450	-0.4006	-0.4244	-0.3941
C_{m0}	0.0550	0.0511	0.0532	0.0901	0.0532
$C_{m}\alpha$	-0.7290	-0.7147	-0.7134	-0.9724	-0.7060
$C_{m}\alpha^2$	-1.7150	-1.6432	-1.6828	-0.9506	-1.6644
C_{mq}	-16.300	-17.036	-16.0190	-13.8587	-15.7331
$C_{m}\delta_e$	-1.9400	-1.8931	-1.8999	-1.7300	-1.8800
PEEN(%)		4.119	1.3517	14.593	3.1053

(*) The aerodynamic derivatives are expressed in the form of the text for convenience.

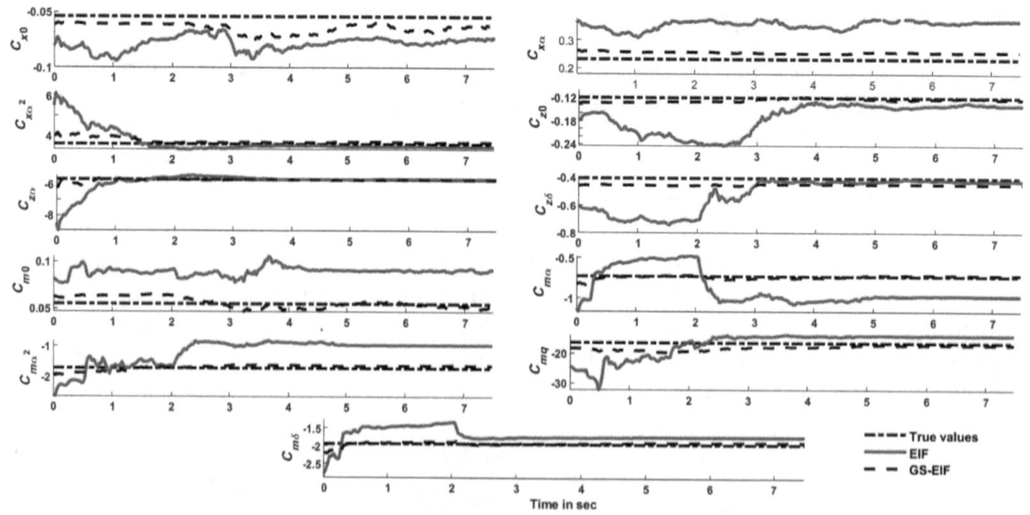

FIGURE 6.1 Time history match of the estimated parameters for Example 6.2.

TABLE 6.2
Comparison of PEENs and Computational Times

		PEEN in %		Comp. Times in Seconds	
	Number of Filters	**EIF**	**GSEIF**	**EIF**	**GSEIF**
Example 6.1	3	4.119	3.2256	0.3281	0.96875
	5		1.3517		1.5156
	6		1.3517		1.875
Example 6.2	20	14.593	4.6785	0.3281	3.5346
	25		3.1053		3.9844
	30		3.1053		4.3594

GSEIF. The estimated parameters in Table 6.1 are those of the average of the last ten data points. The GSEIF offers better estimation than EIF. When three Gaussian components were used, PEEN of GSEIF was 3.23%, though the value is higher than that with five components with PEEN of 1.35%, the estimation accuracy is still better than EIF. When the number of components was increased to six, the PEEN of GSEIF remained the same as that obtained using five components. This shows that five Gaussian components are sufficient for estimation in this case and increasing the number of Gaussian components beyond five will not enhance the performance.

Example 6.2 (far away initial conditions): In this example, the initial state is chosen far (~100 %) from the true values. A greater number of Gaussian components are required in GSEIF, since the initial states are not accurately known and are presumed to be far from the true values. In this example, 30 Gaussian components have been used for GSEIF. A comparison is done by increasing and decreasing the number of components and the results are presented in Table 6.2. As GSEIF uses a number of Gaussian components for estimating, the results are fairly satisfactory and definitely better than those of the EIF. The time histories of the estimated parameters of EIF and GSEIF are shown in Figure 6.1. The comparative CPU time requirements for Examples 6.1

and 6.2 are shown in Table 6.2. The computational time of GSEIF depends upon the number of filters used. As the number of Gaussian components increase the computational load would also increase.

Thus, in this chapter Gaussian-sum extended information filter (GSEIF) has been used for aircraft parameter estimation and the asymptotic stability condition for the same has been derived using Lyapunov–Krasovskii's approach via proposition of a nonlinear observer that uses the gain and the covariance/information matrices/Gramians from the basic EIF and the normalized Lyapunov energy functional. Then, the performances of an EIF and the GSEIF for estimation of aircraft parameters from the realistically simulated flight data have been validated. The number of Gaussian components required for GSEIF depends on how well the initial estimate is known. The computational time increases as the number of components in GSEIF increases. The simulation results obtained proves that GSEIF gives better estimation results than EIF. The results are encouraging and points towards future research aiming the implementation of advanced approximations methods and modelling techniques for more complex nonlinear estimation problems, where the aircraft parameter estimation (APE) is a special case, for which the analytical and numerical convergence results have been established in the present paper.

EXERCISES

Exercise 6.1

What are the main differences between the Kalman filter (KF) and the information filter (IF)?

Exercise 6.2

What is the primary goal of the Gaussian-sum extended information filter (GSEIF), and how does it differ from extended information filter?

Exercise 6.3

Describe the Lyapunov stability analysis approach used to assess the stability of GSEIF. What are the convergence conditions derived from this?

Exercise 6.4

What is the primary focus of the evaluation conducted using the Gaussian-sum extended information filter (GSEIF) in the context of the aerodynamic parameter-estimation problem?

Exercise 6.5

What real-world applications or domains can benefit the most from the Gaussian-sum extended information filter (GSEIF), considering its ability to handle complex, nonlinear systems with uncertain dynamics?

REFERENCES

1. Assimakis, N., Adam, M., and Douladiris, A., Information filter and Kalman filter comparison: Selection of the faster filter, *International Journal of Information Engineering*, Vol. 2, No. 1, pp. 1–5, 2012.

2. Sara, M.G., *Approximations in Nonlinear Filtering: Some Novel Approaches.* Ph.D. Thesis, Electrical and Electronics Engineering Sciences, Visvesvaraya Technological University (VTU), Jnana Sangama, VTU Main Road, Machhe, Belagavi, Karnataka, 2023.
3. Raol, J.R., Girija, G., and Singh, J., *Modelling and Parameter Estimation of Dynamic Systems*, The IEE/IET Control Series, Vol. 65, London, UK, 2004.
4. Garcfa-Velo, J., and Walker, B.K., Aerodynamic parameter estimation for high performance aircraft using extended Kalman filtering, *Journal of Guidance, Control, and Dynamics, USA*, Vol. 20, No. 6, 1997, pp. 1257–1259.
5. Raol, J.R., Gopalratnam, G., and Twala, B., *Nonlinear Filtering: Concepts and Engineering Applications*, CRC Press, Taylor & Francis, Boca Raton, 2017.
6. Muthambara, A.O., *Decentralized Estimation and Control for Multisensor Systems*, CRC Press, Boca Raton, 1998.
7. Alspach, D.L., and Sorenson, H.W., Nonlinear Bayesian estimation using Gaussian sum approximations, *IEEE Transactions on Automatic Control, USA*, Vol. 17, No. 4, pp. 439–448, Aug. 1972.
8. Terajanu, G., Singla, P., Singh, T., and Scott, P.D., Adaptive Gaussian sum filter for nonlinear Bayesian estimation, *IEEE Transactions on Automatic Control, USA*, Vol. 56, No. 9, pp. 2151–2156, 2011.

7 Image Centroid Tracking with Square-Root Kalman Filters

7.1 INTRODUCTION

The key component in defence systems and aerospace applications like air traffic control and/or management (ATC/ATM) is target recognition. Whenever there is an expected danger, automatic-target recognition (ATR) plays an important role in safety and warning. Identification and tracking of objects are the steps involved here, before which the sequence of images are processed. Imaging sensors capture a number of continuous frames which when put together form the required image sequence. Therefore, identity estimation and tracking algorithms will involve the following: (i) detection of target, (ii) segmentation, (iii) calculation of appropriate features like centroid, and (iv) tracking using different tracking algorithms. Performance evaluation and enhancing the efficiency of centroid detection and tracking using fusion strategies is the main focus in this chapter.

For detection of a moving object, there are a number of techniques like template matching [1], which recognizes the segment with the highest correlation as the target, image features, and the object center distance by giving weights. The centroid method is, perhaps, best as it does not depend on the size of the object and it is least sensitive to the target intensities; and before implementing the tracking algorithm, spatial filtering, converting images from grey to binary and formation of cluster of pixels having the same intensities using the nearest neighbourhood criterion need to be performed [2].

7.1.1 Target Tracking and Fusion

Target tracking and fusion are very crucial aspects in the following: (i) military command, (ii) control, (iii) counter-terrorism, (iv) remote sensing, (v) autonomous navigation, (vi) robots, (vii) airfield surveys and (viii) ATC/ATM; also, multi-sensor/source information fusion (MSIF) is very useful for such applications; the latter is defined as a method of combining data from several sources and sensors to produce the specific and single decision about the object under test. This would obtain better accuracy and more direct inferences than could be obtained by use of only one sensor [3]. There are three important levels of sensor data fusion: (i) kinematic fusion, (ii) image fusion and (iii) decision fusion. While there are many tracking algorithms available in the literature, like KF, the information filter (IF) is a quite straightforward technique to handle the data in MSIF including image fusion [3].

DOI: 10.1201/9781003620211-7

7.1.2 Importance and Application of Tracking and Fusion

Some of the research aspects in image-target tracking, in general target tracking, and information fusion are the following: (i) if an incorrect measurement is associated with a track, the track could divert or cause the other tracks to diverge and thus the tracking could be terminated, (ii) combination of multiple data to improve situational awareness would be very useful, (iii) joint/simultaneous localization and mapping (SLAM) in robotics and for moving vehicles would also need tracking, (iv) track-before-detect strategies for exploiting image tracking/fusion would be useful, e.g. as an extended tracking, (v) image processing/computer vision methods for automatic target identification and (vi) programs for enhancing the range and quality of images via passive and active imaging would be needed.

Image tracking and fusion technology has got numerous applications: (a) satellite, (b) seismic and (c) weather forecasting. By fusion of information from radars, scanning lasers and stereo cameras, the depth and velocity observations of targets are tracked. In many computer-vision applications fusion is used in feature tracking, and cluster tracking of the objects of interest. Autopilot applications include the determination of the attitude and heading of the autonomous systems. The brain-computer interface, structural health monitoring and human sensor motor processing require tracking. In the field of economics, like macroeconomics, time series and econometric tracking and fusion could be very useful.

7.1.3 Image Processing for Tracking Applications

Image processing-cum-fusion ensures the quality of a picture by adjusting it with a specific end goal to either (a) upgrade its data content for interpretation and examination, or (b) make it more reasonable for the machine recognition. The main applications of the image processing is in image tracking, which is extensively studied in the present work: (i) it is fundamentally used as a part of remote sensing, satellite imaging fields for the best possible view of images caught by satellites, (ii) it has been used in medical imaging for diagnosing disease by analyzing the captured images from different modalities (computer tomography, CT; and MRI, magnetic resonance imaginary), (iii) it is used in the military for detecting threats by providing the information for target recognition and (iv) in the robotics field, tracked images are used to examine the frequency variations in the surrounding objects.

7.1.4 Filtering and Image Fusion

There is always considerable interest in many military and civilian applications to deploy an automated system for video-based observation/surveillance, and the target-image tracking is one of the very important aspects of locating moving objects in real time using some online and appropriate filtering algorithm. Such an algorithm utilizes each image frame that arrives in a sequence, and outputs the location of moving target/s, and it has two basic steps: (a) detection of moving objects in each frame, and (b) filtering/tracking of the thus detected object/s in each consecutive frame [4]. There are several algorithms that could be utilized for target/image tracking, however, the square-root-type algorithms should be more suitable for the reasons that they are computationally more efficient, accurate and stable (compared to the conventional KF, and IF algorithms); and especially these aspects are very important for online/real-time applications. In several such situations, the acquired image would be often cluttered, dim, spurious and/or noisy, and at times, this might be due to the fact that the distance to the target from the sensing sites/centers is relatively great. The target-image tracking problem involves processing of measurements from deployed sensors/radars for the target of interest, producing at every scan, the estimate of the target's current position, velocity, acceleration states. The uncertainties that are mostly present, are considered as additive stochastic noise in the measured values; as well as the associated uncertainties in the target's states. In addition, more uncertainty with respect to the origin of the observed data, may not have actual observations from these targets; in most cases these are due to unwanted clutter (or false alarms). This aspect requires data association, the question can be asked: which measurement originated from which target? As a result, detection and tracking of a moving object is a reasonably difficult problem in forward-looking infrared (FLIR) image sequences, and more because of: (i) low signal-to-noise (SNR) ratio, (ii) low (intensity) contrast, (iii) presence of background clutter/false alarms and/or (iv) a partial occlusion of target [5]; hence, there is a need for efficient, accurate and numerically stable filtering algorithms for image centroid tracking. Also, an associated idea is the concept of image fusion to enhance the useful information content.

7.1.5 Filtering Algorithms for Tracking Applications

There are several algorithms that could be utilized for target/image tracking, but here, the concentration is on square-root-type algorithms [5]. An image centroid tracking filtering algorithm (CTA) determines the moving object's position and velocity from a noisy time series of images captured by image sensors; and this constitutes a statistical estimation problem. Conventionally, covariance-based KF is used. At times, programming of KF on in finite-word computing machine could pose some problems; and the effect would be largely reduced by implementing it in factorized form; such factorization retains symmetry, and assures non-negativity of the covariance matrix P [5]. One such widely used form is the U-D factorization (UDF) filtering algorithm. The U and D, are the factors of P in KF; U is a unit-upper-triangular matrix (1's on diagonal) and D is diagonal matrix. The main merit in UDF is that the square-root-type algorithm processes square roots of the covariance matrix and it uses half the word length normally needed by the KFs; in UDF, covariance updates KF formulas and estimations are rearranged; specifically, recursions of U/D factors of $P = UDU^T$.

7.1.6 Merits of the Square-Root Information Filter

Information filtering (IF) is the more direct way of dealing with the target tracking and MSDF situations compared to covariance-based KF; it has a direct interpretation for track observation. But, if IF is programmed as is, it would be susceptible to computing round-off, and this can degrade tracking performance. This is important if the algorithm is employed in target tracking in a real-time, online mode. The square-root filtering algorithm is one solution to the problem of numerical accuracy/stability [6]. The implementation of a square-root information filter (SRIF) for tracking a centroid of a target image of interest would have better efficiency and stability and will be very useful for real-time applications.

7.1.7 Image Fusion

The process of consolidating at least two pictures or image data refers to image fusion. Images may be obtained from the similar sensors or different types of sensors. The fused images are handy when time, cost and safety are important parameters. In order to achieve high spatial resolution and high-quality spectral content, multi-level fusion algorithms can be designed, which give more efficient results. Combining pictures from various modalities would enhance the total information, compared to individual source images. The image-fusion process extracts the productive data from each of the source images and incorporates them into a single resultant image [7]. Three distinct levels of image-fusion techniques are as follows: (a) pixel, (b) feature and (c) decision.

Tracking is basically the estimation of a present state of a target using noisy observations. Because of target-model uncertainties and observations uncertainties, the problem is involved for tracking of a single target. This complexity increases when multiple targets are being tracked from observations from more sensors. KF uses these observations that contain random noise and estimates the states of the underlying dynamic system, which is actually a target model.

7.2 COMPUTATION OF AN IMAGE CENTROID

Since, KF is used as a basic tracking algorithm, its equations are given as

State (X, centroid), and error covariance (P) time evolution:

$$\tilde{X}_{(k+1)} = \Phi \hat{X}_{(k)} \tag{7.1}$$

$$\tilde{P}_{(k+1)} = \Phi \hat{P}_{(k)} \Phi^T + GQG^T \tag{7.2}$$

State and error covariance update:

$$K_{(k+1)} = \tilde{P}_{(k+1)} H^T \left[H \tilde{P}_{(k+1)} H^T + R \right]^{-1} \tag{7.3}$$

$$\hat{X}_{(k+1)} = \tilde{X}_{(k+1)} + K_{(k+1)} \left[Z_{(k+1)} - H \tilde{X}_{(k+1)} \right] \tag{7.4}$$

$$\hat{P}_{(k+1)} = \left[\mathrm{I} - \mathrm{K}_{(k+1)} \mathrm{H} \right] \tilde{P}_{(k+1)} \tag{7.5}$$

The image centroid is the state $X(.)$ of the KF; H is the observation model matrix/vector, 'phi' is the state-transition matrix by which the state (the centroid of the tracked image) moves from time t to t+1, Q is process covariance matrix, and R is observations' covariance matrix. K is Kalman gain, and $Z(.)$, the centroid measurement vector.

The object of interest has to be detected in the given frame of an image. There are different parameters to be considered while identifying the target, i.e. the object in an image. An image frame can have a single object or multiple objects. Hence, there are different steps involved for both single and multiple object detection. In the current scenario, only a single object is considered, which has to be tracked. The image centroid is one such important parameter used for detecting the object of interest or target in the image. In an image CTA, particle segmentation is used for separation of a target from the background, if the target is not completely visible, then pixel intensities are discretized into 256 grey bands: (i) the grey-level image is transformed into a binary image using boundary limits of the target, which are ascertained using pixel-intensity histograms from the target and its surrounding; and (ii) these detected pixels are collected into clusters with the nearest-neighbour technique. The clusters are represented by a vector, which may not be a member of the data set. The grey image Im(i, j) is transformed into a binary image of intensity β (i, j) by a hard limit on this intensity:

$$\beta(i,j) = \begin{cases} 1 & I_L \le \mathrm{Im}(i,j) \le I_U \\ 0 & \text{otherwise} \end{cases} \tag{7.6}$$

where I_L and I_U are lower and upper bounds of the target intensity. The detection probability of the pixel (i, j) is

$$\begin{aligned} P\{\beta(i,j) = 1\} &= p(i,j) \\ P\{\beta(i,j) = 0\} &= 1 - p(i,j) \end{aligned} \tag{7.7}$$

$$p(i,j)=\frac{1}{\sigma\sqrt{2\pi}}\int_{I_L}^{I_U} e^{\frac{-(x-\mu)^2}{2\sigma^2}}dx$$

In eqn. (7.7), the image $I(i,j)$ has a Gaussian distribution with mean μ and variance σ^2. The binary image is grouped into clusters with the nearest-neighbour method. A pixel is taken as belonging to the cluster, in case the distance between the pixel and at least one other pixel is less than the proximity distance d_p:

$$\sqrt{\frac{1}{p_t}}<d_p<\sqrt{\frac{1}{p_v}} \tag{7.8}$$

In eqn. (7.8), p_t and p_v are detection probabilities, respectively, of target and noise pixels. The centroid of the group is taken as [2]

$$(x_e,y_e)=\frac{1}{\sum_{i=1}^{n}\sum_{j=1}^{m}I_{ij}}\left(\sum_{i=1}^{n}\sum_{j=1}^{m}iI(i,j),\sum_{i=1}^{n}\sum_{j=1}^{m}jI(i,j)\right) \tag{7.9}$$

where, I_{ij} is the intensity of the pixel and n and m are the dimensions of the cluster.

7.2.1 Synthetic Image Generation

The model of FLIR (forward-looking infrared sensor) for generating synthetic images is briefly described [3]:

Let the 2D of pixels be given as

$$m=m_\xi\times m_\eta \tag{7.10}$$

In eqn. (7.10), each pixel is signified by a single index $\boldsymbol{i}=\boldsymbol{1.....m}$ and the pixel intensity is

$$I_i=s_i+n_i \tag{7.11}$$

where the target intensity is $\boldsymbol{s}_i$ and $\boldsymbol{n}_i$ is the noise intensity, that is Gaussian with zero mean and covariance σ^2. The total target intensity

$$s=\sum_{i=1}^{m}s_i \tag{7.12}$$

If the number of target pixels is $\boldsymbol{m}_s$, then, the average target intensity is

$$\mu_s=\frac{s}{m_s} \tag{7.13}$$

The pixel SNR is

$$r'=\frac{\mu_s}{\sigma} \tag{7.14}$$

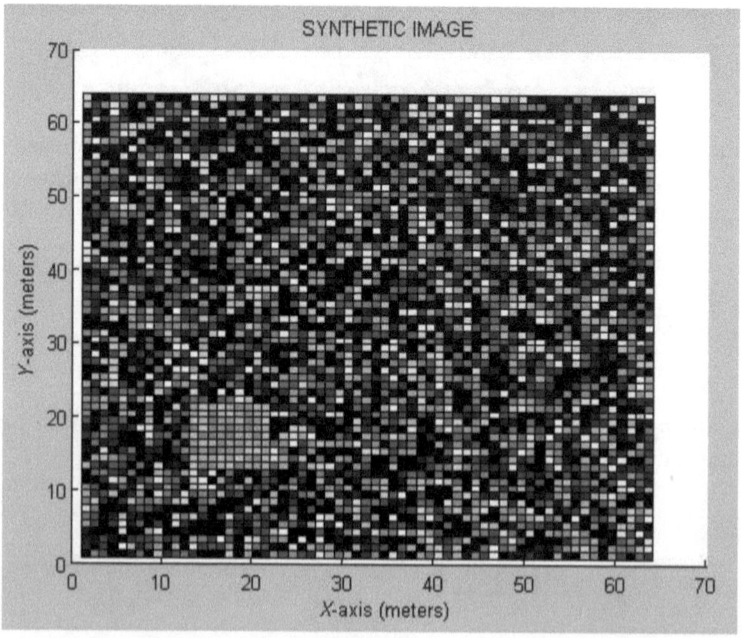

FIGURE 7.1 Synthetic image: target intensity is $N(100, 10)$; and background intensity is $N(50, 50)$.

Constant velocity model is used for generating the data that determine the position of the target at every sampling time. The image frame produced has a 2D array of 64 × 64 pixels, which is considered as the background image, and it is modelled as the white Gaussian field with mean μ_n and variance σ^2_n, i.e. $N(\mu_n, \sigma^2_n)$. A 2D array of pixels, that is the white Gaussian field with a mean μ_t and variance σ^2_t, i.e. $N(\mu_t, \sigma^2_t)$, is used for simulating a target of size (9×9). Here, 50 such frames for the purpose of tracking are generated. In this simulation model, background in the image has parameters N (50, 50) and target in the image is with the parameter N (100, 10). Figure 7.1 shows one of the frames of the synthetically generated image.

7.2.2 Image Centroid Tracking using KF

The image frames produced usually get contaminated by a variety of noise resources: (a) sensor inaccuracies in a camera, (b) variation in the detection sensitivity, (c) environmental variations, (d) nature of radiation, (e) dust covering the lenses, (f) quantization error or (g) during a point-to-point transmission of an image. These noises are removed by passing them through spatial filters. Then the processed image frames are used for tracking. A centroid detection and tracking algorithm (CTA) involves both motion recognition and object recognition in a given scenario of a target. In general, certain features of the target obtained by motion recognition or pattern-recognition techniques are used in associating images to the target. Motion-recognition features of a target involves location, velocity, acceleration, which are determined using the data from successive images, generated synthetically. Whereas object recognition features of a target involves its geometrical structure, i.e. size and shape, different grey levels in the image are obtained using image data. The features of the image considered are intensity and the size of the group. The pixel intensity is discretized into layers of grey-level intensities. It is assumed that sufficient target pixel intensities lie within the limits of certain target layers. The CTA, Figure 7.2, implementation requires the conversion of the data from the image into a binary image by applying limits for the target layers. The binary target image is converted to groups using the nearest-neighbour criterion. For the target of known size, the

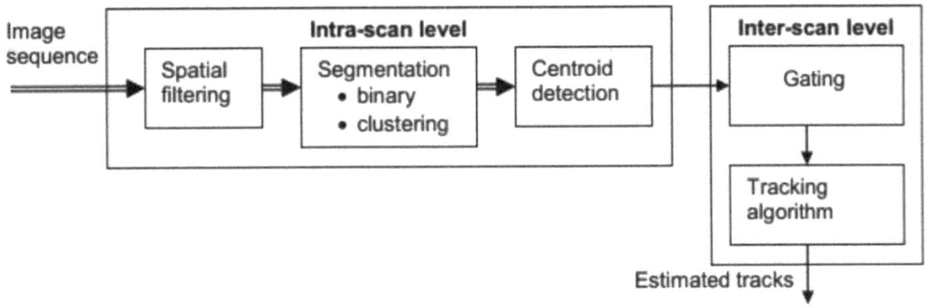

FIGURE 7.2 Image centroid tracking, CTA information flow.

information is used to set limits for removing the groups that differ in a sufficient way from the size of the target group. The centroid of the group is computed for tracking the target.

7.2.3 Performance with Target-Image Noise Variances for CTKF

The two broad methods for fusion of image data are as follows: (i) state-vector fusion (SVF) and (ii) measurement-level fusion (MLF). In SVF, the filter algorithm is applied independently for each image sensor. The centroid estimate is obtained by processing the observations from each image sensor; then fusion is done by combining two independent state estimates. The weighting factors are the corresponding covariance matrices from the KF. The fused estimate and fused covariance matrix are obtained as

$$\hat{x}_f = \hat{x}_1 + \hat{P}_1\left(\hat{P}_1 + \hat{P}_2\right)^{-1}\left(\hat{x}_2 - \hat{x}_1\right) \tag{7.15}$$

$$\hat{P}_f = \hat{P}_1 - \hat{P}_1\left(\hat{P}_1 + \hat{P}_2\right)^{-1}\hat{P}_1^T \tag{7.16}$$

Here, $\hat{x}_1$ and $\hat{x}_2$ are the estimated states from CTKF1 and CTKF2 (CT KF tracking algorithm) obtained using the observations from image sensor 1 and image sensor 2, respectively; $\hat{P}_1$ and $\hat{P}_2$ are the corresponding estimated error covariances from CTKF1 and CTKF2. In the MLF approach, image sensor observations are directly fused in one observation model and uses only one CTKF to obtain the estimate of the fused state vector. CTKF1 and CTKF2 represent the filter performance at two different target noise standard deviations and background noise standard deviations. The performance metrics of image fusion using KF indicate that MLF is better than SVF as seen from Table 7.1. In Figure 7.3 these metrics are plotted.

7.2.4 Performance with Target-Image Noise Variances for CTUDF

At times, implementation of KF on finite-word computing machine could pose a problem; and the effects would be greatly reduced by implementing it in a factorized form; such a factorization preserves symmetry and assures non-negativity of the covariance matrix *P* [5]. Such a requirement would be very useful for online/real-time implementation for the centroid tracking algorithm; because for a large-scale problem, and heavy computational needs, the KF might diverge. One such widely used form of the algorithm is the *U-D* factorization filter; *U* and *D*, are matrix factors of *P* of KF; *U* is unit upper-triangular (1's on diagonal elements) and *D*, a diagonal matrix. The merit of the *U-D* filter (UDF) is that, this square-root-type algorithm processes the square root of *P*, requiring only half the word-length usually used by KFs; i.e. essentially, the covariance update formulas of

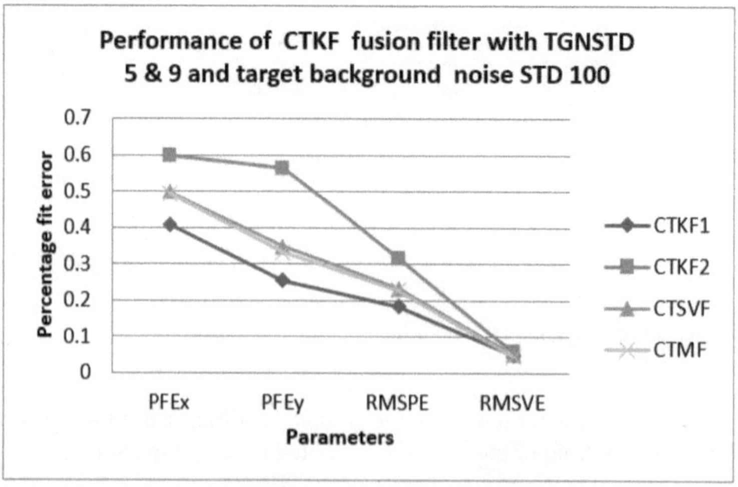

FIGURE 7.3 Performance of CTKF filter/fusion with (TGNSTD) target noise standard deviation 5 and 9 (units in the MATLAB script) and target background (TGBG) std 100.

TABLE 7.1
Performance of CTKF Fusion with Varying Target Noise Standard Deviations (stds)

	Target Noise std 1 & 5*				Target Noise std 1 & 9*				Target Noise std 5 & 9*			
Filter	PFEX	PFEY	RMSPE	RMSVE	PFEX	PFEY	RMSPE	RMSVE	PFEX	PFEY	RMSPE	RMSVE
CTKF1	0.3794	0.2641	0.1772	0.0467	0.3794	0.2641	0.1772	0.0467	0.4076	0.2549	0.1842	0.0493
CTKF2	0.4076	0.2549	0.1842	0.0493	0.5989	0.5644	0.3154	0.0592	0.5989	0.5644	0.3154	0.0592
CTMLF	**0.3883**	**0.2491**	**0.1766**	**0.0420**	**0.4794**	**0.3031**	**0.2172**	**0.0464**	**0.4944**	**0.3314**	**0.2279**	**0.0476**
CTSVF	0.3934	0.2567	0.1800	0.0480	0.4836	0.3196	0.2221	0.0529	0.4985	0.3484	0.2330	0.0542

* The mentioned numbsers are the stds in the MATLAB scripts for the noise intensities.

KF, eqn. (7.8), and eqn. (7.12); and the estimation equations are rearranged. One uses U and D factors of $P = UDU^T$.

The UDF is in two parts like the KF.

7.2.4.1 Time Propagation

The covariance matrix propagation from KF recursion is [7]

$$\tilde{P}(k+1 \mid k) = \varphi \hat{P}(k) \varphi^T + GQG^T \tag{7.17}$$

Let $\hat{P} = \hat{U}\hat{D}\hat{U}^T$ (*a priori* factors are assumed known), and Q is the process noise covariance matrix, the time propagated $\tilde{U}$ and $\tilde{D}$ factors are obtained by utilizing the modified Gram–Schmidt orthogonalization method: let $V = \left[\varphi \hat{U} \mid G \right]$ and $\bar{D} = \text{diag}[\hat{D}, Q]$, and $V^T = [v_1, v_2, ..., v_n]$, then P is $\tilde{P} = \tilde{V}\tilde{D}\tilde{V}^T$; then new U and D components of $\tilde{V}\tilde{D}\tilde{V}^T$ are obtained by the following recursions; for $j = 1, \ldots, n$; we evaluate [5]

$$\tilde{D}_j = < v_j, v_j >_{\bar{D}}; \qquad \tilde{U}_{ij} = (1/\tilde{D}_j) < v_i, v_j >_{\bar{D}} \qquad i = 1, \ldots, j-1 \tag{7.18}$$

$$v_i = v_i - \tilde{U}_{ij} v_j \tag{7.19}$$

In eqn. (7.19), $< v_i, v_j >_{\bar{D}} = v_i^T \bar{D} v_j$; the time-evolution algorithm efficiently produces the new U and D factors, and has the effect of previous U and D factors; transition matrix, and Q, and preserves symmetry of P.

7.2.4.2 Measurement/Data Update

This data update process combines an *a priori* estimate $\tilde{x}$ and $\tilde{P}$ (obtained from the time propagation) with, say a scalar data $z = cx + v$ to obtain the updated estimate/covariance:

$$s = c\tilde{P}c^T + r;\ K = \tilde{P}c^T / s;\ \hat{x} = \tilde{x} + K(z - c\tilde{x}); \tag{7.20}$$

$$\hat{P} = \tilde{P} - Kc\tilde{P} \tag{7.21}$$

In eqn. (7.21), $\tilde{P} = \tilde{U}\tilde{D}\tilde{U}^T$; c is the scalar observation factor, r is measurement noise variance, z is the vector of noisy observations; here, processing is done in a scalar manner to avoid direct matrix inversion as occurring in eqn. (7.5). Kalman gain K, and updated factors $\hat{U}$ and $\hat{D}$ are obtained as [7]

$$g = \tilde{U}^T c^T; g^T = (g_1, \ldots, g_n); w = \tilde{D}g \tag{7.22}$$

$$\hat{d}_1 = \tilde{d}_1 R / s_1;\ s_1 = R + w_1 g_1 \tag{7.23}$$

For $j = 2, \ldots, n$, compute

$$s_j = s_{j-1} + w_j g_j;\ \hat{d}_j = \tilde{d}_j s_{j-1} / s_j$$

$$\tilde{u}_j = \tilde{u}_j + \lambda_j K_j;\ \lambda_j = -g_j / s_{j-1}$$

$$K_{j+1} = K_j + w_j \tilde{u}_j; \tilde{U} = [\tilde{u}_1, \ldots, \tilde{u}_n] \tag{7.24}$$

Then, the Kalman gain is given by

$$K = K_{n+1} / s_n \tag{7.25}$$

In eqn. (7.23), $\tilde{d}$ and $\hat{d}_j$ are the predicted and updated elements of the D matrix, respectively. The time evolution and data update for the state vector, x, are just same as in KF.

7.2.4.3 Performance Evaluation of the CTUDF

Target image tracking with synthetically generated images are evaluated for CTKF and CTUDF filter with different target noise std 1, 3, 5. The performance metrics are given in Table 7.2, and the plot is shown in Figure 7.4. The results show that CTUDF performance better than CTKF.

7.3 SQUARE-ROOT INFORMATION FILTER

In this section, the concept of square-root information filter (SRIF) is discussed. However, the basic aspects of the information filter are first described.

TABLE 7.2
Performance Metrics of CTKF and CTUDF

Filter	Target Noise std = 1*				Target Noise std = 3*				Target Noise std = 5*			
	PFEx	PFEy	RMSPE	RMSVE	PFEx	PFEy	RMSPE	RMSVE	PFEx	PFEy	RMSPE	RMSVE
CTKF	0.5893	0.5467	0.3080	0.0611	0.6181	0.3119	0.2653	0.0651	0.5664	0.4171	0.2696	0.0651
CTUDF	**0.2639**	**0.3715**	**0.1746**	0.1022	**0.2740**	**0.2620**	**0.1453**	0.0925	**0.2767**	**0.3185**	**0.1617**	0.0962

* The mentioned numbsers are the stds in the MATLAB scripts for the noise intensities.

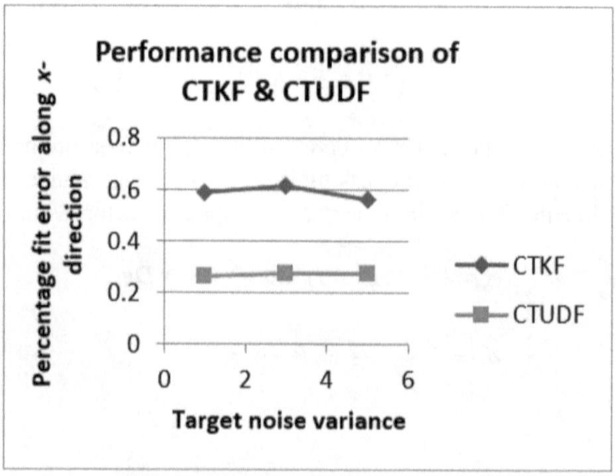

FIGURE 7.4 Performance of CTKF and CTUDF for TGNSTD as 1,3,5.

7.3.1 Information Filter

Information filtering (IF) is an alternative manner of implementing KF equations (or equivalent of), which has better numerical stability properties particularly for applications where the initial estimation uncertainty is large. IF uses the propagation of the inverse of the covariance matrix called the information matrix, instead the covariance matrix itself. Algebraically, the two forms are equivalent (for linear systems), but the IF has the advantage of the ability to handle a singular initial covariance matrix at $t = 0$, which is the case when the *a priori* information is known precisely. IF is found to be particularly advantageous in decentralized fusion networks [3] wherein there are multiple interconnected fusion nodes and no node has global knowledge of the network topology. In these networks, a direct interpretation of node measurement and contribution are provided by the IF in terms of information. IF enables interpretation of the filter performance using information theoretic concepts. The advantage of this formulation is that data update equations are computationally simpler than those of the KF, however, the prediction part is a bit more complex in this form and it relies on a propagation matrix, which is independent of observation made. Information filtering is a more direct way of dealing with the target tracking and multi-sensory data-fusion aspects than KF. It has an advantage in tracking algorithms, since IF provides a direct contribution as additional information. However, if IF is implemented as is, it would be susceptible to computer round-offs, which would affect the tracking. This aspect is very important for target tracking in real-time/online situations.

7.3.2 Square-Root Information Filter (SRIF)

Detection and tracking of a moving target is a challenging problem in forward-looking infrared (FLIR) image sequences because of: (i) low SNR, (ii) low contrast, (iii) background clutter and (iv) partial occlusion of the target. For such a purpose, an efficient and numerically stable centroid tracking algorithm is required. The square-root filter algorithm could provide better numerical accuracy, stability and efficiency. The state-space equations for a dynamic system are given by

$$x(k+1) = \phi x(k) + Bu(k) + Gw(k) \tag{7.26}$$

$$z(k+1) = Hx(k) + \mathrm{v(k)} \tag{7.27}$$

In eqn. (7.26), x is a state vector that contains the centroid coordinates of a target, w is WGN with zero mean and covariance Q, z is the observation data vector, H is the observation model and v is the WG observation noise with zero mean and covariance R.

7.3.2.1 Measurements/Data Update

Consider a linear algebraic measurement model in a simplified form for the sake of illustration:

$$z = Hx + v \tag{7.28}$$

The LS solution of x is generally obtained by minimizing the observation error cost function:

$$J(x) = (z - Hx)^T(z - Hx \tag{7.29}$$

The idea is to estimate x (it could be a parameter vector). Also, assume that *a priori* unbiased estimate $\tilde{x}$ of x and *a priori* information matrix P^{-1} (inverse of the covariance matrix of the KF, P) forms an *a priori* state–information matrix pair $(\tilde{x}, \tilde{P}^{-1})$. The cost function J can be modified by inclusion of the *a priori* information pair to obtain

$$J_a(x) = (z - Hx)^T(z - Hx) + (x - \tilde{x})^T\,\tilde{\mathrm{P}}^{-1}(x - \tilde{x}) \tag{7.30}$$

The matrix P^{-1} (it is the square of some quantity) can be factored as

$$\tilde{P}^{-1} = \tilde{R}^T\tilde{R} \tag{7.31}$$

Inserting eqn. (7.31) in eqn. (7.30) one gets

$$J_a(x) = (z - Hx)^T(z - Hx) + (x - \tilde{x})^T\,\tilde{R}^T\tilde{R}(x - \tilde{x}) \tag{7.32}$$

The second term in the eqn. (7.32) can be simplified as

$$(x - \tilde{x})^T\,\tilde{R}^T\tilde{R}(x - \tilde{x}) = (\tilde{R}\tilde{x} - \tilde{R}x)^T(\tilde{R}\tilde{x} - \tilde{R}x) \tag{7.33}$$

With this, one gets

$$\begin{aligned} J_a(x) &= (z - Hx)^T(z - Hx) + (\tilde{R}\tilde{x} - \tilde{R}x)^T(\tilde{R}\tilde{x} - \tilde{R}x) \\ J_a(x) &= (z - Hx)^T(z - Hx) + (\tilde{z} - \tilde{R}x)^T(\tilde{z} - \tilde{R}x) \end{aligned} \tag{7.34}$$

In eqn. (7.34) $\tilde{z} = \tilde{R}\tilde{x}$; the second term of eqn. (7.34) is given by

$$\tilde{z} = \tilde{R}x + \tilde{v} \quad (7.35)$$

From eqn. (7.34) and using eqn. (7.35), the composite system can be equivalently represented as

$$\begin{bmatrix} \tilde{z} \\ z \end{bmatrix} = \begin{bmatrix} \tilde{R} \\ H \end{bmatrix} x + \begin{bmatrix} \tilde{v} \\ v \end{bmatrix} \quad (7.36)$$

This *a priori* information forms the data equation similar to the observation eqn (7.28). Hence, it can be considered as an additional measurement, and the composite problem can be solved in an efficient manner in a LS sense. This forms the basis for SRIF. The SRI pair and the existing (means new) observations are arranged in the following form and an orthogonal transformation is used to obtain

$$T\begin{bmatrix} \tilde{R}(k) & \tilde{z}(k) \\ H & z(k+1) \end{bmatrix} = \begin{bmatrix} \tilde{R}(k+1) & \hat{z}(k+1) \\ 0 & e(k+1) \end{bmatrix} \quad (7.37)$$

This formulation has been proven to be less sensitive to computer round-off [7]. In eqn. (7.37) T is the Householder orthogonal transformation (HHOT) matrix. Here, $k = 1,2,\ldots,N$, with $e(.)$ as the residuals. Thus, the updated pair $(\hat{z}(k+1), \widehat{R}(k+1))$ is obtained. This processing can be continued with the inclusion of new observation $z(k+1)$ at each scan. If required, the usual state is obtained as

$$\hat{x} = \hat{R}^{-1}\hat{z} \quad (7.38)$$

This is done since in a real-life situation one uses and sees only states in the covariance form! Thus, the data-equation aspect of the information pair and the HHOT gives an elegant and simple solution for the tracking problem. Eqn (7.37) is the data update part of the tracking filter. The dynamic state eqn. (7.26) is introduced in the complete development of SRIF.

7.3.2.2 Time Propagation

The set of equations for the time propagation are given as follows [5]:

$$\begin{bmatrix} \hat{R}_w(k+1) & \hat{R}_{wx}(k+1) & \hat{z}_w(k+1) \\ 0 & \hat{R}(k+1) & \hat{z}(k+1) \end{bmatrix} = T(k+1)\begin{bmatrix} \hat{R}_w(k) & 0 & \hat{z}_w(k) \\ -\hat{R}(k)\phi^{-1}G & \hat{R}(k)\phi^{-1} & \hat{z}(k) \end{bmatrix} \quad (7.39)$$

In eqn. (7.39), the terms on the right-hand side are provided as the initial conditions (to start with), but subsequently these are available from the previous scan of the data update. The left-hand-side terms result after the application of the HHOT to the right-hand-side block matrix. Here, $R(.)$ is the square root of the information matrix $P - 1$. The following is the state time-evolution equation

$$\hat{x}(k+1) = \phi\hat{x}(k) \quad (7.40)$$

Next, the measurement update equation

$$\begin{bmatrix} R^*(k+1) & z^*(k+1) \\ 0 & e(k+1) \end{bmatrix} = T(k+1/k+1)\begin{bmatrix} \widehat{R}(k+1) & \hat{z}(k+1) \\ H & z(k+1) \end{bmatrix} \quad (7.41)$$

The terms in the right-hand side of eqn. (7.41) come from the corresponding terms of the left-hand side of eqn. (7.39). Assign the following equivalence:

$$\begin{aligned} &R^*(k+1) \Rightarrow \hat{R};\ z^*(k+1) \Rightarrow \hat{z}; \\ &\hat{R}_w(k+1) \Rightarrow \hat{R}_w(k+1);\ \hat{z}_w(k+1) \Rightarrow \hat{z}_w \end{aligned} \tag{7.42}$$

Finally, we have the state update equation as

$$\hat{x}(k+1/k+1) = \hat{R}^{-1}\hat{z} \tag{7.43}$$

7.3.3 Image Centroid Tracking using CTSRIF-Performance Evaluation

The set of image frames that is used in the tracking technique is generated synthetically, as done earlier. This image consists of an object and its surrounding along with a noise, which is uniformly

TABLE 7.3
Performance Metrics for CTSRIF

Performance Metrics	TGNSTD = 1	TGNSTD = 3	TGNSTD = 5
PFEx	0.4960	0.5657	0.5661
PFey	0.5719	0.5891	0.6179
RMSPE	0.2895	0.3123	0.3204
RMSVE	0.0092	0.0093	0.0098

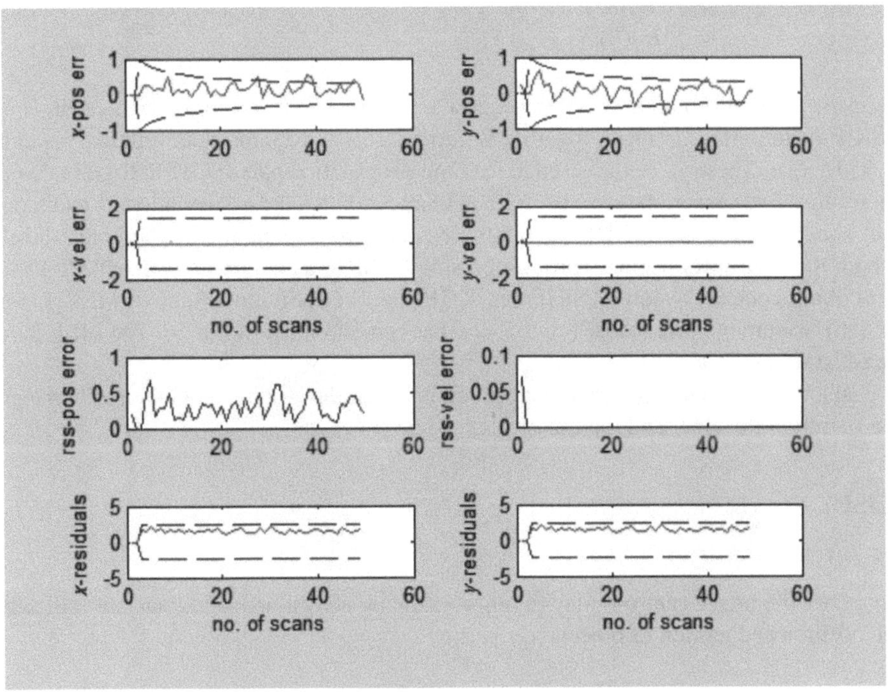

FIGURE 7.5 Performance of CTSRIF: x-y position and velocity errors, RSS position; velocity errors and x-y residuals plots.

TABLE 7.4
Performance Metrics for CTSRIF for TGNSTD = 1 and 5

	% FEs with target noise STD 1 & 5, TGBN std = 50			
Filter (std = *, in the MATLAB script)	**PFEx**	**PFEy**	**RMSPE**	**RMSVE**
CTSRIF1 (std = 1)	0.3778	0.6652	0.2879	0.0261
CTSRIF2 (std = 5)	0.4145	0.4957	0.2417	0.0277
CTSRIMLF	0.3895	0.5738	0.2604	0.0277

distributed. The image generated will have intensity in the range 0 to 255. The target intensity value and its background have a certain mean and variance. The performance metrics are given in Table 7.3 for different target image noise STDs (TGNSTD). Figure 7.5 shows the time histories of the position/velocity states.

As seen from Table 7.3, although the performance of the CTSRIF gets affected with the increasing variance of the target noise, it is still very good and acceptable for the case studied. Also, error time histories are within their theoretical bounds. Hence, the overall performance of the image centroid tracking SRIF algorithm is acceptable.

7.3.4 Image Centroid Fusion using the SRIF Algorithm

A system with two sensors' models H1 and H2 is used and their measurements can be fused at ML. This results in the state estimate due to the fusion of the two-sensor data:

$$T(k)\begin{bmatrix} \tilde{R}_f(k-1) & \tilde{y}_f(k-1) \\ H_1(k) & z_1(k) \\ H_2(k) & z_2(k) \end{bmatrix} = \begin{bmatrix} \hat{R}_f(k) & \hat{y}_f(k) \\ 0 & e(k) \end{bmatrix}; \; k = 1,..., \tag{7.44}$$

Here, $\hat{y}_f$ is the fused information state. Since, now y_f is available, this R_f and y_f are used in eqn. (7.44) for CTSRIF fusion. Measurement updating is carried out, and x_f (for each timestep) is calculated using $x_f = \text{inv}(R_f)\, y_f$. These values are used in the time-propagation part of CTSRIF to obtain optimum tracking results. Two sensor images with different target noise standard deviations 1 and 5 (and with the target background noise, TGBN std as 50) are considered. Using the measurement-level fusion (MLF) both the centroids of images are fused using CTSRIF as in eqn (7.44). Table 7.4 shows the metrics of image centroid fusion CTSRIF with TGBN std = 50. Figure 7.6 shows MLF performance with CTSRIF for target noise std as 1 and 5, and background noise std of 50. The MLF is found to be quite satisfactory.

This chapter sets the stage for further investigations in the development of factorization algorithms for image centroid tracking, and the use of fuzzy logic.

EXERCISES

Exercise 7.1

What role does the image centroid play in object-tracking algorithms, and how is it utilized to estimate the position and motion of objects in a scene?

Exercise 7.2

What are the steps involved in image centroid tracking using the Kalman filter?

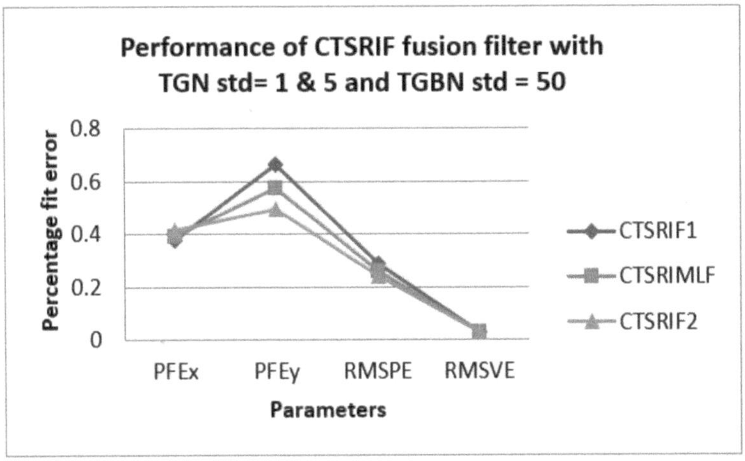

FIGURE 7.6 Performance of CTSRIF.

Exercise 7.3

How can the Kalman filter be used for object-tracking applications?

Exercise 7.4

What factors influence the determination of target image noise variances in digital imaging, and how do these variances impact the quality and processing of images in various applications?

Exercise 7.5

What is an information filter in the context of state estimation, and how does it differ from conventional Kalman filters?

Exercise 7.6

How is the information filter better than the Kalman filter?

Exercise 7.7

What is a square-root information filter (SRIF) in the context of state estimation?

Exercise 7.8

How does SRIF differ from other filtering techniques such as Kalman filters and information filters?

Exercise 7.9

How can image centroid tracking be implemented using a Kalman filter, and what are the advantages of using this approach in computer vision applications?

Exercise 7.10

How can UD factorization be applied to improve image centroid tracking algorithms, particularly in scenarios with high-dimensional or noisy image data?

Exercise 7.11

What is the *U-D* factorization filter?

Exercise 7.12

What is image centroid fusion, and how does it enhance the accuracy and robustness of object tracking in multi-sensor systems?

Exercise 7.13

How can square-root information filters be applied to fuse centroid measurements from multiple images?

Exercise 7.14

What considerations should be taken into account when incorporating synthetic images into the tracking process?

REFERENCES

1. Sue Bae, J., and Song, T.L., Image tracking algorithm using template matching and PSNF-m, *International Journal of Control, Automation and Systems*, Vol. 6, No. 3, pp. 413–423, June 2008.
2. Parimala, P., *Image Tracking and Fusion using Square Root Information Filtering*. Ph.D. Thesis, Electronics and Communications Engineering, Jain University, Bangalore, India, 2017.
3. Raol, J.R., *Data Fusion Mathematics – Theory and Practice*, CRC Press, Taylor & Francis, Florida, 2015.
4. Naidu, V.P.S., *Kinematic and Image Data Fusion with Application to Tracking of Moving Target*. Ph.D. Thesis, University of Mysore, India, 2008.
5. Bierman, G.J., *Factorization Methods for Discrete Sequential Estimation*, Academic Press, New York, 1977.
6. Raol, J.R., and Girija, G., Sensor data fusion algorithms using square root information filtering, *IEE Proceedings-Radar Sonar Navigation, UK*, Vol. 149, No. 2, pp. 89–93, April 2002.
7. Naidu, V.P.S., and Raol, J.R., Pixel-level image fusion using wavelets and principal component analysis, *Defence Science Journal, India*, Vol. 58, No. 3, pp. 338–352, May 2008.

8 Image Centroid Tracking with Fuzzy-Logic-Augmented Filters

8.1 INTRODUCTION

Fuzzy logic (FL) has several applications: (i) in many signal- and image-processing tasks, (ii) control and (iii) data fusion; and is one of the constituents for building artificial intelligent (AI) systems, and of the soft computing technology [1,2].

In this chapter, three FL-based image centroid tracking algorithms (CTAs) are presented: (a) ANFIS (adaptive neuro-fuzzy inference system), (b) FL-function augmented KF (FLCTKF) and (c) FL-based SRIF. These CTAs are evaluated using MATLAB-based simulated synthetic image data.

8.1.1 Fuzzy-Logic System

FL is a multi-valued logic, like in a trapezoidal membership function (MF); and there are many types of MFs; a MF gives any continuous value between '0' zero and '1' to the input variable, that is to be fuzzified. This is the gradation/grad-value/s. The x-axis variable, say pressure is now fuzzified, and its belongingness to a fuzzy set; low, medium or high pressure would be a value within the MF. The fuzzy inference system (FIS) operates over MFs and the rule-base by using fuzzy implication functions (FIFs) [2], see Figure 8.1.

The defuzzification obtains crisp outputs to connect to the real world. The If…Then… rules are given by the human expert who has a lot of practical experience of operation/analysis/design of classical and traditional systems.

In the present, FL is used directly, via the act of modulating the residuals of the KF, in target-image CTA to deal with uncertainty in image data. Such a use of a FLS (FL system) requires the following to be done:: (i) select appropriate MFs – a fuzzification process, (ii) gather the rule-base, with the help of a design expert for IO mapping, (iii) select appropriate fuzzy operators, (iv) select FIF and aggregation approaches and (v) select suitable defuzzification technique.

A FLS permits a partial membership of a member of a set; a FS (fuzzy set) A on a universe of discourse (UOD) U (the elements u) is expressed as

$$A = \int \{\mu_A(u)/u\} \quad \forall\, u \in U \tag{8.1}$$

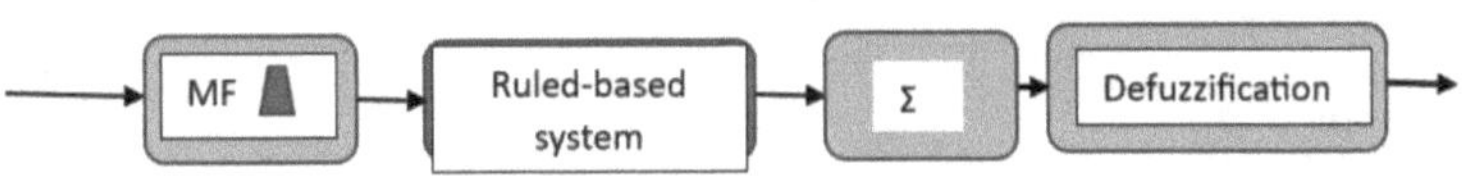

FIGURE 8.1 Fuzzy inference system (FIS).

DOI: 10.1201/9781003620211-8

$$\text{or } A = \sum \{\mu_A(u)/u\} \; \forall \; u \in U \tag{8.2}$$

Here, $\mu_A(u)$ is a MF function value of u on the set A and obtains a mapping of the UOD U the closed interval [0, 1]. The classical logic operations AND, OR and NOT have some new interpretations in FL. For FL the corresponding operators specified are *min*, *max*, and *complement* defined as

$$\mu_{A \cap B}(u) = \min[\mu_A(u), \mu_B(u)] \quad \text{(intersection, AND)} \tag{8.3}$$

$$\mu_{A \cup B}(u) = \max[\mu_A(u), \mu_B(u)] \quad \text{(union, OR)} \tag{8.4}$$

$$\mu_{\bar{A}}(u) = 1 - \mu_A(u) \qquad \text{(complement, NOT)} \tag{8.5}$$

In FL, first the rules are specifically defined for a given application; and obtained from an expert or can be derived from the data of the system. A rule is 'If u is A, Then v is B'. The FIS via FIF maps the input FSs into the output FSs. The outputs of all the fired rules are then aggregated, i.e. to get a single FS. The fuzzifier-MF takes input values and assigns the degree to which these numbers belong to each of the FSs. The steps in FIS are as follows: (a) the inputs u, and v are fuzzified using MF ($\mu^i(u)$, and $\mu^i(v)$) for ith rule; the appropriate MFs are specified; (b) since antecedent (If…) part of each rule has more than one clause, the FL operator resolves the antecedent to a one number between 0 and 1; (c) FIF shapes the consequent part (Then…), based on the antecedent; (d) then, aggregation, Σ, as in Figure 8.1 is performed; and (e) a crisp value of the output variable, say w, is obtained by the defuzzification method.

8.2 FUZZY-LOGIC SYSTEM: ADAPTIVE NEURO-FUZZY INFERENCE SYSTEM – ANFIS

The ANFIS system can be incorporated in a filtering scheme to improve the performance of the combined system; it can be used for image centroid tracking and fusion also. It employs the rules to represent the system's behaviour when an accurate model of the system is not available [3]. Using a collection of *I/O* data, one can build a FIS that would approximate the data and use the rules to get the result, Figure 8.2.

The ANFIS consists of MFs and If…Then… rules with adjustable coefficients, which are chosen to adapt the MFs to the data; the MF parameters that specify the shape of the MF are now tuned to the variations in the *I/O* data. In ANFIS, the artificial neural network (ANN)-based learning mechanism is used to learn information from the data. This facilitates determination of suitable MFs; their defining coefficients that would allow the FIS to track *I/O* data; thus, the ANFIS is an adaptive network and is functionally equivalent to FIS. It utilizes a hybrid learning method to estimate the coefficients, which are tuned using a BP (back-propagation) steepest descent method or its combination with the LS method. The MFs are determined by using ANN and the *I/O* data of a given system. The components of an ANFIS are as follows: (i) if u_1 is A_1 and u_2 is B_1, Then $y_1=c_{11}u_1 + c_{12}u_2 + c_{10}$; (ii) if u_1 is A_2 and u_2 is B_2, Then $y_2=c_{21}u_1 + c_{22}u_2 + c_{20}$; u_1, u_2 are nonfuzzy inputs, and y as the desired output; (iii) some internal computations using product operators/normalization, etc.; and (iv) computation of output, $\bar{w}_i = \dfrac{w_i}{w_1 + w_2}$, $\bar{w}_i y_i = \bar{w}_i(c_{i1}u_1 + c_{i2}u_2 + c_{i0})$, $i = 1,2.$, and $y_p = \bar{w}_1 y_1 + \bar{w}_2 y_2$, where y_p is the output.

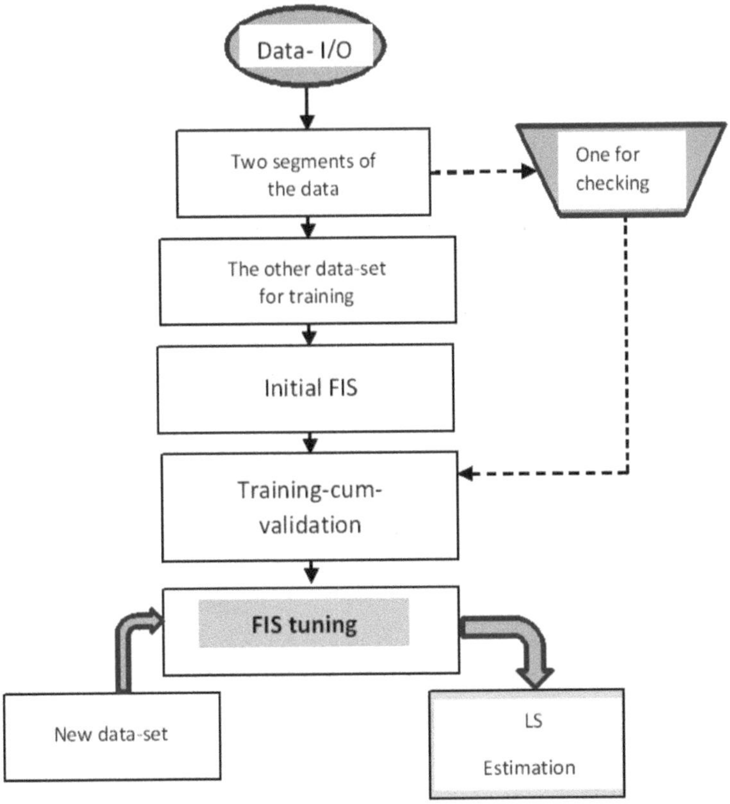

FIGURE 8.2 ANFIS procedure for image centroid tracking.

The output is obtained as

$$\begin{aligned} y_p &= \bar{w}_1 y_1 + \bar{w}_2 y_2 = \bar{w}_1(c_{11}u_1 + c_{12}u_2 + c_{10}) + \bar{w}_2(c_{21}u_1 + c_{22}u_2 + c \\ &= (\bar{w}_1 u_1)c_{11} + (\bar{w}_1 u_2)c_{12} + \bar{w}_1 c_{10} + (\bar{w}_2 u_1)c_{21} + (\bar{w}_2 u_2)c_{22} + \bar{w}_2 c_{20} \end{aligned} \tag{8.6}$$

The hybrid algorithm would adjust the consequent parameters in a forward pass of the algorithm, and the premise parameters are adjusted in a backward pass. The MATLAB-ANFIS programming steps and system are used for centroid tracking.

8.3 FUZZY-LOGIC-BASED KF

FL can be used for tuning a conventional KF, and CTA can be developed by augmenting KF with FL; in such a system, FL aids soft decision making [2], via fuzzy 'IF… Then' rules. In CTA, FL is used to modulate the residual of KF, i.e. at the data update level. The FLKF is the same KF, except eqn. (8.4):

$$\hat{X}(k+1) = \tilde{X}(k+1) + K f_{FL}(k+1) \tag{8.7}$$

In eqn. (8.7), *f(k+1)* as an output of the FL-based process, in general, is a nonlinear function of the residuals '*e*' (and their derivative). One also has the *x-y* axes position observations of the target. The final *f* vector consists of modified residuals for *x* and *y* axes

$$f_{FL}(k+1) = \left[f_{FLx}(k+1) \quad f_{FLy}(k+1) \right] \tag{8.8}$$

The innovation vector '*e*' is split into its *x* and *y* components, e_x and e_y, the target motion in each axis being independent. The vector consists of e_x and $\dot{e}_x$, and one output $f_{FLx}(k+1)$, with $\dot{e}_x$ computed as

$$\dot{e}_x = \frac{\{e_x(k+1) - e_x(k)\}}{T} \tag{8.9}$$

In eqn. (8.9), T is the sampling interval in seconds, and eqn. (8.9) can be used for y and z axes, if required. Appropriate MFs are used and the technique is evaluated with the MATLAB-based FL logic tool box.

8.4 FUZZY-LOGIC-BASED SRIF

The fuzzy-logic-based SRIF algorithm for target image centroid tracking is presented. It combines certain merits of FL and SRIF. A data-fusion study for image centroid tracking is presented with parametric studies. The performance metrics are evaluated by utilizing the sequence of synthetic images. In the measurement/data update part, it does not need the computation of the Kalman gain, since now it uses the measurement/data update part of the SRIF, and hence, it is the gain-free filter; the main reason is that the information state y_f is now directly available from the orthogonal transformation. It retains the merits of the CTSRIF. Two sensors' models H1 and H2 are used and their measurements can be fused at data level. Here, $\hat{y}_f$ is the fused information state.

8.5 PERFORMANCE EVALUATION OF THE FL-BASED CT ALGORITHMS

8.5.1 Mathematical Model and Data Generation

The constant-velocity target motion model is used for the generation of data for centroid tracking:

$$X(k+1) = \begin{bmatrix} 1 & T & 0 & 0 \\ 0 & 1 & 0 & 0 \\ 0 & 0 & 1 & T \\ 0 & 0 & 0 & 1 \end{bmatrix} X(k) + \begin{bmatrix} \frac{T^2}{2} & 0 \\ T & 0 \\ 0 & \frac{T^2}{2} \\ 0 & T \end{bmatrix} w(k) \tag{8.10}$$

In eqn. (8.10), $X(k) = \left[x \quad \dot{x} \quad y \quad \dot{y} \right]^T$ is a state vector; $w(k)$ is zero-mean white Gaussian noise (WGN) with co-variance matrix $Q = \begin{bmatrix} \sigma_{w1}^2 & 0 \\ 0 & \sigma_{w2}^2 \end{bmatrix}$.

The measurement model is given as

$$z(k+1) = \begin{bmatrix} 1 & 0 & 0 & 0 \\ 0 & 0 & 1 & 0 \end{bmatrix} X(k+1) + v(k+1) \tag{8.11}$$

Here, $v(k)$ is the centroid observation noise, and is zero-mean WGN, and has the diagonal covariance R. A 2D array of 64 × 64 pixels is used as the background image, and is generated by using white Gaussian noise as $N(\mu_n, \sigma_n^2)$. One more 2D array of pixels is the target of size (90 × 90) generated by using white Gaussian as $N(\mu_t, \sigma_t^2)$. The total scan number is 50 and the sampling frequency is 1 frame/sec; $T = 1$. The initial conditions are $X = \begin{bmatrix} x & \dot{x} & y & \dot{y} \end{bmatrix}^T = \begin{bmatrix} 10 & 1 & 10 & 1 \end{bmatrix}^T$. The synthetic image is then transformed into binary image using the bounds $I_U = 110$; $I_L = 90$ of a target layer; grouped into clusters by the nearest-neighbour data-association technique with the proximity distance $d_p = 2$. If the background is too noisy, the algorithm would produce more clusters and centroids. This would require a nearest-neighbour KF (NNKF) to correlate the true observation to the target. The metrics are as follows: % fit errors (for the observations and states), and root mean square, rms values of position and velocity errors are computed. The state – and observation – errors and residuals are computed to evaluate the performance of these CTAs.

The performance evaluation using simulated synthetic images is presented next [4].

8.5.2 Performance with ANFIS-CTA and FLKF-CTA

Table 8.1 gives the values of the metrics for three values of the stds of the noise in the target image itself. There is somewhat upward trends in the metrics with an increase in the stds of the noise in the target images; yet, the metrics have lower values in the case of FLKF-CTA compared to ANFIS-CTA. The ANFIS needs a two-step process: training and evaluation. For training, the error and its derivatives are used as input, and the target output is used as the output. Figures 8.3 and 8.4 show the time histories of state errors, and observations' residuals for the CTFLKF; which are within the theoretical bounds, and the performance is considered very satisfactory.

Similar plots were generated for the ANFIS based CTA, but not produced here. Thus, it is seen that the FL-based CTAs perform very satisfactorily and can be further utilized for image fusion and building AI-based data fusion and target tracking.

Although, the ANFIS-CTA performed slightly worse than that of the FLKF, the overall performance of both the algorithms is found to be very satisfactory based on the numerical values of the performance metrics. The time histories of the errors, are all found to be within the respective theoretical bounds. These tracking algorithms can be implemented for real-time applications and can

TABLE 8.1
Metrics for Two FL-Augmented CTAs

% Fit Errors; Background Noise (BN) Std = 50	TGN Std = 1*	TGN Std = 3	TGN Std = 5
		ANFIS- CTA	
PFE-x	0.859	0.917	1.304
PFE-y	0.909	1.071	1.160
RMSPE	0.479	0.540	0.668
RMSVE	0.206	0.225	0.250
		FLKF-CTA	
PFE-x	**0.5697**	**0.631**	**1.061**
PFE-y	**0.648**	**0.792**	**0.866**
RMSPE	**0.331**	**0.388**	**0.525**
RMSVE	**0.0745**	**0.0781**	**0.122**

* These std values are entered in the MATLAB script to generate noise.

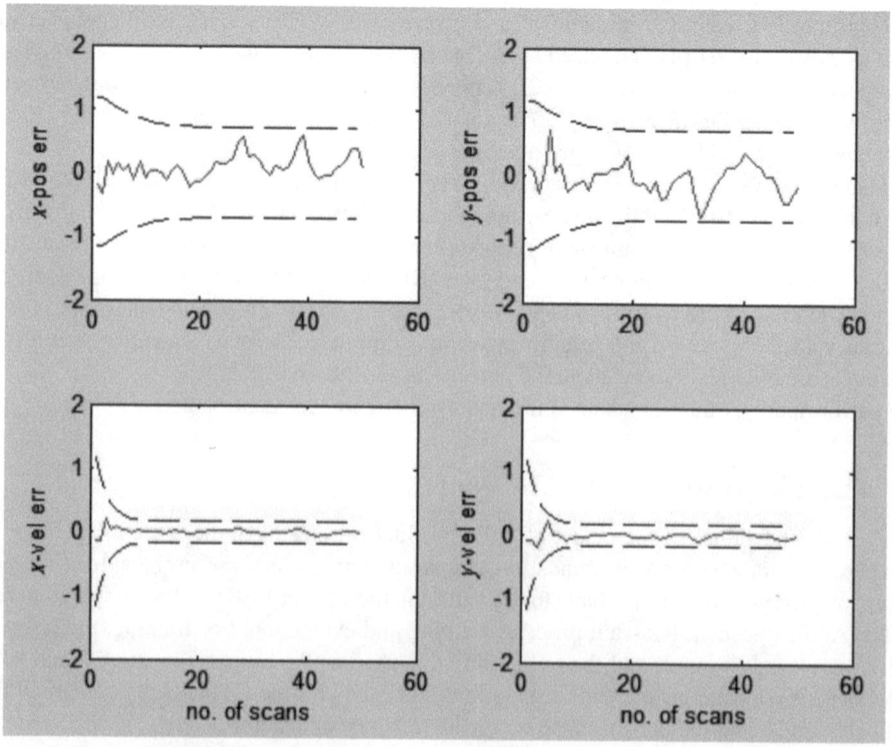

FIGURE 8.3 State errors with bounds: FLKF.

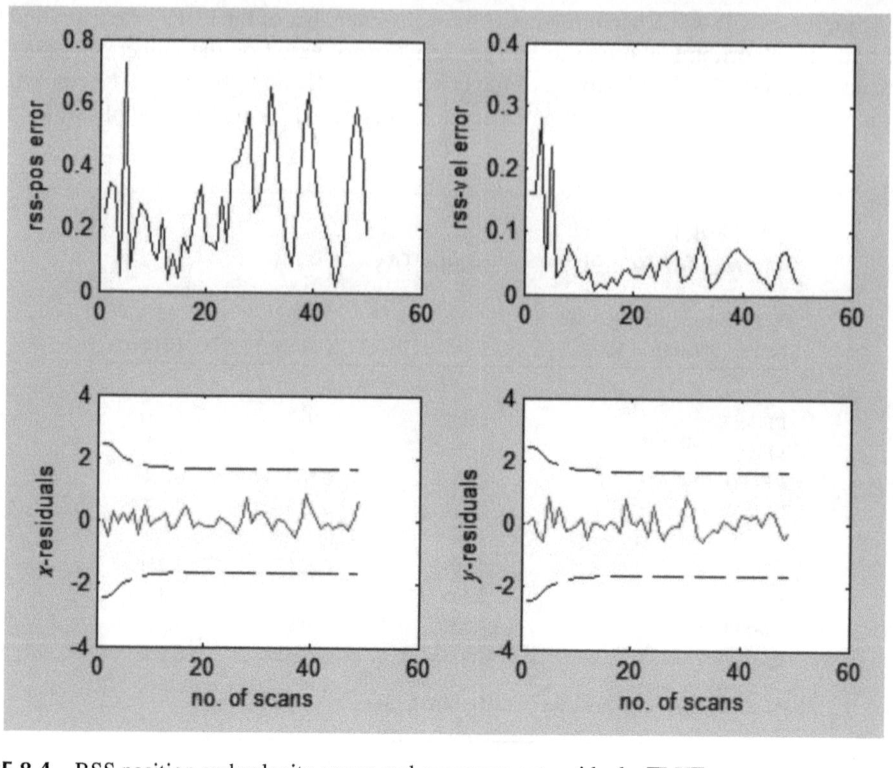

FIGURE 8.4 RSS position and velocity errors and measurement residuals: FLKF.

TABLE 8.2
Metrics for CTSRIF and CTFSRIF: Case 1

	Fit error with TGNstd = 5, BGNstd= 50			
Filter	**PFEx**	**PFEy**	**RMSPE**	**RMSVE**
CTSRIF	0.6130	0.7113	0.3512	0.0315
CTFSRIF	**0.5918**	**0.6851**	**0.3469**	0.0315

TABLE 8.3
Metrics for CTFSRIF and CTFSRIF: Case 2

	Fit error with TGNstd = 5, BGNstd = 100			
Filter	**PFEx**	**PFEy**	**RMSPE**	**RMSVE**
CTSRIF	0.4145	0.4957	0.2417	0.0277
CTFSRIF	**0.3822**	**0.4833**	**0.2361**	0.0277

be tested first using FPGA (field programmable gate arrays) and DSP (digital signal processing) hardware.

8.5.3 Performance Evaluation of FL-SRIF-Based CTA

Tables 8.2 and 8.3 give the performance metrics of CTSRIF and CTFSRIF tracking filters for different background-target-image-noise standard deviation (BGNstd) and different target-image-noise standard deviation (TGNstd).

It is seen that there is some trend of the performance metrics w.r.t. the standard deviations. It is established that CTFSRIF performs somewhat better than CTSRIF in a similar centroid tracking task.

8.5.4 Performance Evaluation of FL-SRIF-Based CTA for Image Fusion

Next, two sensor images with different target-noise standard deviations (TGNstds) = 1 and 5; and with the target-background-noise standard deviation, BGNstd=100 are considered. Using the MLF, both the centroids of images are fused using CTSRIF, and CTFSRIF. Tables 8.4 shows the performance metrics of image centroid fusion CTSRIF and CTFSRIF with BGNstd=100. Figure 8.5 shows MLF for target-noise std as 1 and 5 and background-noise std of 100. The MLF is found to be quite satisfactory, and CTFSRIF is slightly better than CTSRIF.

This chapter establishes that the FL-augmented image centroid tracking algorithms perform better than those without fuzzy logic.

TABLE 8.4
Metrics of Image Fusion using CTFSRIF for TGNstd = 1 and 5

	Fit error with TGNstd = 1 and 5, BGNstd = 100			
Filter	**PFEx**	**PFEy**	**RMSPE**	**RMSVE**
CTFSRIF1 (std = 1)	0.3658	0.6552	0.2876	0.0261
CTFSRIF2 (std = 5)	0.3822	0.4833	0.2361	0.0277
CTFSRIMLF-Fusion	0.3775	0.5638	0.2600	0.0277

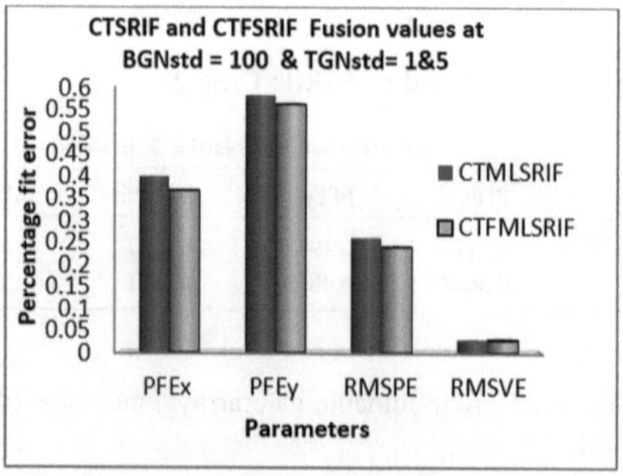

FIGURE 8.5 Fusion: PFEs of CTSRIF and CTFSRIF in MLF.

EXERCISES

Exercise 8.1

What is a fuzzy-logic system, and how does it facilitate decision-making in uncertain or complex environments?

Exercise 8.2

How do adaptive neuro-fuzzy inference systems (ANFIS) combine fuzzy logic and neural networks to address complex modelling and decision-making tasks?

Exercise 8.3

What are the key principles behind the integration of fuzzy logic with the Kalman filter to enhance the accuracy and robustness of state estimation in dynamic systems?

Exercise 8.4

How does the integration of fuzzy logic with the square-root information filter enhance the accuracy and reliability of image centroid tracking, particularly in scenarios with uncertainty or noise?

Exercise 8.5

How does fuzzy-logic impact the performance of centroid tracking algorithms in various tracking scenarios?

REFERENCES

1. Rodjito, P., *Position Tracking and Motion Prediction Using Fuzzy Logic*. Honours Thesis, Dept. of Computer Science, Paper 520. http://digitalcommons. colby.edu/ honorstheses/520, accessed July 2016.
2. Raol, J.R., *Multisensor Data Fusion with MATLAB*, CRC Press, Florida, 2010.
3. Jang, J.-S., ANFIS: Adaptive-network based fuzzy inference system, *IEEE Transactions on Systems, Man, and Cybernetics, USA*, Vol. 23, No. 3, May-June 1993.
4. Parimala, P., *Image Tracking and Fusion using Square Root Information Filtering*. Ph.D. Thesis, Electronics and Communications Engineering, Jain University, Bangalore, India, pp. 1–85, Aug. 2017.

9 Image Centroid Tracking-cum-Fusion using New Factorization Filtering Algorithms

9.1 INTRODUCTION

Parametric studies and performance results of image centroid tracking-cum-fusion using SRIF and SRVD (square root-VD, eigenfactor) algorithms are presented in this chapter [1]. Both the algorithms are used, perhaps for the first time, for the said application. The direct fusion of the centroids of two images, being tracked, using SRIF and the square-root eigenfactor filter (SRVD) is considered. The SRIF algorithm has been studied in the previous chapter.

Also, three new factorization filtering algorithms are presented for image centroid tracking. An observer based on the nonlinear function of residuals is also presented, and its asymptotic stability results are established. This is a novel work in the area of image-centroid tracking [1].

9.2 EIGENFACTOR SQUARE-ROOT FILTER SRVD: TIME PROPAGATION AND DATA UPDATE

A singular value decomposition-based (SRVD) filtering algorithm for image centroid tracking and fusion is proposed. Such an SVD-based algorithm might be of some importance in certain applications where continuous monitoring of the eigenfactors is necessary in order to reveal singularities that might occur during the running of the algorithm; this would be an added merit in large-scale multi-dimensional fusion-processing systems. This also helps to identify the states that are nearly dependent [2].

Here, V is eigenvector matrix and D is the matrix with the diagonal elements as the singular values of the given original matrix.

9.2.1 *V-D* Discrete-Time Measurement Update

The idea is to obtain the *a posteriori* eigenfactors given the *a priori* eigenfactors, the latter can also be called square-root factors. Given the time propagated factors $V(k + 1/k)$ and $D^{1/2}(k + 1/k)$ of $P(k + 1/k)$ (or the factors from initial conditions), the measurement matrix $H(k + 1)$, and the measurement covariance matrix $R_m(k + 1/k)$, we define the augmented matrix [2] as

$$A(k+1) \triangleq [V(k+1/k)D^{\frac{-1}{2}}(k+1/k)\ H^T R_m^{\frac{-T}{2}}] \tag{9.1}$$

Then perform a SVD (singular value decomposition) of $A(.)$ to obtain

$$A(k+1) = Y(k+1)[S(k+1)\ 0]Z^T(k+1) \tag{9.2}$$

DOI: 10.1201/9781003620211-9

In eqn. (9.2), $Y(.)$, and $Z(.)$ are the eigenvector matrices, with the columns as the respective eigenvectors. Then, the measurement updated spectral factors (V and D are so called) are obtained as

$$\begin{aligned} V(k+1/k+1) &= Y(k+1) \\ D^{\frac{1}{2}}(k+1/k+1) &= S^{-1}(k+1) \end{aligned} \tag{9.3}$$

In eqn. (9.3), $Y(.)$ is the an $n \times n$ orthogonal matrix, and $S(k+1)$ is $n \times n$ diagonal matrix with elements as the singular values of $A(k+1)$, i.e. the positive square roots of the eigenvalues of matrix $A(k+1)\,A^T(k+1)$ [2]. The Kalman gain is obtained by defining the $M(./.)$ matrix as

$$M(k+1/k) = W^T(k+1/k)H^T(k+1) \tag{9.4}$$

$$W(k+1/k) = V(k+1/k)D^{\frac{1}{2}}(k+1/k) \tag{9.5}$$

$$K(k+1) = W(k+1/k)M(k+1/k)[M^T(k+1/k)M(k+1/k)+R_m]^{-1} \tag{9.6}$$

9.2.2 *V-D* Time Propagation

Given the measurement updated factors $V(k/k)$ and $D^{1/2}(k/k)$ of $P(k/k)$, the state-transition matrix, the input gain matrix $G(k)$, and $Q(k)$; define the composite matrix as

$$A(k) = [\phi(k)V(k/k)D^{\frac{1}{2}}(k) \quad G(k)Q^{\frac{1}{2}}(k)] \tag{9.7}$$

Then, perform the SVD to decompose eqn. (9.7) into

$$A(k) = Y(k)[S(k) \quad 0]Z^T(k) \tag{9.8}$$

Then, obtain the time-propagated factors as

$$\begin{aligned} V(k+1/k) &= Y(k) \\ D^{\frac{1}{2}}(k+1/k) &= S(k) \end{aligned} \tag{9.9}$$

These square-root factors (V, D are also called so) are used in the measurement part of the filtering algorithm, in eqn. (9.1), and then the complete filtering algorithm is obtained.

9.3 PERFORMANCE EVALUATION OF THE SRIF FUSION

A system with two sensor models H1 and H2 is used and their measurements can be fused at data level (MLF). Satisfactory tracking performance of SRIF has led to its application to image centroid fusion. The target background is set at (TGBN/BN/SB: mean = 50, std = 50, 100, 150). Two images are considered with the target image set as (TGIM: mean = 100, std = 10), and with variation in the target-noise standard deviation, TGNstd = 1, (image 1, CTSRIF1), and TGNstd = 3 (image 2, CTSRIF2); these images are considered two-at-a-time for centroid tracking-cum-fusion. The performance metrics are shown in Table 9.1. Two of such results are plotted in Figure 9.1.

Also, the fusion results with TGNstd = 1 and 5; and TGNstds = 5 and 9 have been obtained for the centroid fusion of two images using SRIFs; however, these extensive results are not presented here; however, most of the trends of these results were also found to be similar to those in Table 9.1 and Figure 9.1 with minor variations.

TABLE 9.1
Metrics of Image Centroid Tracking-cum-Fusion using SRIF

Filter	TGNstds = 1 & 3; (BN/SB = 50)			TGNstds = 1 & 3; (BN/SB = 100)			TGNstds = 1 & 3 (BN/SB = 150)		
	PFEx	PFEy	RMSPE	PFEx	PFEy	RMSPE	PFEx	PFEy	RMSPE
CTSRIF1 (std = 1)	0.5837	0.7373	0.3604	0.3678	0.6552	0.2879	0.3386	0.5880	0.2600
CTSRIF2 (std = 3)	0.6116	0.6173	0.3330	0.3862	0.3599	0.2023	0.3558	0.2765	0.1727
CTSRIMLF	0.5986	0.6459	0.3375	0.3805	0.4869	0.2368	0.3472	0.4076	0.2052

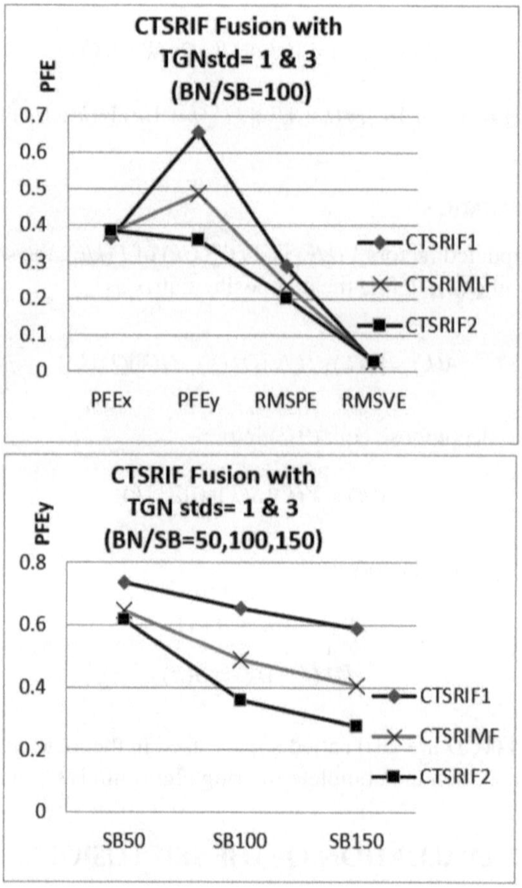

FIGURE 9.1 Fusion performance of SRIF.

9.4 PERFORMANCE EVALUATION OF THE SRVD ALGORITHM

A similar procedure as in the previous section has been used here also. The results with TGNstds = 1 and 3, TGNstds = 1 and 5 and TGNstds = 5 and 9 have been obtained for the centroid fusion of two images, using the SRVD filter. These extensive results (as generated for SRIF) have also been generated here, but only a few results are given in Table 9.2. Figure 9.2 depicts one such result for this case.

TABLE 9.2
Metrics from CTSRVD Filter Fusion

	With TGNstds = 1 & 3 (BN/SB = 100)			
Filter	**PFEx**	**PFEy**	**RMSPE**	**RMSVE**
CTSRVD1 (vnn1=1)	0.4651	0.5632	0.2799	0.0509
CTSRVD2 (vnn2=3)	0.3759	0.2141	0.1658	0.0505
CTSRVDMLF	0.4388	0.4752	0.2479	0.0507

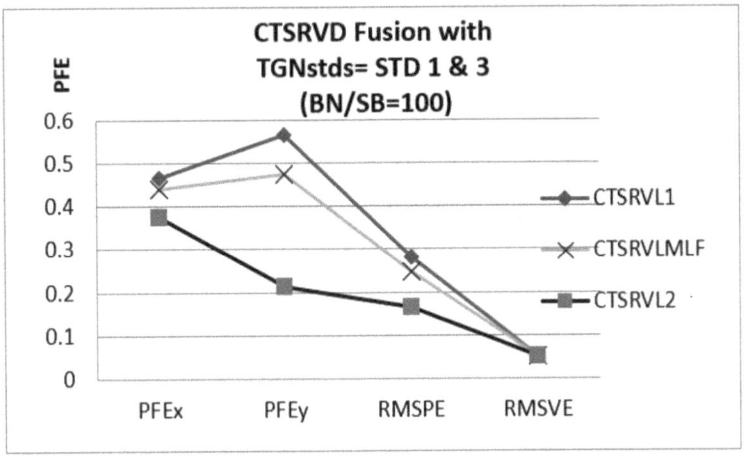

FIGURE 9.2 Fusion with SRVD with TGNstds = 1 and 3 (CTSRVL is CTSRVD).

Here again, the trends of various performance metrics across these fusion combinations (images with TGNstds = 1 and 3; 1 and 5; and 5 and 9) for this SRVD filter were also found to be almost similar with minor variations. In the case of this filter, the variation of PFEs w.r.t. BN/SB, after the value of BN/SB = 100 was found to increase slightly, however, up to BN/SB = 100, it was consistent with that of SRIF.

The inferences from the presented and other results (not presented here) for the image centroid tracking-cum-fusion are as follows: (i) the image centroid tracking performance has been found to be very satisfactory across the square-root-type filters; and (ii) the image centroid tracking-cum-fusion performance also has been found to be in accordance with the well-established theory of general data fusion, for the square-root filtering algorithms studied here.

In the next few sections, we propose three new filtering algorithms for image centroid tracking-cum-fusion:

(i) SRVD-SRIF: eigenvalue-eigenvector factorization-square-root information filtering that is based on the eigenfactor *V-D* filtering (SRVD) and the SRIF. The time propagation part is the same as the *V-D* filter [2] and the measurement/data update part is the same as that of the SRIF, and it can be called eigenfactor-SRIF, or SRVD-SRIF;
(ii) UDF-SRIF: here, the time-propagation part is the same as that of the UDF, and the data update is the same as that of SRIF; and
(iii) FLMO: is based on the concept of fuzzy membership function (MF) operating over the residuals of the linear observer. The Lyapunov energy functional is used to derive the asymptotic stability result for this overall nonlinear observer for centroid tracking.

All three algorithms are evaluated for their performance in image centroid tracking (and fusion where applicable) by simulations carried out in MATLAB.

9.5 A NEW FILTERING ALGORITHM BASED ON SRVD AND SRIF

It is presumed that all the initial values of P, Q, R, ϕ, H, G, R_m are the same as the SRVD filter and the SRIF. For more clarity and simplicity, the algorithmic steps are given for the combined covariance-information domain of filtering and the direct MLF of the centroids of the two input images (in fact more images can be easily fused as in the case of SRIF). Now, since these initial values are known, one proceeds as follows:

Initial part with initial conditions: use some guesstimate of the fused state, the same as the one used in the SRVD, and SRIF filters, and compute the covariance matrix $P(0)$:

$$P(0) = \{x_f(0) - \hat{x}_f(0)\}\{x_f(0) - \hat{x}_f(0)\}^T \tag{9.10}$$

Then, obtain SVD of $P(0)$:

$$[Y, S, Z] = \text{SVD}(P);\ \text{since } P \text{ is a } n \times n \text{ matrix, } S \text{ will be so.} \tag{9.11}$$

Here, $P \rightarrow Y \times S \times Z'$. Then, assign the SVD factors as follows:

$$V = Y;\ D12 = S \tag{9.12}$$

Also, obtain initial

$$R(0)\text{=sqrtm(inv}(V \times D12 \times Z'));\ \rightarrow R(0) = \text{sqrtm(inv}(P)) \tag{9.13}$$

This is the square root of the information matrix, i.e. IM = inv(P). For simplicity $\hat{R}_f$ is denoted as R, $\hat{x}_f$ is denoted as x, and $\hat{y}_f$ is denoted as y in the following. Then, we get

$$y(0) = R(0) \times x(0) \tag{9.14}$$

Thus, at this stage R, and y, the information pair (like SRIF) as required for the time propagation of the state estimate are available.

9.5.1 Time-Propagation Part

The time-propagated information state is obtained as follows:

$$y(\text{new}) = R \times \phi \times \text{inv}(R) \times y(\text{previous}) \tag{9.15}$$

In fact, eqn. (9.15) is the same as $xf(\text{new}) = \phi\, xf(\text{previous})$. In eqn. (9.15) use R from eqn. (9.13), and y from eqn. (9.14) for the start; then, in the next cycle these will be available from the output of the measurement-data! part. Then, form the following A matrix using the values from eqn. (9.14)

$$A = \left[\phi \times V \times D12 \cdot G \times \text{sqrtm}(Q)\right] \tag{9.16}$$

This means that the previous eigenfactors, V, $D12$ are now to be augmented with the new information, i.e. by using, ϕ, G and Q. Then, the new time-propagated factors are obtained as

$$[Y, S1, Z] = \text{SVD}(A); \quad \text{here, } A \rightarrow Y \times [S\ 0] \times Z'; \text{ and } S1 = [S\ 0] \tag{9.17}$$

Then, assign the factors as follows:

$$V = Y; D12 = S \tag{9.18}$$

Now, since A in eqn. (9.16) is a nonsquare matrix, S will also be nonsquare and hence, assign only the first sub-matrix of S that is square here. Next, obtain the R matrix needed in the measurement/data update part of SRIF:

$$R = \text{sqrtm}(\text{inv}(A \times A')) \tag{9.19}$$

In this R matrix, the effects of G and Q are included, by virtue of eqn. (9.16). Hence, matrix R is the time-propagated factor. So, at this stage R is obtained from eqn. (9.19), and y from eqn. (9.15), as the time-propagated information pair required for the measurement/data update.

9.5.2 Measurement-Data Update

Since R and y are available as mentioned above from the output of the time-propagation part, say at $k - 1$; form the following composite matrix, modified for the fusion of two inputs (image centroids), here

$$T(k)\begin{bmatrix} \tilde{R}_f(k-1) & \tilde{y}_f(k-1) \\ H_1(k) & z_1(k) \\ H_2(k) & z_2(k) \end{bmatrix} = \begin{bmatrix} \hat{R}_f(k) & \hat{y}_f(k) \\ 0 & e(k) \end{bmatrix}; \; k = 1, \ldots, \tag{9.20}$$

In eqn. (9.20), one has direct MLF of the centroids of the two input images. By applying the orthogonal transformation as in eqn. (9.20), the updated R and y (fused only) information pair is obtained. These are used in eqn. (9.19). At this stage, x_f (for each timestep) can be computed from

$$x_f = \text{inv}(R_f)y_f \tag{9.21}$$

Now, the V, $D12$ factors are required in time-propagation part in eqn. (9.15), hence, formulate 'A' using V and $D12$ factors from eqn. (9.18) as follows:

$$A = [V \times \text{inv}(D12)\ H' \times \text{inv}(\text{sqrtm}(R_{m'}))]; \tag{9.22}$$

In eqn. (9.22), R_m is the covariance matrix of the measurement noise. Then obtain the SVD of A as

$$[Y, S1, Z] = \text{SVD}(A); \quad A \rightarrow Y \times [S\ 0] \times Z'; \text{ here, } S1 = [S\ 0] \tag{9.23}$$

Then, assign the factors as

$$V = Y; \quad D12 = \text{inv}(S) \tag{9.24}$$

Use eqn. (9.24) in eqn. (9.22) and repeat the cycle by going to the time-propagation part.

TABLE 9.3
Metrics of Image Centroid Tracking using VDSRIF

CTVDSRIF	TGNstd = 1			TGNstd = 3			TGNstd= 5		
Filter	PFEx	PFEy	RMSPE	PFEx	PFEy	RMSPE	PFEx	PFEy	RMSPE
BN/SB = 50	0.2087	0.5923	0.2404	0.2190	0.2615	0.1306	0.1679	0.4414	0.1808
BN/SB = 100	0.1454	0.5689	0.2248	0.1525	0.2289	0.1053	0.1444	0.3831	0.1567

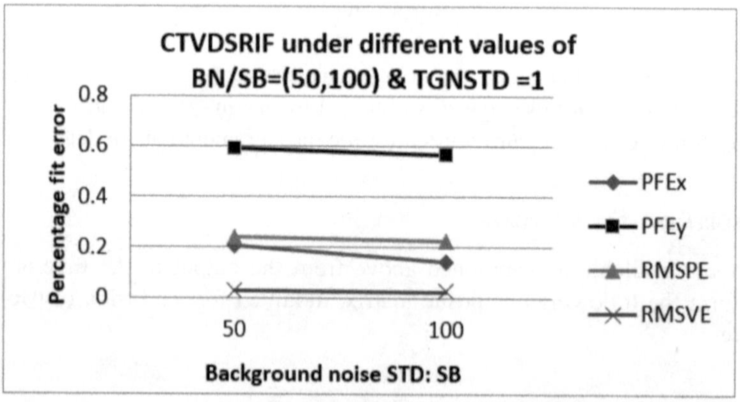

FIGURE 9.3 Tracking performance of VDSRIF with TGNstd = 1.

9.5.3 Performance Evaluation of the VDSRIF Algorithm

Similar procedure as used for the SRIF has been used here also. Since, this is the new filter implemented for image centroid tracking its tracking performance has been evaluated first. One such result from Table 9.3 is plotted in Figure 9.3.

The trend of the PFE w.r.t. BN/SB variation is like that of with SRIF. Now, for the fusion, the results with TGNstds = 1 and 5; and 5 and 9 have also been obtained for fusion of centroid of two images using the new VDSRIF filter. The extensive results (as generated for SRIF and SRVL/VD) have also been generated here for VDSRIF, but are not tabulated for brevity. However, Table 9.4 shows one such result, and a plot is given in Figure 9.4.

Here again, the trends of various performance metrics across these fusion combinations (images with TGNstds = 1 and 5; and 5 and 9) for this filter were also found to be almost similar with minor variations, and hence the fusion performance is found to be satisfactory.

9.5.3.1 Inferences Drawn from the VDSRIF Filtering Algorithm

i. In the time-propagation part, it does not need inversion of the state-transition matrix as required in the SRIF.
ii. It does not need specification of the information state y_w, related to the process noise. This is because it uses the eigenfactors/SVDs of the composite matrix.
iii. In the measurement/data update part, it does not need the computation of the Kalman gain, since, now it uses the measurement/data update part of the SRIF, and hence, it is the gain-free filter. The main reason is that the information state y_f is now directly available from the orthogonal transformation.

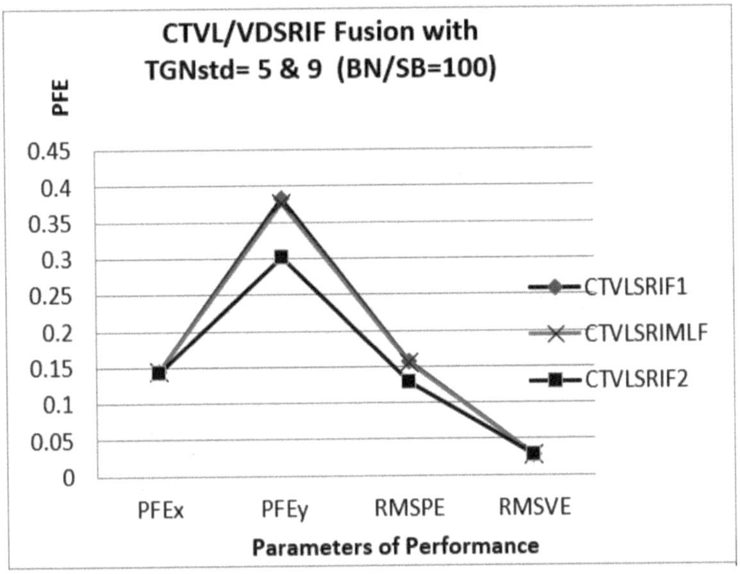

FIGURE 9.4 Fusion using VDSRIF with TGNstd = 5 and 9.

TABLE 9.4
Metrics of Image Centroid Tracking-cum-Fusion using VDSRIF

	TGNstd = 1 & 5 (BN/SB = 50)			TGNstd = 1 & 5 (BN/SB = 100)		
Filter	**PFEx**	**PFEy**	**RMSPE**	**PFEx**	**PFEy**	**RMSPE**
CTVDSRIF1 (std = 1)	0.2087	0.5923	0.2404	0.1554	0.5689	0.2248
CTVDSRIF2 (std = 5)	0.1679	0.4414	0.1808	0.1444	0.3831	0.1567
CTVDSRIFM	0.2027	0.5466	0.2232	0.1520	0.5139	0.2052

iv. It is the hybrid algorithm based on the eigenfactors and SRIF, a combination of the covariance and the information filter (SRIF).
v. It retains the merits of the two filters: (a) SRVD filter, and (b) SRIF.
vi. The new filter eliminates some demerits of both the individual filters.

9.6 A NEW FILTERING ALGORITHM BASED ON UDF AND SRIF

This filter combines the merits of the factorized *U-D* filtering (UDF) and SRIF algorithms for tracking as well as image centroid fusion, and eliminates some demerits of the CTUDF and CTSRIF algorithms [3]. Thus, by capturing the merits of UDF and SRIF algorithms, a new filtering algorithm UDSRIF using the time-propagation part of the *U-D* filter and the measurement update part of the SRIF filter is proposed and validated. It can be used for fusion of image centroids, of any number, at measurement level. All initial values of P, F, G, Q, H, R_m are like that taken in the previous algorithm, and P, F are $n \times n$ matrices. H and R_m are individually composite matrices. R_m is the covariance matrix of the measurement noise. Since, these values are known, steps for designing this new filter are as follows:

9.6.1 Initial Part with Initial Conditions

$$P(0) = \{x_f(0) - \hat{x}_f(0)\}\{x_f(0) - \hat{x}_f(0)\}^T \tag{9.25}$$

Now, take only diagonal elements of $P(0)$ and assign to the

$$D = \text{diag}\{P\};\ U = \text{eye}(n, n) \tag{9.26}$$

Thus, the initial U-D factors of $P(0)$ are known; $P(0) \rightarrow$ UDUT, approximately, and the off-diagonal elements of $P(0)$ are neglected. Now, the initial $R(0)$ matrix as the square root of the information matrix is needed and is given as

$$R(0) = U^{-T}D^{-1/2} \tag{9.27}$$

For simplicity $R_f(0)$ is denoted as $R(0)$, $x_f^{(0)}$ is denoted as $x(0)$, $y_f^{(0)}$ as $y(0)$ in the following. Then,

$$y(0) = R(0) \times x(0) \tag{9.28}$$

Thus, at this stage R, and y as required for the time propagation of the state estimate are available.

9.6.2 Time-Propagation Part

For the time-propagation part of the CTUDSRIF filter, the time-propagated information state is obtained as

$$y_f(\text{new}) = R \times F \times \text{inv}(R) \times y_f(\text{previous}) \tag{9.29}$$

In eqn. (9.29), use R from eqn. (9.27), and y from eqn. (9.28) for the start; then, in the next cycle these will be available from the output of the measurement/data part. Then, form the following matrices using the values from eqn. (9.23), and known F, G and Q:

$$V = [F \times U\ G];\ Dc = \text{diag}[D, Q] \tag{9.30}$$

This means that the previous U, D factors are to be augmented with the new information, i.e. by using F, G and Q. Then, call the time-propagation part (UDTP) of the normal U-D filter with the inputs as in eqn. (9.27). Next, obtain the R matrix needed in the measurement/data update part of SRIF by using the updated U, D factors obtained by the time propagation of (9.30)

$$R = U^{-T}D^{-1/2} \tag{9.31}$$

In this R matrix, the effects of F, G and Q are included by virtue of eqn. (9.30). Hence, matrix R is time-propagated factor. So, at this stage we have R from eqn. (9.31), and y from eqn. (9.29), i.e. the information pair for us to go to SRIF measurement/data update part.

9.6.3 Measurement-Data Update

Since, now R, y are available from the output of the time-propagation part, form the composite matrix

$$T(k)\begin{bmatrix} \tilde{R}(k-1) & \tilde{y}(k-1) \\ H & z(k) \end{bmatrix} = \begin{bmatrix} \hat{R}(k) & \hat{y}(k) \\ 0 & e(k) \end{bmatrix} \tag{9.32}$$

From eqn. (9.32), get the estimated $\hat{R}$ and $\hat{y}$. Since now updated y is available, the Kalman gain is not needed. These $\hat{R}$ and $\hat{y}$ are used in eqn. (9.29). At this stage, since x (for each timestep) is needed, use

$$\hat{x} = \text{inv}\left(\hat{R}\right)\hat{y} \tag{9.33}$$

The U, D factors required in time-propagation part, in eqn. (9.31) are obtained as follows:

First, covariance matrix P is formulated as

$$P = \text{inv}(R^2); \tag{9.34}$$

Then, with the P as in eqn. (9.34), call the following subroutine, assuming P as $(n \times n)$ matrix:

$$\begin{gathered}
\text{for } j = n{:}2; \\
D(j) = P(j,j);\ a = 1./D(j); \\
\text{for } k = 1{:}\,j-1; \\
b = P(k,j);\ U(k,j) = a \times b; \\
\text{for } i = 1{:}k; \\
P(i,k) = P(i,k) - b \times U(i,j); \\
\text{end} \\
\text{end} \\
j = j-1; \\
\text{end} \\
D(1) = P(1,1);
\end{gathered} \tag{9.35}$$

Since, now the new U, D factors are available from eqn. (9.35), substitute in the time-propagation part and repeat the cycle.

The merits of the proposed filtering algorithm, UDSRIF are the following:

- In the time-propagation part, it does not need inversion of the state-transition matrix as required in the SRIF.
- It does not need specification of the information state y_w, related to the process noise, as is required in the case of SRIF, because it uses the square root of the information matrix formed from the time-propagated *U-D* factors of the composite matrix.
- In the measurement-data update part, it does not need the computation of the Kalman gain, since now it uses the measurement-data update part of the SRIF, and hence, it is the gain-free filter; the main reason is that the information state y_f is now directly available from the orthogonal transformation, and the covariance state if required can be easily computed.
- It is the hybrid algorithm based on the *U-D* factors and SRIF, a combination of the covariance and the information filter (SRIF).
- It retains the merits of the two filters: (i) *U-D* filter, and (ii) SRIF; and the new algorithm eliminates some demerits of both the filters.

9.6.4 Performance Evaluation of the UDSRIF Algorithm

First the image centroid tracking performance is evaluated, see Table 9.5 and Figure 9.5.

It is seen that there is not much of a trend of the performance metrics with respect to the stds. However, it was established earlier that CTUDSRIF performed somewhat better than CTKF in a similar centroid tracking task, and that the position and velocity state errors (time histories) were found to lie within their theoretical bounds as predicted by the CTUDF.

For a system with two sensors H1 and H2, their measurements can be fused at data level. The least-squares solution is obtained by applying the method of orthogonal transformation. This method of solving is likely to be less susceptible to computer round-off errors. This process results in the state estimate of the two-sensor data fused.

$$T(k)\begin{bmatrix} \tilde{R}_f(k-1) & \tilde{y}_f(k-1) \\ H_1(k) & z_1(k) \\ H_2(k) & z_2(k) \end{bmatrix} = \begin{bmatrix} \hat{R}_f(k) & \hat{y}_f(k) \\ 0 & e(k) \end{bmatrix};\ k = 1,..., \tag{9.36}$$

Here $\hat{y}_f$ is the information state. The updated R_f and y_f (fused only). Since, now y_f is available, we do not need the Kalman gain. These R_f and y_f are used in eqn. (9.29). At this stage, since x_f (for each timestep) is needed, use

$$x_f = \text{inv}(R_f)\, y_f \tag{9.37}$$

The fusion is MLF. The target background is set at (BN:mean = 50, std = 50, 100). Two images are considered with target image set as (mean = 100, std = 10), and with variation in the TGNstd = 1 (image 1, CTUDSRIF1), and TGNstd = 5 (image 2, UDSRIF2); these images are considered two-at-a-time for centroid tracking-cum-fusion. The performance metrics are shown in Table 9.6; and two of such results are shown in Figure 9.6.

TABLE 9.5
Metrics of Image Centroid Tracking using CTUDSRIF

Parameter (*) Metrics-> (%fit errors)	TGNstds 1	5	9
PFEx	0.2255	0.1961	0.1989
PFEy	0.7927	0.5986	0.4782
RMSPE	0.3158	0.2414	0.1985
RMSVE	0.0290	0.0315	0.0316

(*: $Q = 0.00001$; BN/SB std = 50)

TABLE 9.6
Metrics for Fusion using CTUDSRIF

	TGNstds = 1 & 5 (BN/SB = 50)				TGNstds= 1 & 5 (BN/SB = 100)			
Filter	PFEx	PFEy	RMSPE	RMSVE	PFEx	PFEy	RMSPE	RMSVE
CTUDSRIF1(std = 1)	0.2255	0.7927	0.3158	0.0290	0.1554	0.6551	0.2580	0.0261
CTUDSRIF2(std = 5)	0.1961	0.5986	0.2414	0.0315	0.1596	0.4354	0.1777	0.0277
CTUDSRIFMLF	0.2226	0.6459	0.2616	0.0315	0.1644	0.5933	0.2357	0.0277

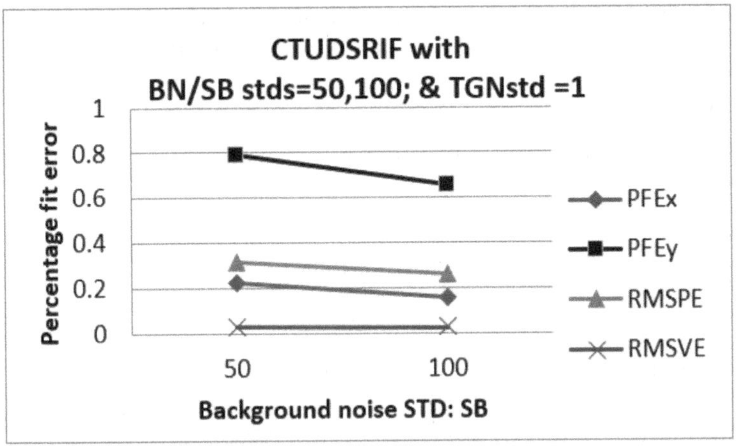

FIGURE 9.5 Tracking results with UDSRIF.

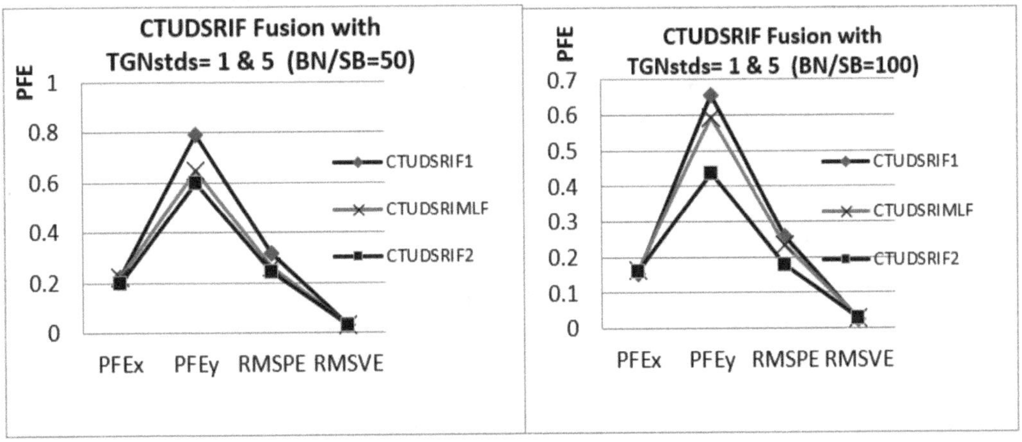

FIGURE 9.6 Performance of CTUDSRIF (TGNstds = 1 and 5).

Also, the target-image fusion results with TGNstds = 5 and 9 have been obtained for centroid fusion of two images, using UDSRIF. The trends of these results were found to be like those in Table 9.6, and Figure 9.6. Based on the performance metrics and plots, it has been found that this algorithm gives satisfactory performance in tracking and fusion. The performance of the square-root-type filtering algorithm UDSRIF has remained nearly robust despite there are certain, and small, trends in the performance metrics w.r.t. target noise variance and background noise variance.

9.7 OBSERVER BASED ON THE FUZZY NONLINEAR FUNCTION OF RESIDUALS

An application of the observer theory to image centroid tracking is considered, and especially the fuzzy membership function (FMF) is used as a nonlinear operator on the observer residuals. The latter is a new aspect in the domain of centroid tracking as well as in the theory of observers. In estimation/filtering theory, the objective of an observer is to reconstruct the state of a dynamic system using the knowledge of the system's data/signals. For linear systems, the system's states are usually estimated using the Luenberger observer (LO) or KF.

The asymptotic result for the observer error dynamics, which are influenced with (any) FMF is derived. The performance of the proposed observer is illustrated with implementation in MATAB.

9.7.1 Linear System and Observer Error Dynamics

Let a linear system be given as

$$\begin{aligned} \dot{x}(t) &= Ax(t) + Bu(t) \\ y(t) &= Hx(t) \end{aligned} \tag{9.38}$$

In a normal situation a linear observer for the system of eqn. (9.38) can be proposed

$$\begin{aligned} \dot{\hat{x}}(t) &= A\hat{x}(t) + Bu(t) + L(t)(y(t) - \hat{y}(t)) \\ \hat{y}(t) &= H\hat{x}(t) \end{aligned} \tag{9.39}$$

However, now it is proposed to use fuzzy logic in the observer design, by way of using FMF of the residuals:

$$\begin{aligned} \dot{\hat{x}}(t) &= A\hat{x}(t) + Bu(t) + L(t)f_m(r(t)) \\ \dot{\hat{x}}(t) &= A\hat{x}(t) + Bu(t) + L(t)f_m(He(t)) \\ \hat{y}(t) &= H\hat{x}(t) \end{aligned} \tag{9.40}$$

In eqn. (9.40), the residuals and the state errors are

$$r(t) = y(t) - \hat{y}(t); \;\; e(t) = x(t) - \hat{x}(t) \tag{9.41}$$

$L(t)$ is an observer gain matrix given as

$$L(t) = P(t)H^T R^{-1} \tag{9.42}$$

From eqn. (9.42), it is seen that the observer gain is obtained from the KF/EKF theory. This is done because the KF and the linear observer belong to the same class of the state estimators and both can be easily applied to the same or similar dynamic systems. In the case of the observer of eqn. (9.39), R in eqn. (9.42) is some positive definite matrix, and is regarded as the weighting matrix. In eqn. (9.39), f_m is an FMF that operates on input residuals $r(t)$, (in turn on $e(t)$) and obtains the output that is used in the observer structure; this provides a nonlinear function of the residuals effect because any FMF is inherently nonlinear. In eqn. (9.42) $P(t)$ is obtained as the solution of the observer Riccati differential (ORD) equation

$$\dot{P}(t) = P(t)A^T(t) + A\ (t)P(t) - P(t)H^T R^{-1} HP(t) + Q \tag{9.43}$$

In eqn. (9.43), Q is considered as some weighting matrix, whereas in KF/EKF theory it is an intensity or a covariance matrix of the process noise. By subtracting eqn. (9.40) from eqn. (9.38), the following observer error dynamics are obtained:

$$\begin{aligned} \dot{e}(t) &= A(t)e(t) - L(t)f_m(r(t)) \\ &= A(t)e(t) - L(t)f_m(He(t)) \end{aligned} \tag{9.44}$$

9.7.2 Asymptotic Stability Analysis

The following conditions for the local asymptotic behaviour of the observer error dynamics of eqn. (9.44) are considered:

1. The solution of the ORD eqn. (9.43) is bounded [4,5]:

$$p_l I \leq P(t) \leq p_u I \tag{9.45}$$

with $p_l, p_u > 0$ as positive constants ($P(t)$ is theoretically positive definite matrix), and are the lower and upper bounds, respectively; I is the identity matrix.
2. The nonlinearity $f_m(.)$ is bounded

$$\left\| f_m(.) \right\| \leq \rho \left\| x(t) - \hat{x}(t) \right\|^2 \tag{9.46}$$

With the bounding constant equal to or greater than zero. Since in the present case the nonlinear function is an FMF, the overall bound on the nonlinearity is $0 \leq \left\| f_m(.) \right\| \leq 1$.

Then, the nonlinear observer error dynamics in eqn. (9.44) are locally asymptotically stable, if conditions 1 and 2 are satisfied, and the time derivative of the LE functional is negative definite.

First, consider the normalized LE functional to establish the asymptotic stability of the error dynamics (6.35):

$$V(t) = e^T(t)Y(t)e(t) \tag{9.47}$$

Matrix $Y(t)$ is the normalizing matrix and is recognized as an information matrix $\{Y(t) = P^{-1}(t)\}$; in the case of KF, $P(t)$ is considered as the covariance matrix of the state errors. The matrix $Y(t)$ is also called the information Gramian.

The LE functional, eqn. (9.47) is positive definite because it is governed by condition 1, the inequality of eqn. (9.45):

$$\frac{1}{p_u}\left\| e(t) \right\|^2 \leq e^T(t)Y(t)e(t) \leq \frac{1}{p_l}\left\| e(t) \right\|^2 \tag{9.48}$$

The point in obtaining the asymptotic result is that the time derivative of the LE functional, eqn. (9.47); under the constraints governed by error dynamics, the specified bounds, and the gain/covariance dynamics; should be negative definite. This would ensure the local asymptotic stability of the error dynamics, and hence the convergence of the observer. Next, this time derivative of $V(t)$ is obtained as

$$\begin{aligned} \dot{V}(t) &= e^T \dot{Y} e + e^T Y \dot{e} + \dot{e}^T Y e \\ &= e^T(t)\dot{Y}(t)e(t) + e^T(t)\{A^T(t)Y(t) + Y(t)A(t)\}e(t) \\ &\quad - 2e^T Y(t)L(t) f_m(He(t)) \end{aligned} \tag{9.49}$$

Substitute for $\dot{Y}(t) = -Y(t)\dot{P}(t)Y(t)$, (since $Y(t)P(t) = I$), and eqns. (9.42) and (9.43) in eqn. (9.49) to obtain

$$\dot{V}(t) = -e^T(t)Y(t)QY(t)e(t) + e^T(t)H^T R^{-1} He(t) - 2e^T(t)H^T R^{-1} f_m(He(t)) \tag{9.50}$$

In obtaining eqn. (9.50), because of the substitution of eqn. (9.42), and eqn. (9.34), several common terms cancel out. In eqn. (9.50), assume that

$$\|R^{-1}\| \leq \frac{1}{R_r}; \|H^T H\| \leq h^2$$

(R_r and h being positive constants); and $\|e(t)\|^2 \leq \varepsilon^2$, and since, these are known and pre-specified quantities, or should be finite, the following condition is obtained

$$\dot{V}(t) \leq -e^T(t)Y(t)QY(t)e(t) + \frac{h^2\varepsilon^2}{R_r} - 2e^T(t)H^T R^{-1} f_m(He(t)) \tag{9.51}$$

Using the inequality from eqn. (9.45), one obtains

$$\dot{V}(t) \leq -\frac{1}{p_u^2} q_s \|e(t)\|^2 + \frac{h^2}{R_r}\|e(t)\|^2 - 2\frac{h}{R_r} e^T(t) f_m(He(t)) \tag{9.52}$$

Hers, q_s is the smallest (positive) eigenvalue of the matrix $Q(t)$, that is positive definite. Using bound for the FMF, one obtains

$$\dot{V}(t) \leq -\left(\frac{1}{p_u^2} q_s - \frac{h^2}{R_r}\right)\|e(t)\|^2 - 2\frac{h}{R_r} e^T(t)\rho\|e(t)\|^2 \tag{9.53}$$

For $\|e(t)\| \leq \varepsilon$, one has the following condition from eqn. (9.35)

$$\dot{V}(t) \leq -\left(\frac{1}{p_u^2} q_s - \frac{h^2}{R_r} + 2\frac{h\rho}{R_r}\varepsilon\right)\|e(t)\|^2 \tag{9.54}$$

$$\dot{V}(t) \leq -\left(\frac{p_u^2 h^2 - q_s R_r}{2h\rho p_u^2}\right)\|e(t)\|^2 \tag{9.55}$$

One needs to ensure that the factor appearing in eqn. (9.54) is positive,$\{p_u^2 h^2 > q_s R_r\}$, since all the other individual coefficients are anyway positive. Since, the LE functional is positive definite and its time derivative is locally negative definite as in eqn. (9.55), the observer error dynamics of the newly proposed nonlinear observer for the system with FMF as the nonlinear function (of the residuals) is locally asymptotically stable. The derived result is quite general, since it is applicable even when any FMF is used to modulate the observer residuals, because the magnitude of all the FMFs are bounded between 0 and 1.

9.7.3 Performance Evaluation of an FL-Modulated Observer for Centroid Tracking

The tracking dynamics are represented as in eqn. (9.38); the centroid coordinates are represented as a state x in the observer dynamics. FL can be used for tuning observer dynamics, and in the present application FMF is used for modulating the observer residuals, that means for the purpose of centroid tracking, FL augments the observer at the measurement level:

$$\begin{aligned}\dot{\hat{x}}(t) &= A\hat{x}(t) + L(t) f_m(r(t)) \\ \hat{y}(t) &= H\hat{x}(t)\end{aligned} \tag{9.56}$$

Here, $f_m(r(t))$ is regarded as an output of the FL-based process and is a nonlinear function of inputs, which are the residuals 'r' (and often derivative of 'r') of the observer. It is assumed that position in x-y axes-measurements of the target are available. The f vector consists of the modified residuals' sequence for x and y axes, expressed in the discrete time domain, for the sake of clarity as

$$f_m(k+1) = \left[f_{mx}(k+1) \quad f_{my}(k+1) \right] \tag{9.57}$$

To determine eqn. (9.57), the residual vector 'r' is first separated into its x and y units, r_x and r_y. The f vector for the x direction is computed, it can then include the y axis. This vector consists of two inputs, r_x and $\dot{r}_x$, and single output $f_m(k+1)$, where $\dot{r}_x$ is computed by

$$\dot{r}_x = \frac{\{r_x(k+1) - r_x(k)\}}{T} \tag{9.58}$$

Here, T is sampling interval in seconds, and eqn. (9.58) can be easily extended to y direction. Appropriate FMF/FIS (fuzzy inference system) has been used and the technique is evaluated using the MATLAB-based fuzzy logic tool box, where necessary (evalfis: Type= sugeno; NumInputs=2; NumOutputs=1; NumRules=4; AndMethod='prod'; OrMethod='max'; ImpMethod='prod'; AggMethod='max'). It is ascertained here, that the asymptotic result derived using FMF applied only to residual term $r(t)$, eqn. (9.55) is also equally applicable when it is also applied to the derivative of $r(t)$, eqn. (9.58). The value of R (or 'R_r') appearing in eqn. (9.43) is used as a tuning parameter. In order to implement the algorithm, one should solve the matrix ORD eqn. (9.44), and for this one can use the method of transition matrix described in previous sections. To simulate the motion of the target in a frame, the kinematic model of target motion is used for which a continuous-time target motion model is used. The state vector is given as $X(k) = \left[x \quad \dot{x} \quad y \quad \dot{y}\right]^T$.

The measurement model is given as

$$z(t) = \begin{bmatrix} 1 & 0 & 0 & 0 \\ 0 & 0 & 1 & 0 \end{bmatrix} X(t)$$

For the background image, a 2D array of 64 × 64 pixels is used, which is a WGN $N(\mu_{bg}, \sigma_{bg}^2)$, with std of the background image (BN) as 50. One more 2D array of pixels, as WGN $N(\mu_{tg}, \sigma_{tg}^2)$ is used for a target of size 9 × 9. The initial state vector of the target in the image frame is $X = \left[x \quad \dot{x} \quad y \quad \dot{y}\right]^T = \left[10 \quad 1 \quad 10 \quad 1\right]^T$. The synthetic image/s are converted into binary image/s (with $I_U = 110$ and $I_L = 90$ limits of a target layer). The performance metrics: % fit errors (PFE) for the observations and states, and rms values of position and velocity errors are computed. The performance of the FMF-based observer is illustrated in Figures 9.7 to 9.10 (for the TGNstd = 5; BNstd = 50); and Table 9.7, which gives the numerical values of the performance metrics; PFE = 100 × norm(error)/norm(true signal).

The rms errors are given in Table 9.7; rsspe = sqrt(xperr.^{2} + yperr.^{2}); and rmspe = sqrt(mean(xperr.^{2} + yperr.^{2})); it is seen that the metrics show some trend when the std of the target image noise is increased, however, in an overall sense the values are quite small and hence, a good centroid tracking has been obtained.

Since, all the errors' time histories are within their theoretical bounds, the performance of the proposed FMF-based observer for target image centroid tracking is considered very satisfactory. Figure 9.10 shows the plot of the eigenvalues of the Gramian matrix P, related to state $x1$, $x2$ (--), and state $x3$ and $x4$ (.); showing the convergence of the observer.

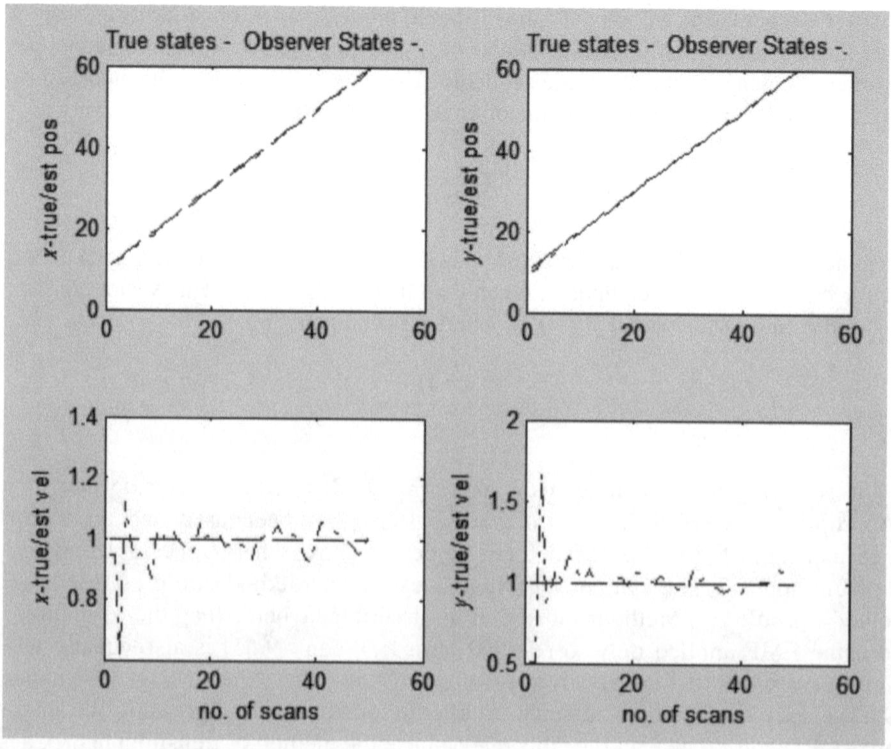

FIGURE 9.7 Time history match of the true (-) and observer states (-.).

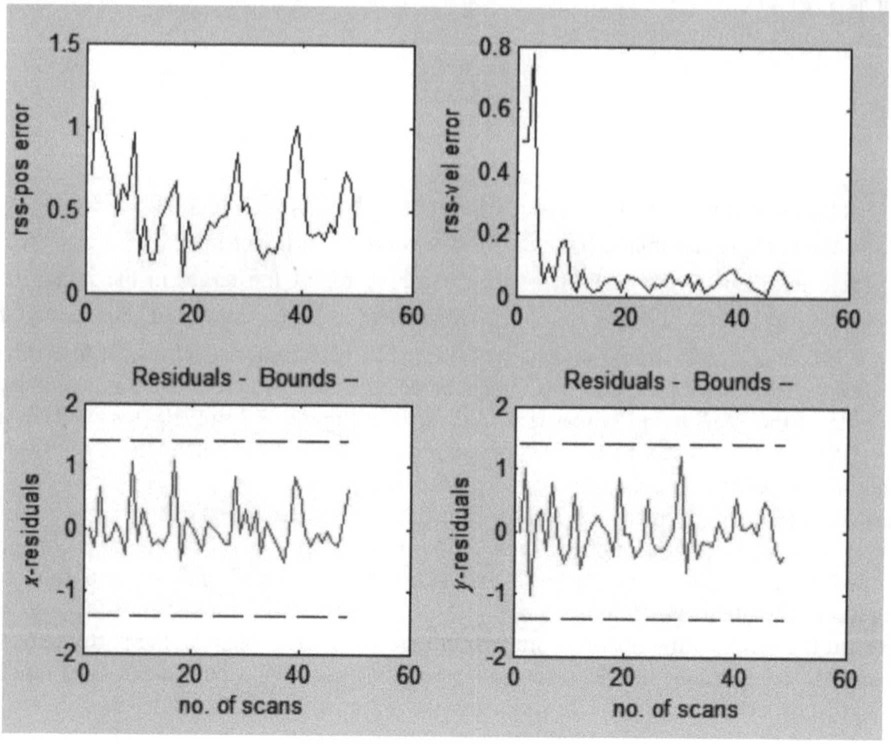

FIGURE 9.8 RSS position and velocity errors and residuals (-), bound (--) (obtained from the Gramian, *R*).

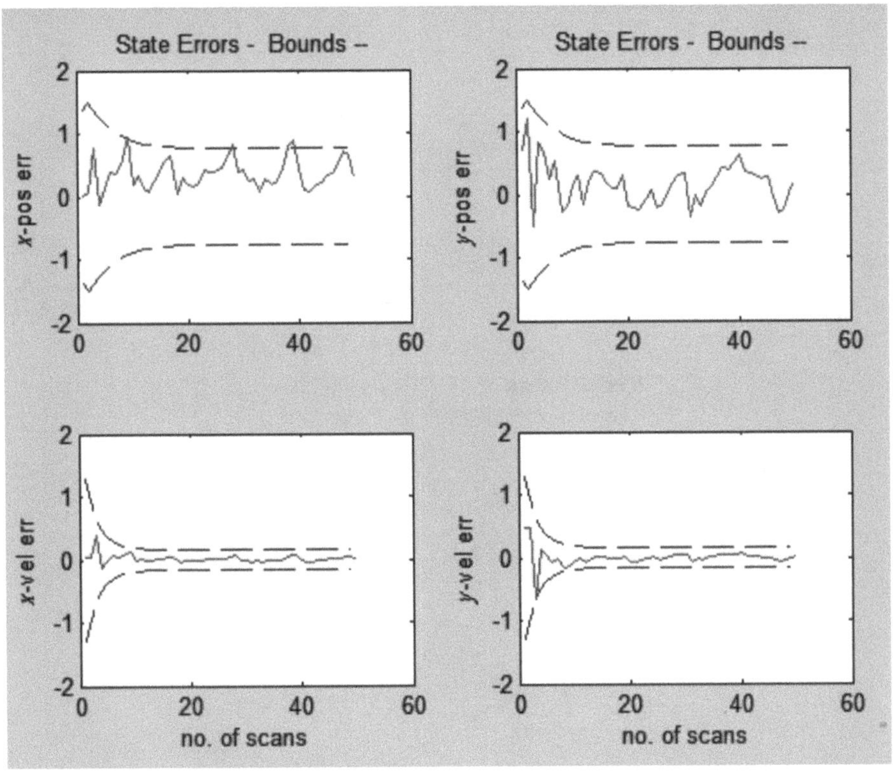

FIGURE 9.9 Observer state errors with bounds: errors (-), bounds (--) (obtained from the Gramian, P of the solution of the observer Riccati equation).

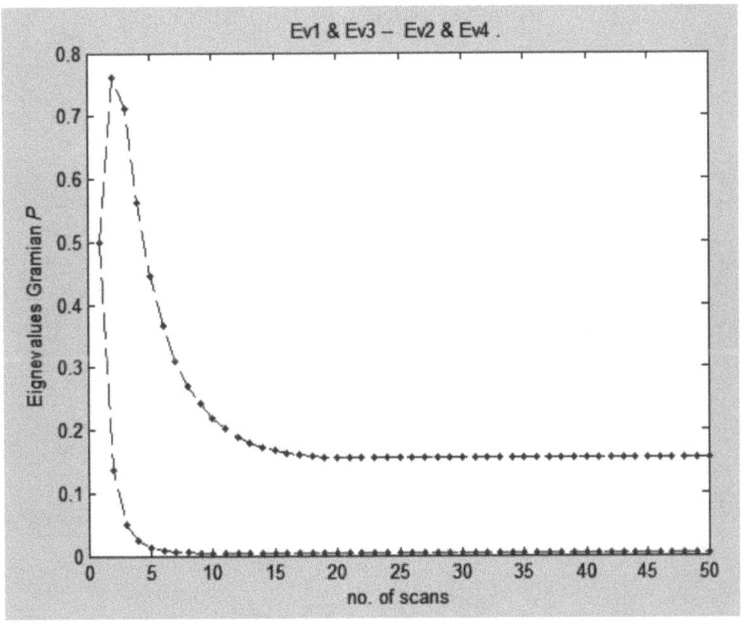

FIGURE 9.10 Eigenvalues of the Gramian matrix P (related to state $x1$, $x2$ (--), and state $x3$ and $x4$ (.)).

TABLE 9.7
Metrics from FMF-Based Observer

Standard deviation of target noise →	Std = 1	Std = 3	Std = 5
PFE-x	0.6620	0.7236	1.1135
PFE-y	0.7483	0.8857	0.9590
RMSPE	0.3829	0.4383	0.5631
RMSVE	0.1058	0.1102	0.1608

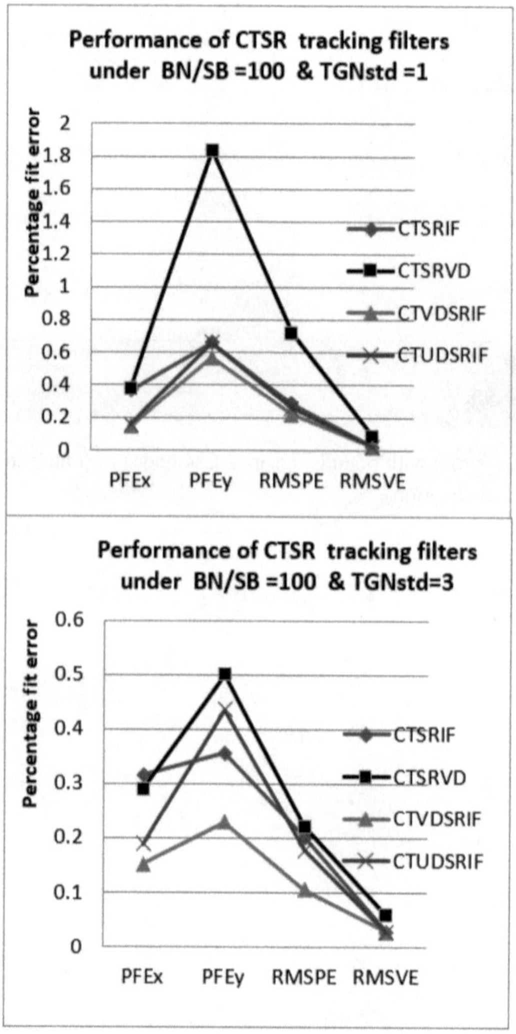

FIGURE 9.11 Tracking performance for BN/SBstd = 100, TGNstd = 1 and 3.

TABLE 9.8
CTSR Tracking Filters' Performance (BNstd = 100; TGNstd1 = 1; TGNstd2 = 3)

	With Target Noise STD 1 BN: Background Noise std 100				With Target Noise STD 3 BN: Background Noise std 100			
Tracking Filter	PFEx	PFEy	RMSPE	RMSVE	PFEx	PFEy	RMSPE	RMSVE
CTSRIF	0.3678	0.6552	0.2879	0.0261	0.3162	0.3559	0.2023	0.0280
CTSRVD	0.3818	1.8338	0.7178	0.0871	0.2903	0.4999	0.2215	0.0580
CTVDSRIF	0.1454	0.5689	0.2248	0.0261	0.1525	0.2289	0.1053	0.0280
CTUDSRIF	0.1554	0.6551	0.2580	0.0261	0.1886	0.4354	0.1777	0.0277

TABLE 9.9
Metrics of CTSR Fusion with BN/SBstd = 100 and TGNstds = 1 and 3

	With Target noise stds = 1 & 3 BN/SB noise std = 100			
Fusion Filter	PFEx	PFEy	RMSPE	RMSVE
CTSRIF	0.3805	0.4869	0.2368	0.0280
CTSRVD	0.3600	1.5131	0.5960	0.0832
CTVDSRIF	0.1549	0.4301	0.1750	0.0280
CTUDSRIF	0.1748	0.5046	0.2045	0.0280

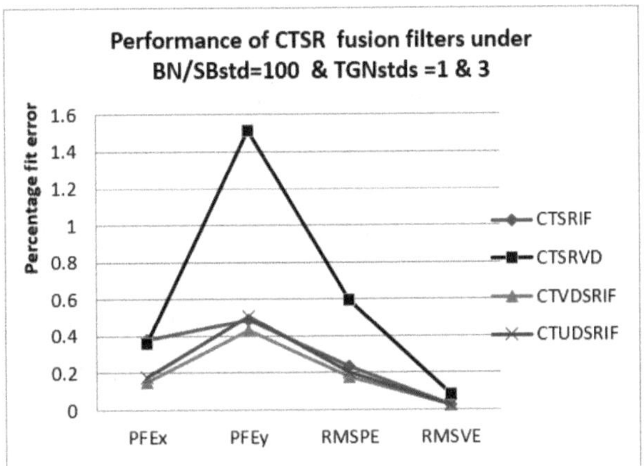

FIGURE 9.12 Fusion performance with BN/SBstd = 100, TGNstds = 1 and 3.

9.7.4 Overall Performance of the Square-Root-Type Image Centroid Tracking Algorithms

The overall performance of certain square-root centroid algorithms (CTSR) is presented in Table 9.8 and Figure 9.11 for image centroid tracking and in Table 9.9 and Figure 9.12 for the centroid fusion with TGNstds= 1 and 3; separately and background-noise BNstd=100.

It is inferred from the presented results, that the CTVDSRIF (one of the newly proposed algorithms) performs better than others in tracking as well as in the fusion task; while CTSRIF (new application to the image centroid problem) and CTUDSRIF (newly proposed algorithm) have similar performances. The CTSRVD algorithm (new application to the image centroid problem) performs reasonably well.

Thus, in this chapter, three new algorithms have been proposed and evaluated for image centroid tracking, and fusion where applicable. The merits of these algorithms have been stated. For the proposed new FL-based nonlinear observer/estimator, the asymptotic stability result has been derived using the LE functional. The performance of the observer is validated by the behaviour of the eigenvalues of the observer Gramian matrix. The performance of the proposed FMF-based observer is found to be very satisfactory and the analytical results corroborate the theoretical result presented. These results can be extended for measurement/state vector-level image fusion, in comparison with conventional filtering algorithms used for data fusion.

Overall performance of two newly proposed and two newly applied image centroid tracking-cum-fusion algorithms has also been evaluated and has been found to be greatly acceptable.

EXERCISES

Exercise 9.1

How does the eigen factor square-root filter differ from conventional filtering techniques in improving tracking accuracy and robustness?

Exercise 9.2

How does the eigen factor square-root filter handle time propagation and data update processes, and what advantages does this approach offer over traditional filtering methods?

Exercise 9.3

What is an observer-based fuzzy nonlinear function of residuals?

Exercise 9.4

How is an observer-based fuzzy nonlinear function of residuals different from other methods?

Exercise 9.5

How can we combine the *U-D* filter with the SRIF filter for object tracking?

Exercise 9.6

What is asymptotic stability analysis in the context of dynamical systems?

REFERENCES

1. Parimala, P., *Image Tracking and Fusion Using Square Root Information Filtering*. Ph.D. Thesis, Electronics and Communications Engineering, Jain University, Bangalore, India, pp. 1–85, Aug. 2017.
2. Oshman, Y., and Bar-Itzhack, I.Y., Square root filtering via covariance and information eigenfactors, *Automatica, Pergamon Journals Ltd., IFAC, UK*, Vol. 22, No. 5, pp. 599–604, 1986.

3. Bierman, G.J., *Factorization Methods for Discrete Sequential Estimation*, Academic Press, New York, 1977.
4. Raff, T., and Allgower, F., An EKF-based observer for nonlinear time-delay systems. In *Proc. of the American Control Conference, Minneapolis, MN, USA*, July 2006, Vol. 4, pp. 3130–3133. http://bibbase.org/network/publication/raff-allgwer-anekfbasedobserverfornonlineartime delay-systems-2006, accessed January 2016.
5. Reif, K., Sonnemann, F., and Unbehauen, R., An EKF-based nonlinear observer with a prescribed degree of stability, *Automatica*, Vol. 34, pp. 1119–1123, Sept. 1998, www.researchgate.net/publication/ 220157426, accessed January 2017.

10 H-Infinity Fuzzification Filtering and Target Tracking

10.1 INTRODUCTION

In this chapter, the major emphasis is on utilization of some HI filtering algorithms. Some of these algorithms are augmented by fuzzy logic [1] and thus the joint algorithms are used for tracking of maneuvering and nonmaneuvering targets, and this is a novel approach for estimation and target tracking. The main advantage of the HI-based tracking and fusion algorithms is that these do not need any specification of statistics of the process (state-) noise and the measurement noise, which are often not known. In the HI filtering approach, the disturbances are usually assumed to be arbitrary energy-bounded signals, and an HI filter is designed to minimize the maximum of the gain from the input disturbance signals to the (output) estimation errors. This gain is defined in terms of HI norm, and the related IO signals are represented in terms of the variances of: (i) differences in the initial conditions, (ii) input state and measurement noise processes and (iii) the errors in the estimation. In HI theory the random processes are considered as generalized random variables, and their variances are known as the corresponding Gramians, and not the covariances. HI norm-based estimators give robust state estimates, but not the optimal estimates, and often it is important for several practical systems to obtain robustness rather than optimality. The estimators are obtained by the process of minimization of the magnitude of the HI norm, and keeping it below some threshold, signifying the fact that despite the presence of various noise and error processes, the HI norm is bounded, a basic requirement of robustness of an estimator, or a controller. These estimators offer alternative solutions and still carry the flavour of conventional Kalman filtering approaches, the former being robust, and the latter being optimal. At least for linear systems, both are equivalent, and look similar in form and structure.

First, the *a posteriori* HI-based sensor fusion algorithm [2] is given. Then, FL-based adaptive tuning approaches of Kalman/HI filtering algorithms [3] are given, one of which is used in scheme 3, described below.

Subsequently, some new schemes based on FL-augmented HI filter are presented; the outline of which is as follows:

(i) Scheme 1: basic HI fuzzification with fuzzy correlation method-FLHIF.
(ii) Scheme 2: further fuzzification of FLHIF for determining the weights for sensor fusion using SVF-FLHIFF.
(iii) Scheme 3: use of adaptive FL-based degree of matching (DoM)-AHIFDOM.
(iv) Scheme 4: a finite memory-sliding window-based adaptive HI filter-AHIF.

These new schemes are evaluated for data fusion, and monitoring maneuvering and non-maneuvering targets.

DOI: 10.1201/9781003620211-10

10.2 FUZZY-INFINITY FILTER FUZZIFIATION (FLHIFF) – SCHEME 1

A basic scheme makes use of the FL correlator in the KF/HI filter to obtain the FL-based filter [1]. In the KF/HI filter, the measurement update equation for the $(k+1)$th scan is defined as

$$\hat{x}(k+1) = F\hat{x}(k) + K(e(k+1)) \tag{10.1}$$

where, $e(k+1) = y(k+1) - HF\hat{x}(k)$ is an innovation sequence vector. For the modified filter measurement update, the equation is redefined as

$$\hat{x}(k+1,k+1) = \tilde{x}(k+1,k) + K(C(k+1)) \tag{10.2}$$

where $\tilde{x}(k+1,k) = F\hat{x}(k)$ and $C(k+1)$ is a fuzzy correlator function (FCF), along x and y axes: $C(k+1) = \left[C_x(k+1) \; C_y(k+1)\right]$. Vector e is split into its x, y components, i.e. e_x and e_y. FCF is initially computed using e_x and later generalized to include e_y. FCF considers vector e_x, $\dot{e}_x$ as inputs and provides modified or correction vector C as an output. With T as the sampling period, input $\dot{e}_x$ is computed using

$$\dot{e}_x = \frac{e_x(k+1) - e_x(k)}{T} \tag{10.3}$$

FCF is based on the Mamdani model; and consists of fuzzification, fuzzy inference system (FIS), and defuzzification units. In fuzzification crisp inputs e_x, $\dot{e}_x$ are converted to fuzzy sets using MFs, which are defined using variables: SN – small negative, SP – small positive, MN – medium negative, MP – medium positive, LN – large negative, LP – large positive and ZE –zero error. MFs adopted for these error inputs in scheme 1 are shown in Figure 10.1(a),(b),(c). Mamdani FIS for the FCF consisting of 49 rule base (from experts' opinions), and the rule matrix is given in Table 10.1.

A sample of 49 rules adopted is in the form: If e_x is SN and $\dot{e}_x$ is SN THEN C_x is MN. Mandani inference used is realized using min as AND operator, max as OR operator, min for fuzzy implication function (FIF) and max for aggregation operation. Defuzzification operation converts the fuzzy values obtained from FIS to crisp output values by using centroid method.

10.3 FL-BASED HI FUZZIFIED FILTER (FLHIFF) – SCHEME 2

The FLHIF is further extended to realize FLHIFF to eliminate outlier effects and minimize estimation error during MSDF [2]. Estimation error average from each sensor (from FLHIF) is computed using

$$\bar{e}_s^{\text{FLHIF}}(k) = \frac{\left(e_{s_x}^{\text{FLHIF}}(k) + e_{s_y}^{\text{FLHIF}}(k) + e_{s_z}^{\text{FLHIF}}(k)\right)}{M} \tag{10.4}$$

Sensors number $s = 1,2$; $e_{s_x}^{\text{FLHIF}}(k), e_{s_y}^{\text{FLHIF}}(k)$, and $e_{s_z}^{\text{FLHIF}}(k)$ represent the estimation error using FLHIF [3] along the x,y,z measurement channels and M is the total number of channels. Estimation errors at every scan number is computed by using

$$e_{s_x}^{\text{FLHIF}}(k) = x_{m_s}(k) - \hat{x}_s^{\text{FLHIF}}(k)$$

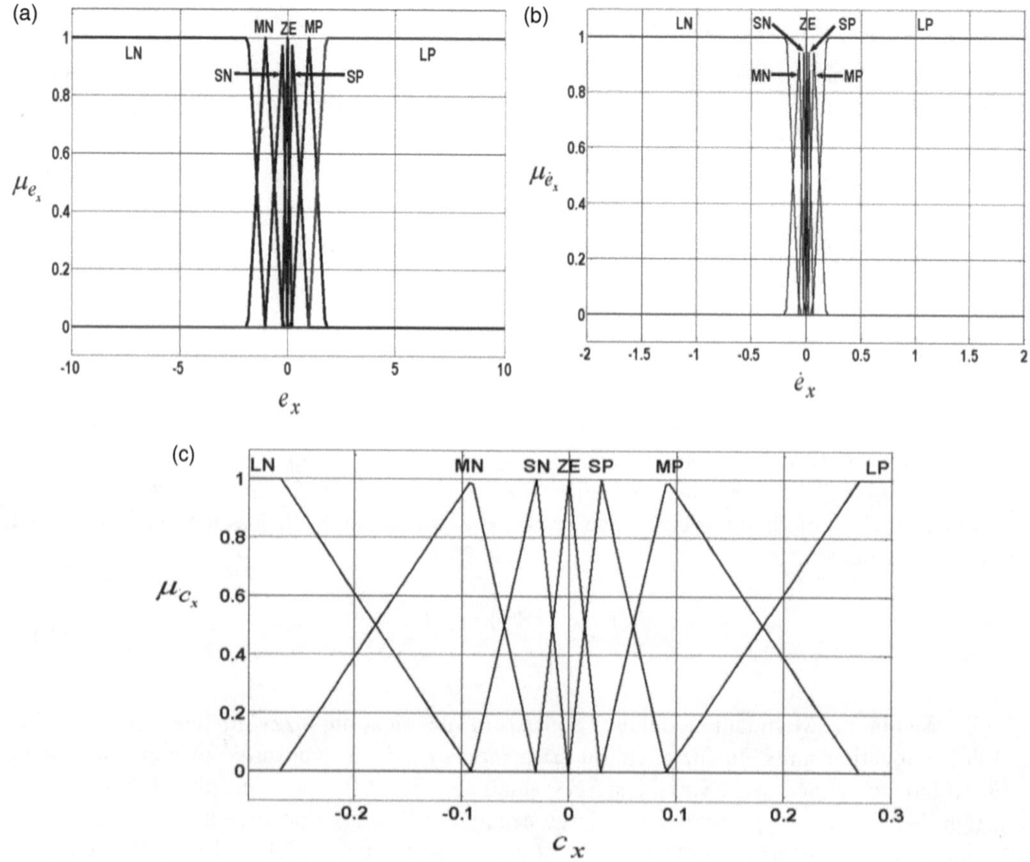

FIGURE 10.1 MFs used in the Mamdani model for scheme 1.

TABLE 10.1
Rules Adopted in HI Fuzzification: Scheme 1

	e_x						
$\dot{e}_x$	**SN**	**SP**	**MN**	**MP**	**LN**	**LP**	**ZE**
SN	MN	ZE	MN	SP	MN	MP	SN
SP	ZE	MP	SN	MP	MN	MP	SP
MN	MN	SN	MN	ZE	LN	SP	MN
MP	SP	MP	ZE	MP	SN	LP	MP
LN	MN	MN	LN	SN	LN	ZE	MN
LP	MP	MP	SP	LP	ZE	LP	MP
ZE	SN	SP	MN	MP	MN	MP	ZE

$$e_{s_y}^{\mathrm{FLHIF}}(k) = y_{m_s}(k) - \hat{y}_s^{\mathrm{FLHIF}}(k) \tag{10.5}$$

$$e_{s_z}^{\mathrm{FLHIF}}(k) = z_{m_s}(k) - \hat{z}_s^{\mathrm{FLHIF}}(k)$$

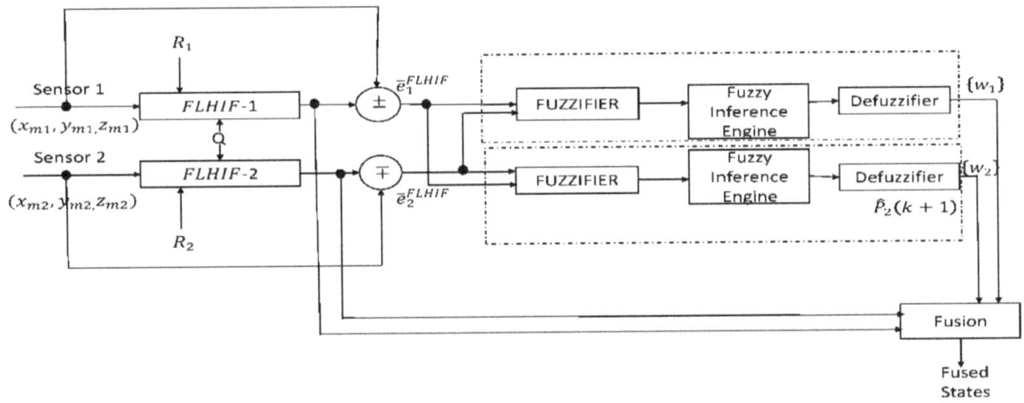

FIGURE 10.2 Schematic of FLHIFF – scheme 2.

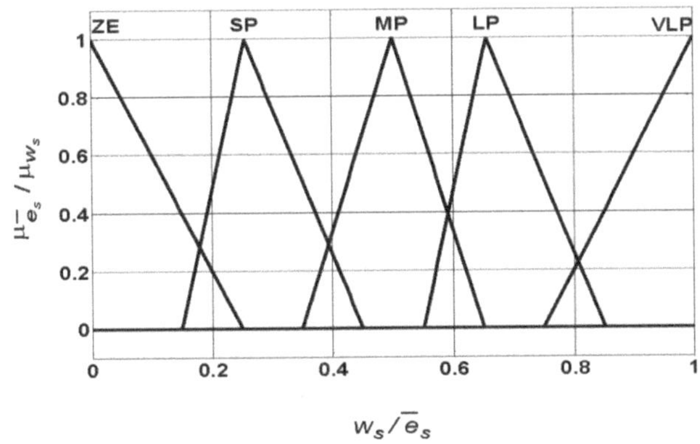

FIGURE 10.3 FMFs used in FLHIFF.

where x_{m_s}, y_{m_s}, z_{m_s} represent the target's position measurements along the x,y,z axis, and their estimated positions using FLHIF is represented using $\hat{x}_s^{\text{FLHIF}}$, $\hat{y}_s^{\text{FLHIF}}$, $\hat{z}_s^{\text{FLHIF}}$. In FLHIFF – scheme 2, Figure 10.2, two FLSs corresponding to two sensors. Error $\bar{e}_s^{\text{FLHIF}}$ is considered as an input to FLS and adaption weights w_1, w_2 are used to represent output sets of FLS; with this SVF estimates are obtained using

$$\hat{X}_f^{\text{FLHIFF}}(k) = \hat{X}_1^{\text{FLHIF}}(k) + w_1(k)\left(w_1(k) + w_2(k)\right)^{-1}\left(\times\left(\hat{X}_2^{\text{FLHIF}}(k) - \hat{X}_1^{\text{FLHIF}}(k)\right)\right) \tag{10.6}$$

where $\hat{X}_1^{\text{FLHIF}}$, $\hat{X}_2^{\text{FLHIF}}$ represent estimated states obtained from FLHIF for sensor 1 and 2.

Each fuzzy logic system of FLHIFF – scheme 2 consists of a fuzzifier, FIS, and defuzzifier; Mamdani-based FLS [4] is considered. The fuzzifier normalizes $\bar{e}_s^{\text{FLHIF}}$ signals between [0,1] using MFs which are defined using variables SP – small positive, MP – medium positive, LP – large positive, VLP – very large positive and ZE-zero error. MFs used are shown in Figure 10.3. Rule matrix for sensor 1 and sensor 2 is depicted in Table 10.2.

The fuzzy inference is realized using min as AND operator, max as OR operator, min for FIF and max for aggregation operation. Fuzzy values obtained from FIS are defuzzified using the centroid

TABLE 10.2
Rules Adopted in Realization of FLHIFF – Scheme 2

	$\bar{e}_2^{FLHIF}$ – Sensor 1					$\bar{e}_2^{FLHIF}$ – Sensor 2				
$\bar{e}_1^{FLHIF}$	**SP**	**MP**	**LP**	**VLP**	**ZE**	**SP**	**MP**	**LP**	**VLP**	**ZE**
SP	MP	MP	LP	LP	MP	MP	MP	SP	SP	MP
MP	MP	MP	MP	LP	SP	MP	MP	MP	SP	LP
LP	SP	SP	MP	MP	ZE	LP	LP	MP	MP	VLP
VLP	ZE	SP	MP	MP	ZE	VLP	LP	MP	MP	VLP
ZE	MP	LP	LP	VLP	MP	MP	SP	SP	ZE	MP

method. Defuzzification provides adaption weight sets $\{w_1\}$, $\{w_2\}$ used in eqn. (10.6). for MSDF-SVF in FLHIFF – scheme 2.

Proposed FLHIFF – scheme 2 is used for MSDF-based tracking for nonlinear systems, i.e. maneuvering targets.

10.4 DEGREE OF MATCHING (DOM) – SCHEME 3

Filter divergence could be observed when more/severe uncertainties are present in process noise and/or measurement noise; to eliminate effects of these uncertainties in real-time tracking applications, an adaptive HI filter is proposed. Adaption is achieved using concept of covariance matching. In AHIFDOM difference in DOM between theoretical HIF values and the innovation sequence vector/residuals (e) covariance is computed and in accordance to it R is tuned to achieve matching. Tuning is achieved using a system designed based on fuzzy logic. To arrive at DOM, theoretical covariance of residuals e in HIF is computed using

$$S(k) = HP_iH' + R(k) \tag{10.7}$$

The actual/measured covariances of the innovation sequence vector e are approximated by its sample covariance computed using averaging methods inside a moving estimation window of size N_w. The covariance of e is represented as $\hat{S}$:

$$\hat{S}(k) = \frac{1}{N_w} \sum_{i=k-N_w+1}^{k} e_i(i)(e_i(i)) \tag{10.8}$$

The DOM at kth scan is a difference defined as

$$\text{DOM}(k) = S(k) - \hat{S}(k) \tag{10.9}$$

The computed DOM is considered as an input to a FLS. The tuning factor represented as ΔR is the output of the fuzzy logic system, the updation of R is defined as

$$R(k+1) = R(k) + \Delta R \tag{10.10}$$

FLS for this scheme 3 is based on the Mamdani model; the crisp inputs of DOM are converted to fuzzy sets using MFs, which are defined using variables NZ – near zero, S – small, M – medium,

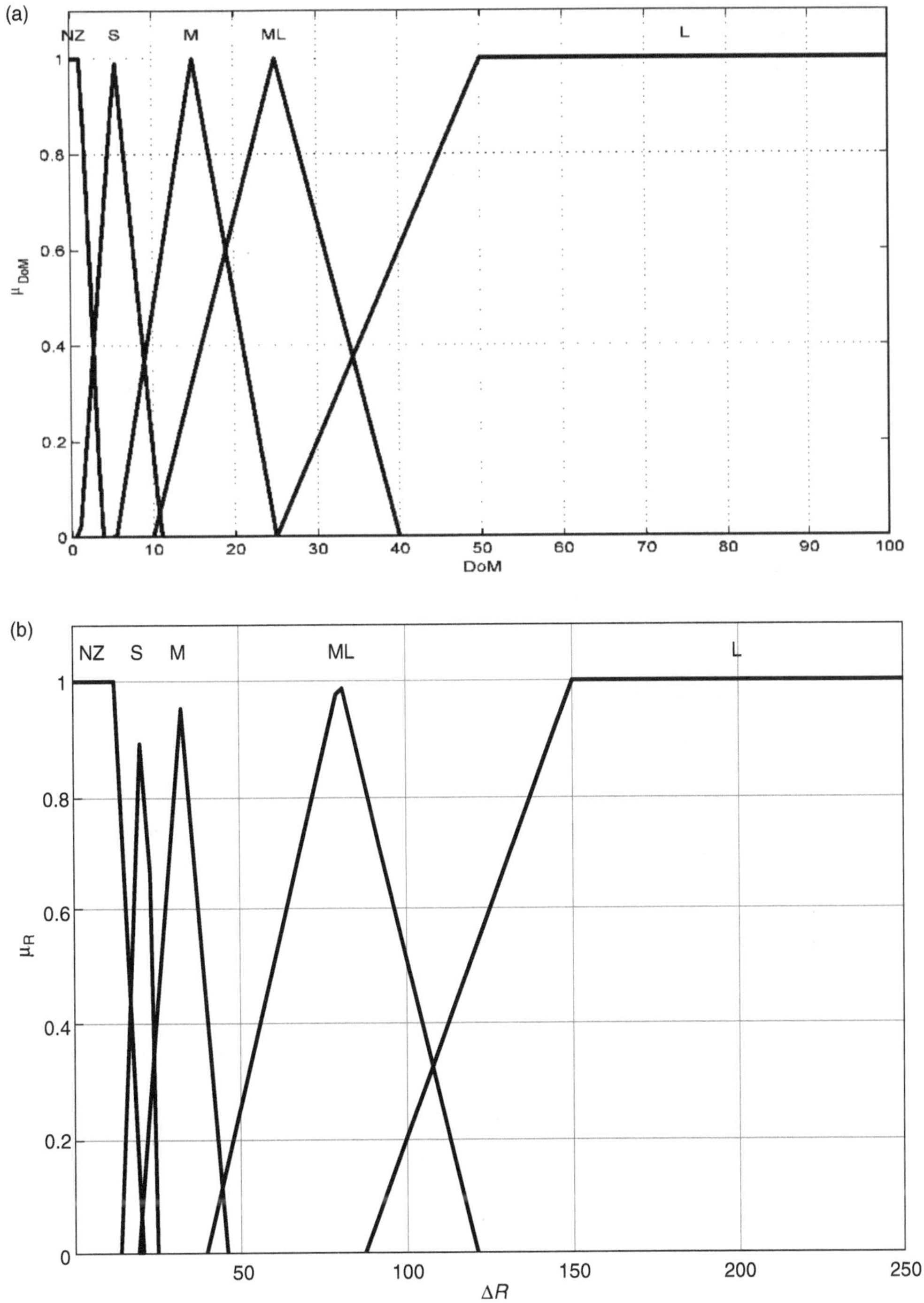

FIGURE 10.4 MFs used in AHIFDOM – scheme 3.

ML – moderatevly large and L – large. MFs used for a value of DOM [5] are shown in Figure 10.4(a), (b).

The rules used are defined as follows:

$$\begin{aligned} &\text{If DOM is NZ then } \Delta R \text{ is L.} \\ &\text{If DOM is S then } \Delta R \text{ is ML.} \\ &\text{If DOM is M then } \Delta R \text{ is M.} \\ &\text{If DOM is ML then } \Delta R \text{ is S.} \\ &\text{If DOM is L then } \Delta R \text{ is NZ} \end{aligned} \tag{10.11}$$

Mamdani inference is realized using min as AND operator, max as OR operator, min for FIF and max for aggregation operation. Defuzzification operation converts the fuzzy values obtained from inference system to crisp output values. For defuzzification, the centroid method is used.

10.5 PERFORMANCE EVALUATION

The proposed algorithms are evaluated using the MATLAB and FL toolbox for mild maneuvering and evasive maneuvering target tracking. Using 3 degrees of freedom (DOF), the kinematic model performance of AHIFDOM is compared with the fuzzy DOM-based adaptive Kalman filter represented as AKFDOM.

10.5.1 Target Tracking for Maneuvering Objects using HIF and FLHIFF

Two sets of maneuvering conditions for target tracking are used to evaluate the performance of FLHIF and FLHIFF (scheme 1 and scheme 2): (i) mild, and evasive maneuvers. Simulated data are generated using kinematic model defined in eqn. (10.1). In the experiments for maneuvering objects sampling time $T = 1$ sec. The total number of scans considered for the simulation is 25. The state transition matrix F and process noise gain matrix G are

$$F = \begin{bmatrix} 1 & T & T^2/2 & 0 & 0 & 0 \\ 0 & 1 & T & 0 & 0 & 0 \\ 0 & 0 & 1 & 0 & 0 & 0 \\ 0 & 0 & 0 & 1 & T & T^2/2 \\ 0 & 0 & 0 & 0 & 1 & T \\ 0 & 0 & 0 & 0 & 0 & 1 \end{bmatrix} \tag{10.12}$$

$$G = \begin{bmatrix} T^2/2 & 0 \\ T & 0 \\ 1 & 0 \\ 0 & T^2/2 \\ 0 & T \\ 0 & 1 \end{bmatrix} \tag{10.13}$$

The measurements are obtained with sensor dynamic matrix H defined as

$$H = \begin{bmatrix} 1 & 0 & 0 & 0 & 0 & 0 \\ 0 & 0 & 0 & 1 & 0 & 0 \end{bmatrix} \tag{10.14}$$

Process noise variance $Q = 0.1$; $Q_{xx} = Q_{yy} = Q; R_{xx} = R_{yy} = R$; and $R = 25$. The initial state of the dynamic system $(x, \dot{x}, \ddot{x}, y, \dot{y}, \ddot{y})$ is $(100, 30, 0, 100, 20, 0)$. To simulate maneuvering targets additional acceleration of (x_{ac}, y_{ac}) at scan number 8 and acceleration of $(-x_{ac}, -y_{ac})$ at scan 15 are introduced. Data simulation for maneuvering targets is carried out with process noise vector $w(2\times 1)$:

$$w(1) = \begin{Bmatrix} \left(\text{Gauss}()\times\sqrt{Q_{xx}}\right) & \forall\, k: k \neq 8,15 \\ \left(\text{Gauss}()\times\sqrt{Q_{xx}}\right) + x_{ac} & k = 8 \\ \left(\text{Gauss}()\times\sqrt{Q_{xx}}\right) - x_{ac} & k = 15 \end{Bmatrix} \tag{10.15}$$

$$w(2) = \begin{Bmatrix} \left(\text{Gauss}()\times\sqrt{Q_{yy}}\right) & \forall\, k: k \neq 8,15 \\ \left(\text{Gauss}()\times\sqrt{Q_{yy}}\right) + y_{ac} & k = 8 \\ \left(\text{Gauss}()\times\sqrt{Q_{yy}}\right) - y_{ac} & k = 15 \end{Bmatrix} \tag{10.16}$$

Mild maneuvering and evasive maneuvering targets are simulated by varying acceleration parameters x_{ac} and y_{ac}.

10.5.2 Target Tracking for Mild Maneuvering Objects using HIF and FLHIFF

Acceleration magnitudes of $x_{ac} = 6$ m/s^2 and $y_{ac} = -6$ m/s^2 are introduced at scan 8. At scan 15 acceleration magnitudes of $x_{ac} = -6$ m/s^2 and $y_{ac} = 6$ m/s^2 are introduced to simulate mild maneuvering targets. Initial states close to true values are considered for both the filters, and initial-state covariance matrices are assumed to be unity.

The true and estimated positions obtained are shown in Figure 10.5(a). Mild maneuvering object position estimates are almost identical to the true values initially, i.e. from scan 1–7 due to system linearity. On introducing acceleration variables, tracking of maneuvering objects is better in the case of FLHIFF when compared to FLHIF. Estimation errors of position, velocity and accelerations are measured using root sum square position error – RSSPE, root sum square velocity error – RSSVE and root sum square acceleration error – RSSAE.

For mild maneuvering target tracking RSSPE, RSSVE results obtained are shown in Figure 10.5(b) and Figure 10.5 (c). Errors observed for position, and velocity estimation are maximum during mild maneuvering phase of targets in FLHIF and lowest in initial linear phase of the system, i.e. nonmaneuvering; and better maneuvering target tracking performance is exhibited by FLHIFF compared to the FLHIF filter.

10.5.3 Target Tracking for Evasive Maneuvering Objects using HIF and FLHIFF

Large acceleration magnitudes introduced at scan 8 are $x_{ac} = 392$ m/s^2 and $y_{ac} = 392$ m/s^2. At scan 15, $x_{ac} = -392$ m/s^2 and $y_{ac} = 392$ m/s^2 are introduced. State covariance matrices are assumed to be unity. A comparison of the *x*, *y* position estimates and true positions of evasive maneuvering targets is shown in Figure 10.6(a). Evasive maneuvering object position estimates for two schemes are almost identical to the true values throughout as the system (with respect to target positions) is almost linear in nature. Results of RSSPE, RSSVE to compare performance of andshown in Figure 10.6(b), and Figure 10.6(c). During nonmaneuvering phase minimal errors are observed for both the schemes. Estimation error of target position, velocity and acceleration increase when evasive maneuvering

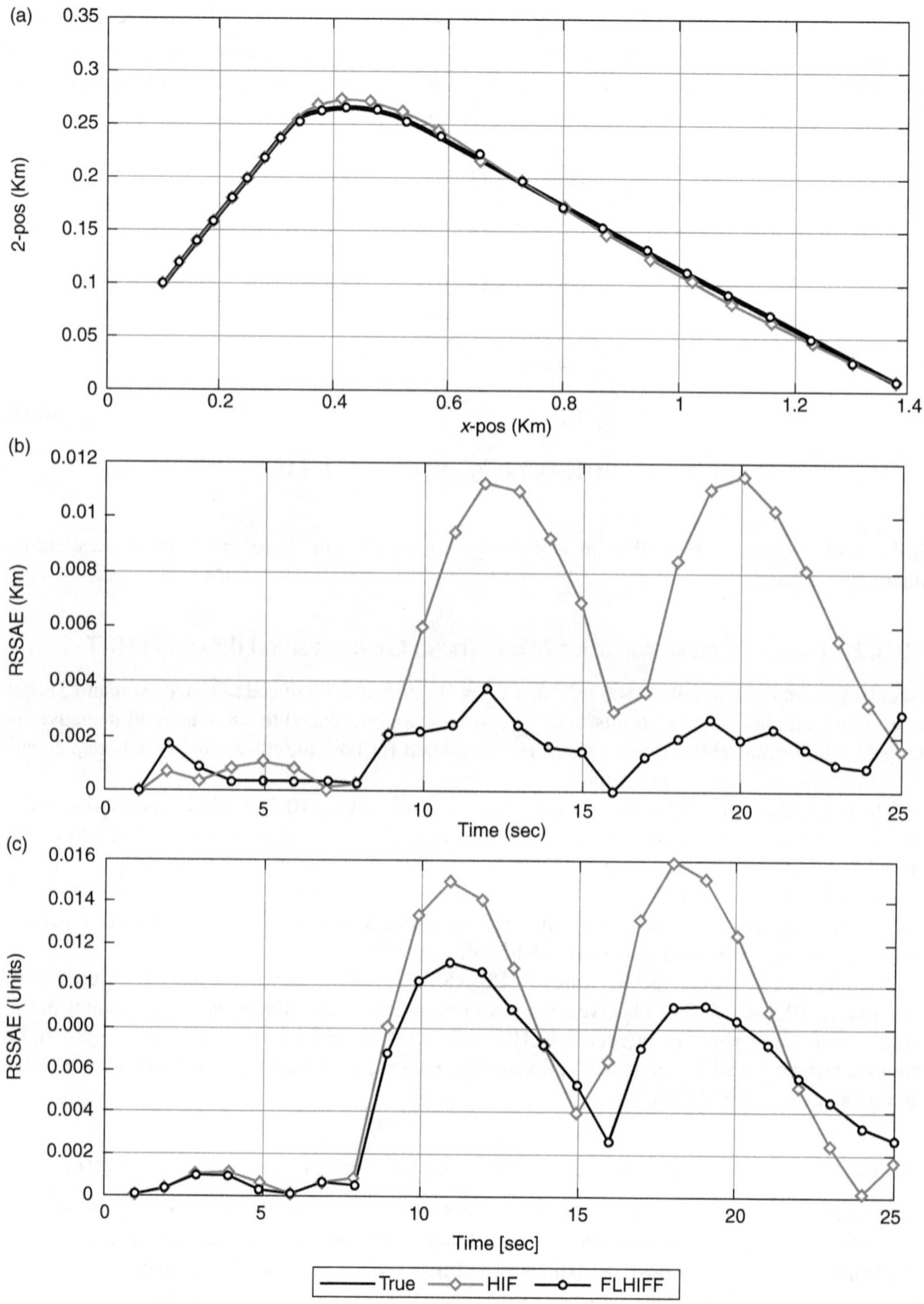

FIGURE 10.5 Mild maneuvering target tracking results for schemes 1 and 2.

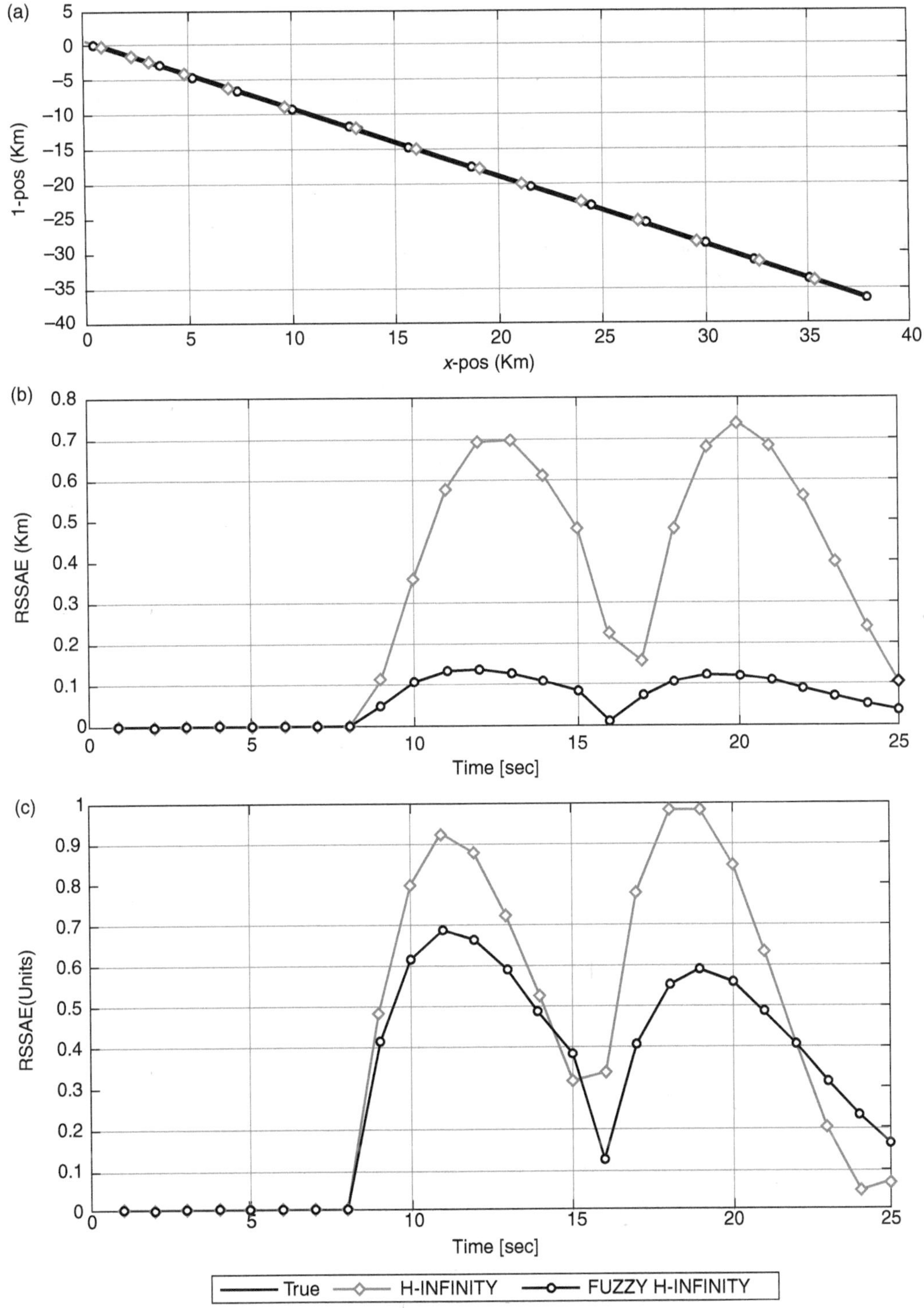

FIGURE 10.6 Evasive maneuvering target tracking results for two schemes.

is initiated during scan 8 or at 8 sec of simulation time. Overall lower evasive maneuver target estimation errors for position, velocity, and acceleration (not shown here) are observed for when compared to

Results presented considering both the cases prove that estimation performance of FLHIFF is superior to HIF estimations. Better tracking performance is attributed to inclusion of dual fuzzy logic systems in FLHIFF, initially to eliminate estimation errors at measurement update level and later for outlier effect elimination at fusion level.

10.5.4 Target Tracking for Nonmaneuvering Objects using AKFFDOM and AHIFFDOM

In this study, kinematic models are used to generate the simulated data for 100 sec. in X, Y and Z Cartesian planes, with $T = 0.25$. State vector x consists of target position, velocity and acceleration data in all three Cartesian planes, i.e. X, Y, Z axes. Hence a total of 400 scans are obtained in this simulation. The matrices F, G and H are defined as follows:

$$F = \begin{bmatrix} 1 & T & T^2/2 & 0 & 0 & 0 & 0 & 0 & 0 \\ 0 & 1 & T & 0 & 0 & 0 & 0 & 0 & 0 \\ 0 & 0 & 1 & 0 & 0 & 0 & 0 & 0 & 0 \\ 0 & 0 & 0 & 1 & T & T^2/2 & 0 & 0 & 0 \\ 0 & 0 & 0 & 0 & 1 & T & 0 & 0 & 0 \\ 0 & 0 & 0 & 0 & 0 & 1 & 0 & 0 & 0 \\ 0 & 0 & 0 & 0 & 0 & 0 & 1 & T & T^2/2 \\ 0 & 0 & 0 & 0 & 0 & 0 & 0 & 1 & T \\ 0 & 0 & 0 & 0 & 0 & 0 & 0 & 0 & 1 \end{bmatrix} \tag{10.17}$$

$$G = \begin{bmatrix} T^3/6 & 0 & 0 \\ T^2/2 & 0 & 0 \\ T & 0 & 0 \\ 0 & T^3/6 & 0 \\ 0 & TT^2/2 & 0 \\ 0 & T & 0 \\ 0 & 0 & T^3/6 \\ 0 & 0 & T^2/2 \\ 0 & 0 & T \end{bmatrix} \tag{10.18}$$

$$H = \begin{bmatrix} 1 & 0 & 0 & 0 & 0 & 0 & 0 & 0 & 0 \\ 0 & 0 & 0 & 1 & 0 & 0 & 0 & 0 & 0 \\ 0 & 0 & 0 & 0 & 0 & 0 & 1 & 0 & 0 \end{bmatrix} \tag{10.19}$$

The initial condition of the states are $[x, \dot{x}, \ddot{x}, y, \dot{y}, \ddot{y}, z, \dot{z}, \ddot{z}] = [200, 2, 0, 200, 10, 0.01, 200, -0.5, 0.001]$. For measurement noise $\sigma = 10$. Simulations are carried out for AKFFDOM and AHIFFDOM. Target estimation results obtained using both the adaptive filters are stored and error matrices are computed. True target positions with estimated tracked target position using AKFFDOM and AHIFFDOM is

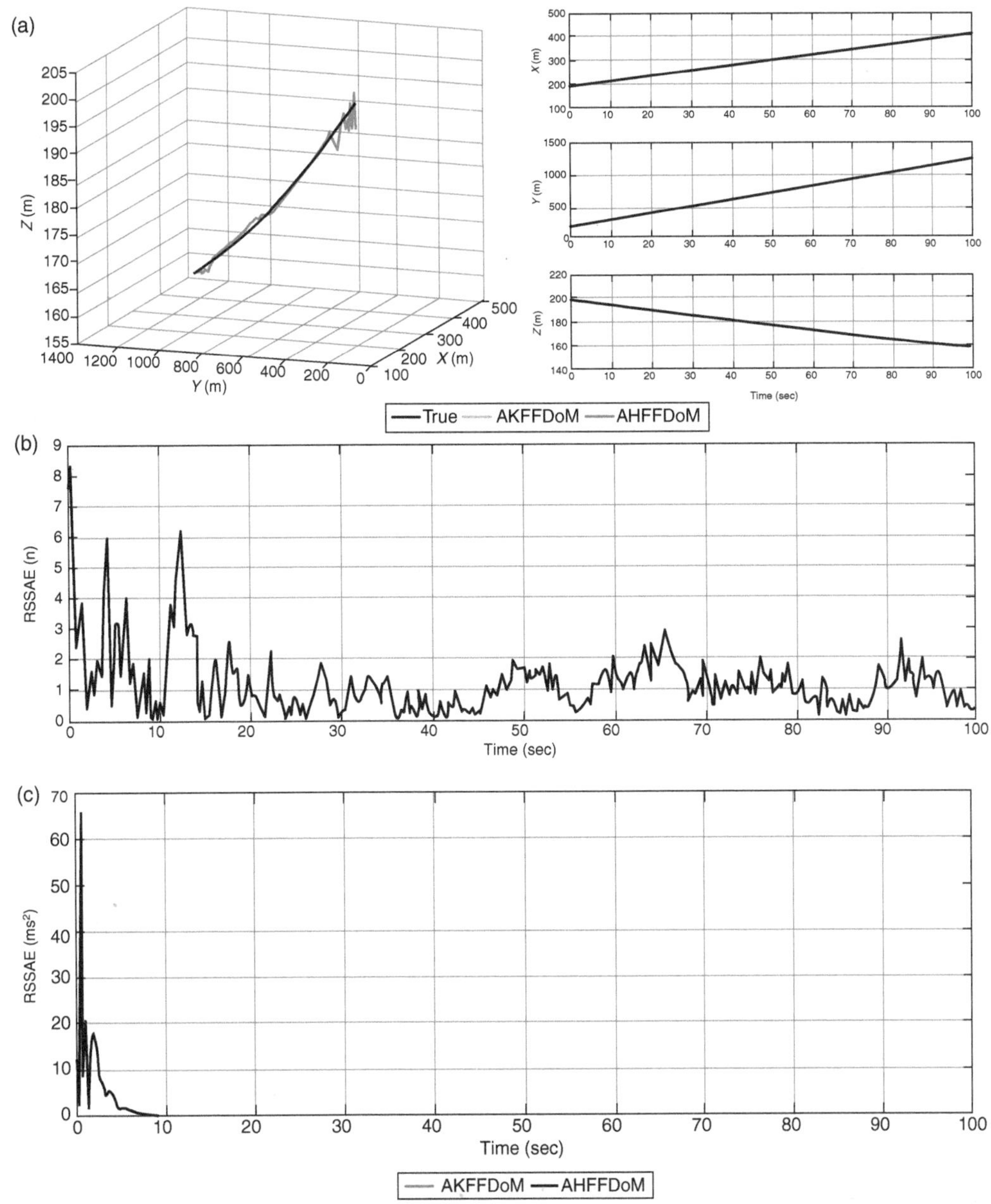

FIGURE 10.7 Nonmaneuvering target tracking using AKFFDOM and AHIFFDOM.

shown in Figure 10.7(a); a 3D-view depicting true and tracked estimated position with 2D estimates along *X*, *Y* and *Z* coordinates for both the adaptive filters is shown.

Results presented prove the good matching of both adaptive filters AKFFDO and AHIFFDO with true values. Matching along *X*,*Y* positions is more robust when compared to *Z* positions. Results obtained for RSSPE, RSSAE to compare the performance of AKFFDOM and AHIFFDOM are shown in Figure 10.7(b), and Figure 10.7(c). Almost identical performance is observed in AKFFDOM and AHIFFDOM. Position estimation is better in AHIFFDOM than AKFFDOM by 0.16%. Low values of RSSPE, and RSSAE observed for both filters indicate good target estimating

TABLE 10.3
PFE in *X,Y,Z* Position for AKFFDOM and AHIFFDOM

Algorithm	PFE – X(%)	PFE – Y(%)	PFE – Z(%)
AKFFDOM	0.9840	0.3912	1.7046
AHIFFDOM	0.9619	0.3823	1.6661

accuracy for nonmaneuvering targets. Percentage fit error (PFE) in position computation is carried out to further compare the performance of AKFFDO and AHIFFDOM

$$\text{PFE} = 100 \times \left(\frac{\text{norm}\left(\hat{x}^a - x^a_{\text{True}}\right)}{\text{norm}\left(x^a_{\text{True}}\right)} \right) \tag{10.20}$$

where variable a is used to represent position axis X,Y,Z, x_{True} represents true trajectories of objects, and $\hat{x}^a$ is estimated trajectories of tracked objects obtained from AKFFDOM, AHIFFDOM, PFEs are shown in Table 10.3. The lowest PFE is reported for Y position estimates in both the filters. Lower PFE is observed in AHIFFDOM compared to AKFFDOM.

10.6 ADAPTIVE H-INFINITY FILTER (AHIF) – SCHEME 4

In existing Kalman filter [6] based solutions, Gaussian noise of sensor measurements with knowledge of their statistical properties is assumed to be present. The ability of HIF to handle high-dimensional sensor data (necessary for MSDF), nondeterministic noise variations, inaccurate system models and applicability to nonlinear systems render it as a robust filter for real-time applications.

An adaptive HIF (AHIF) based on a sliding window is presented, see Figure 10.8.

Here, sliding-window length can be measured as $l_i \geq 0$; and $(x_i\ y_i)$ are referred as data points. Figure 10.8 represents a time-variant sliding video length where we can have a window length $l_i = 1$ at scan i at the first instant. For the following time instant, to alter the sliding-window length as $l_{i+1} = l+1$ we can add another data points as $(x_{i+1}\ y_{i+1})$. Similarly, for next instant, to alter the sliding window length as $l_{i+2} = l+1$ we can add another data points as $(x_{i+2}\ y_{i+2})$. Some other standard patterns of sliding window can be considered. The new time index k can easily replace the old scan i. The time index k possesses some unique properties as time index k is incremented in any condition either data points are added to the window, or discarded from the window. If window length is l_i at any scan i, then the new time index k can be represented as $2i - l_i + 1$. This can be better explained as follows:

(i) The new time index k incremented each time a data point (x_i, y_i) added at scan i, this is defined as,

$$\overline{y}_k = y_i,\ \overline{x}_k = x_i,\ \text{and}\ \overline{R}_k = 1 \tag{10.21}$$

(ii) The new time index k can be incremented even if each time a data point (x_{i-l_i}, y_{i-l_i}) is removed at scan i, this is defined as,

$$\overline{y}_k = y_{i-l_i}, \overline{x}_k = x_{i-l_i},\ \text{and}\ \overline{R}_k = -1 \tag{10.22}$$

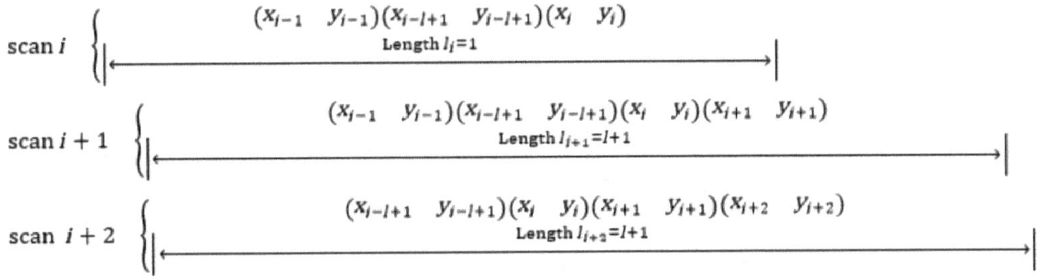

FIGURE 10.8 Sliding window with varying length.

A similar method can be used to build a state-space model, which is partially equal to indefinite quadratic form $\overline{J}_k$; the state model is

$$x(k+1) = Fx(k) + Gw(k) \tag{10.23}$$

$$y(k) = Hx(k) + v(k) \tag{10.24}$$

The sliding-window adaptive filtering iss given as

(a) The data points $(x_i \quad y_i)$ at scan i can be updated as

$$\hat{x}_i(k+1) = F\hat{x}_i(k) + K_i\left(y_i(k+1) - H_iF\hat{x}_i(k)\right) \tag{10.25}$$

where estimate $\hat{x}_i(k+1)$ is stated to enclose all the data of sliding window from scan j to scan i and where

$$K_i = P_i(k+1)H_i^T\left(I + H_iP_i(k+1)H_i^T\right)^{-1} \tag{10.26}$$

$$P_i(k+1) = FP_i(k)F' + GQG' - FP_i(k)\begin{bmatrix} H_i^T & L_i^T \end{bmatrix} R_i^{-1} \begin{bmatrix} H_i \\ L_i \end{bmatrix} P_i(k)F' \tag{10.27}$$

For every sensor $i \in m$ in MSDF, let x represent a vector, its estimation is represented using $\hat{x}_i$ with covariance given as $P_i = \mathrm{cov}(\tilde{x}_i)$. Estimation error is represented using $\tilde{x}_i = \hat{x}_i - x$. Two matrices L_i and H_i are initialized and defined by unity matrix I Covariance time propagation of HIF for the ith sensor is defined as

$$R_i = \begin{bmatrix} I & 0 \\ 0 & -\gamma^2 I \end{bmatrix} + \begin{bmatrix} H_i \\ L_i \end{bmatrix} P_i(k) \begin{bmatrix} H_i^T & L_i^T \end{bmatrix} \tag{10.28}$$

(b) The data points $(x_i \quad y_i)$ at scan i can be down indexed as $\{x_{i-l_i} \ \ y_{i-l_i}\}$:

$$\hat{x}_i(k+1) = F\hat{x}_i(k) + K_i\left(y_{i-l_i}(k+1) - H_{i-l_i}F\hat{x}_i(k)\right) \tag{10.29}$$

$$K_i = P_i(k+1)H_{i-l_i}^T\left(I + H_iP_i(k+1)H_i^T\right)^{-1} \tag{10.30}$$

$$P_i(k+1) = FP_i(k)F' + GQG' - FP_i(k)\begin{bmatrix} H_i^T & L_i^T \end{bmatrix} R_i^{-1} \begin{bmatrix} H_i \\ L_i \end{bmatrix} P_i(k)F' \tag{10.31}$$

$$R_i = \begin{bmatrix} -I & 0 \\ 0 & -\gamma^2 I \end{bmatrix} + \begin{bmatrix} H_{i-l_i} \\ L_{i-l_i} \end{bmatrix} P_i(k)\begin{bmatrix} H_{i-l_i}^T & L_{i-l_i}^T \end{bmatrix} \tag{10.32}$$

$$R_i = \begin{bmatrix} I & 0 \\ 0 & -\gamma^2 I \end{bmatrix} + \begin{bmatrix} H_i \\ L_i \end{bmatrix} P_i(k)\begin{bmatrix} H_i^T & L_i^T \end{bmatrix} > 0 \tag{10.33}$$

$$R_i = \begin{bmatrix} -I & 0 \\ 0 & -\gamma^2 I \end{bmatrix} + \begin{bmatrix} H_{i-l_i} \\ L_{i-l_i} \end{bmatrix} P_i(k)\begin{bmatrix} H_{i-l_i}^T & L_{i-l_i}^T \end{bmatrix} < 0 \tag{10.34}$$

10.6.1 Target Tracking for Maneuvering Objects using HIF and AHIF

The proposed AHIF is evaluated using the MATLAB and FL toolbox, and its performance is compared with HIF for maneuvering target tracking. Mild maneuvering and evasive maneuvering target tracking cases are studied to estimate the exact position and location of the target.

10.6.1.1 System Model

Two maneuvering conditions are used to evaluate performance: mild and evasive maneuvering target tracking [7]. Sampling time $T = 1$ sec., total number of scans $k = 25$. The state transition matrix F, process noise gain matrix G, and H are

$$F = \begin{bmatrix} 1 & T & T^2/2 \\ 0 & 1 & T \\ 0 & 0 & 1 \end{bmatrix} \tag{10.35}$$

$$G = \begin{bmatrix} T^2/2 & T & 1 \end{bmatrix} \tag{10.36}$$

$$H = \begin{bmatrix} 1 & 0 & 0 & 0 & 0 & 0 \\ 0 & 0 & 0 & 1 & 0 & 0 \end{bmatrix} \tag{10.37}$$

Here, $Q = 0.1$, $Q_{xx} = Q_{yy} = Q$ and $R_{xx} = R_{yy} = R = 25$. Initial state of the system $(x, \dot{x}, \ddot{x}, y, \dot{y}, \ddot{y}) = (100, 30, 0, 100, 20, 0)$. To simulate maneuvering targets additional acceleration of (x_{ac}, y_{ac}) is introduced at scan number 8 and acceleration of $(-x_{ac}, -y_{ac})$ at scan number 15; for maneuvering targets we have process noise vector $w(2\times 1)$:

$$w(1) = \begin{cases} \left(\text{Gauss}()\times\sqrt{Q_{xx}}\right) & \forall k : k \neq 8,15 \\ \left(\text{Gauss}()\times\sqrt{Q_{xx}}\right) + x_{ac} & k = 8 \\ \left(\text{Gauss}()\times\sqrt{Q_{xx}}\right) - x_{ac} & k = 15 \end{cases} \tag{10.38}$$

$$w(2)=\begin{Bmatrix} \left(\text{Gauss}()\times\sqrt{Q_{yy}}\right) & \forall k\text{: } k\neq 8,15 \\ \left(\text{Gauss}()\times\sqrt{Q_{yy}}\right)+y_{ac} & k=8 \\ \left(\text{Gauss}()\times\sqrt{Q_{yy}}\right)-y_{ac} & k=15 \end{Bmatrix} \tag{10.39}$$

10.6.2 Target Tracking for Mild Maneuvering Objects using HIF and AHIF

Acceleration magnitudes of $x_{ac} = 6$ m/s^2 and $y_{ac} = -6$ m/s^2 are introduced at scan 8, and at scan 15 acceleration magnitudes of $x_{ac} = -6$ m/s^2 and $y_{ac} = 6$ m/s^2 are introduced to simulate mild maneuvering targets. The true and estimated positions obtained using AHIF are shown in Figure 10.9 considering window lengths: 3, 5 and 7. For mild maneuvering target tracking RSSPE results for different window lengths 3, 5 and 7 are shown in Figure 10.10.

PFE metrics in position are shown in Table 10.4. The lowest PFE is reported for *X* position estimates in AHIF for WL –3. Similarly, the lowest PFE is reported for *Y* position estimates in AHIF for WL –3. Lower PFE is observed in AHIF – 3 compared to HIF and FKF, (HI filter, fuzzy KF).

10.6.2.1 Assertion of Suitable Sliding Window Length in AHIF

To estimate exact position we have tested our model considering different sliding window lengths as WL = 3,5 and 7. On introducing accelerations, tracking of maneuvering objects is better in the case of AHIF for window length 3 compared to AHIF for window length 5 and 7, see Table 10.4. The tracking estimation of maneuvering objects is better when sliding-window length is minimum,

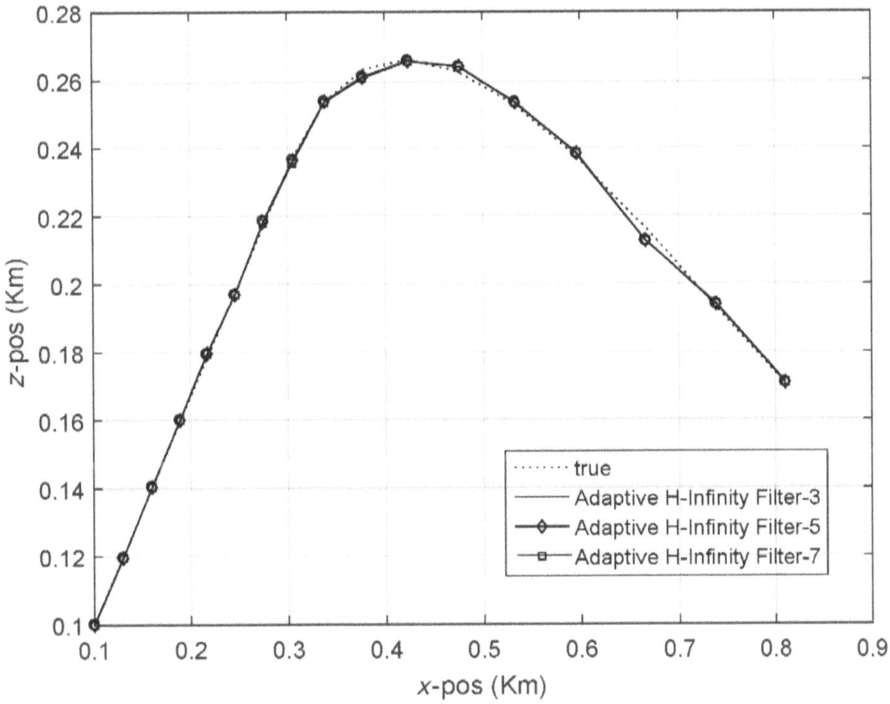

FIGURE 10.9 Mild maneuvering tracking: true and estimated positions with AHIF: window lengths – 3, 5 and 7.

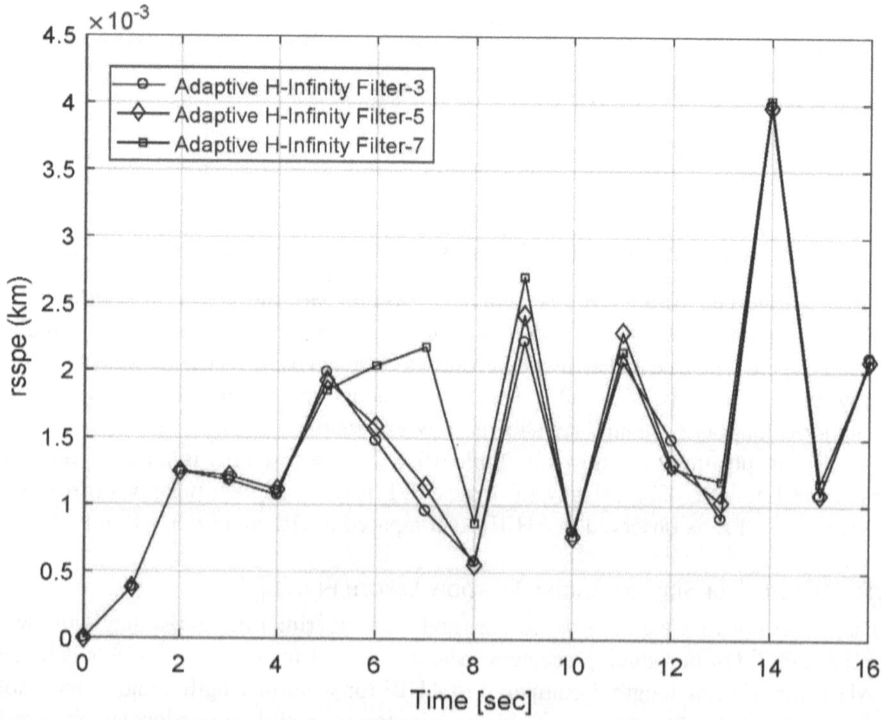

FIGURE 10.10 RSSPE results with AHIF for mild maneuvering target tracking.

due to quick adaptation. The true and estimated positions obtained using HIF, FKF and AHIF for window length 3 are shown in Figure 10.11, errors observed for position estimation are maximum during mild maneuvering phase of targets in HIF and lowest in AHIF for the window length 3, Figure 10.12.

The percentage reductions in the PFE by the AHIF for various window lengths, with respect to HIF, and FKF are also given in Table 10.4.

These results establish that better maneuvering target tracking performance is exhibited by AHIFcompared to the HIF and FKF filter.

10.6.3 Target Tracking for Evasive Maneuvering Objects using HIF and AHIF

Large acceleration magnitudes are introduced: at scan 8, $x_{ac} = 392$ m/s^2 and $y_{ac} = -392$ m/s^2; and at scan 15, $x_{ac} = -392$ m/s^2 and $y_{ac} = 392$ m/s^2. The tracking system is simulated for HIF, FKF and AHIF. The true and estimated positions of evasive maneuvering obtained using AHIF are shown in Figure 10.13 for window lengths 3, 5 and 7. For evasive maneuvering target tracking RSSPE results by AHIF for different window lengths 3, 5 and 7 is shown in Figure 10.14. PFE metrics are shown Table 10.5.

10.6.3.1 Assertion of Suitable Sliding-Window Length in AHIF

The true and estimated positions obtained using HIF, FKF and AHIF for window length = 3 is shown in Figure 10.15.

The RSSPEs for window length = 3 are shown in Figure 10.16.

The percentage reduction in PFE for various window lengths are also given in Table 10.5. The results establish that AHIF exhibits better evasive maneuvering target tracking performance compared to HIF and FKF filters.

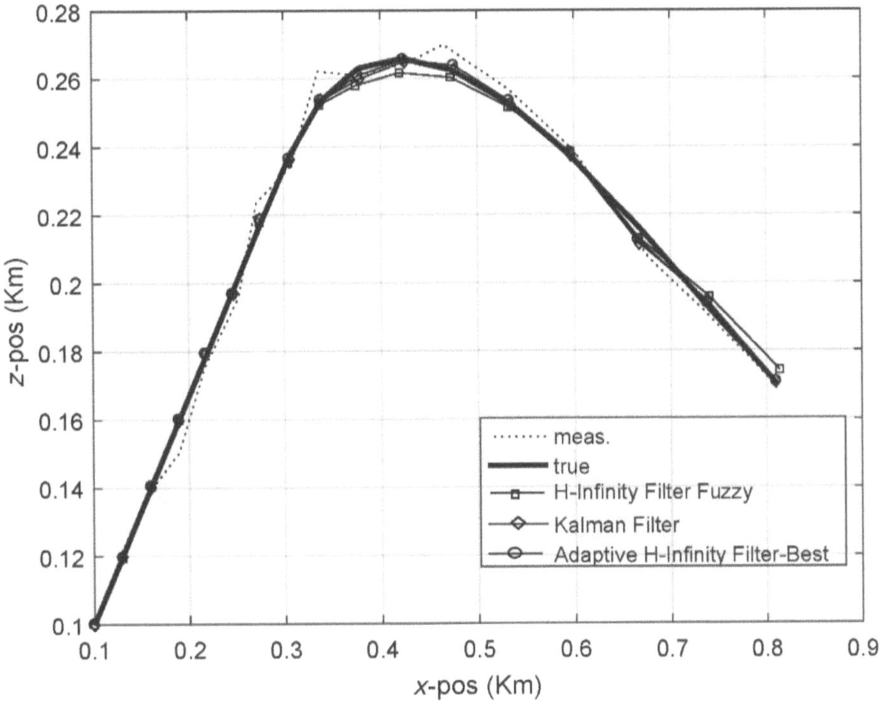

FIGURE 10.11 Mild maneuvering tracking results for true and estimated positions using HIF, FKF, AHIF for window length = 3.

TABLE 10.4
PFE for HIF, FKF, AHIF for Mild Maneuvering – Scheme 4

Algorithm	PFE – *X*	% Reduction in PFE, w.r.t. (HIF/FKF)	PFE – *Y*	% Reduction in PFE, w.r.t. (HIF/FKF)
HIF	0.368054937	–	1.176292633	–
FKF	0.299506842	–	0.78021731	–
AHIF – 3	0.21335349	42.03/28.76	0.6432287	45.31/17.55
AHIF – 5	0.213577993	41.97/28.69	0.662656686	43.66/15.06
AHIF – 7	0.233267274	36.62/22.11	0.701457035	40.36/10.09

10.7 FUZZY-LOGIC-BASED H-INFINITY FILTER (FLHIF)

In this section, the H-infinity filter HIF technique is used for target tracking using fused sensor data from multiple sensors. In target tracking applications, various properties are inherited such as sensor data, various conflicts, miss acquisition and state-vector noise variances. In contrast to KF, HIF is a more robust filtering technique due to its better handling capabilities of uncertainties. MSDF and H-infinity filtering can be achieved through normalization techniques and root mean square (RMS) data, respectively. State-space models are used to track nonmaneuvering targets using HIF.

The schematic representation of FLHIF established is as shown in Figure 10.2. Fuzzy-logic-based HIF(FLHIF) is an element of FLHIFF proposed to reduce errors during state-measurement updates. Outputs of FLHIF are further handled using fuzzy-logic systems to reduce/minimize outlier

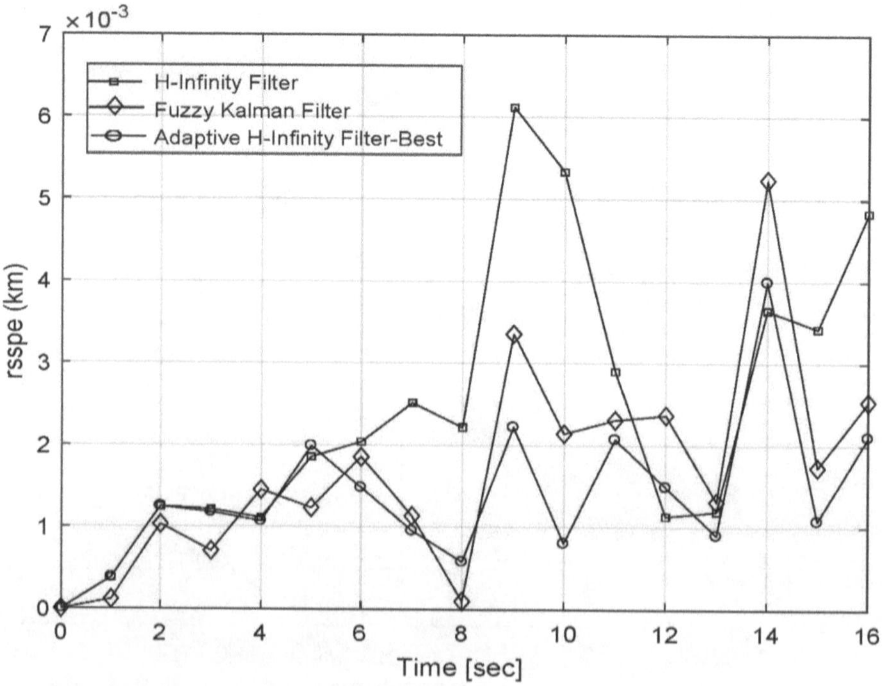

FIGURE 10.12 RSSPE with HIF, FKF and AHIF for mild maneuvering target tracking.

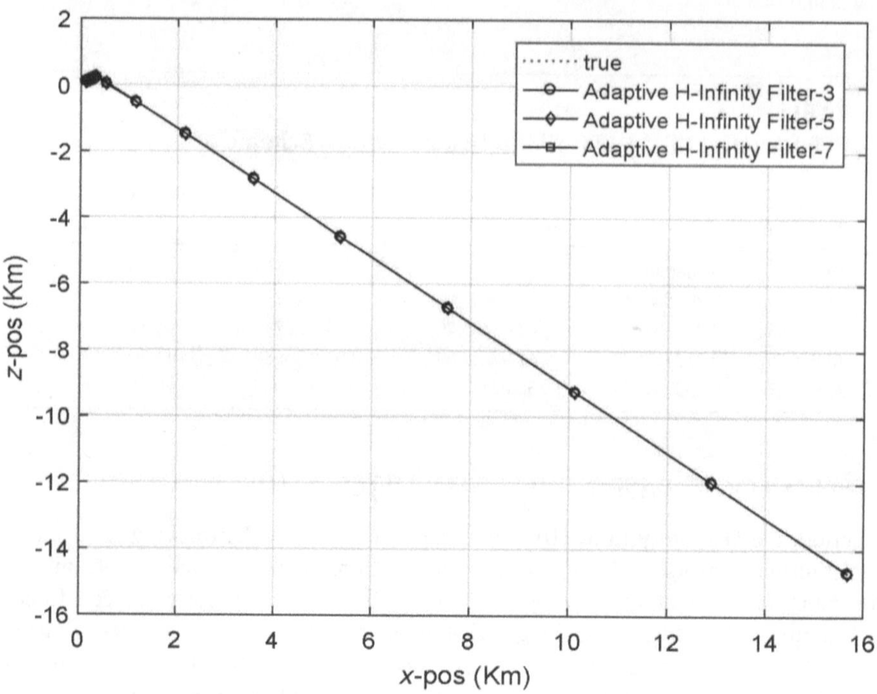

FIGURE 10.13 Evasive maneuvering tracking for true and estimated positions using AHIF for window lengths – 3, 5 and 7.

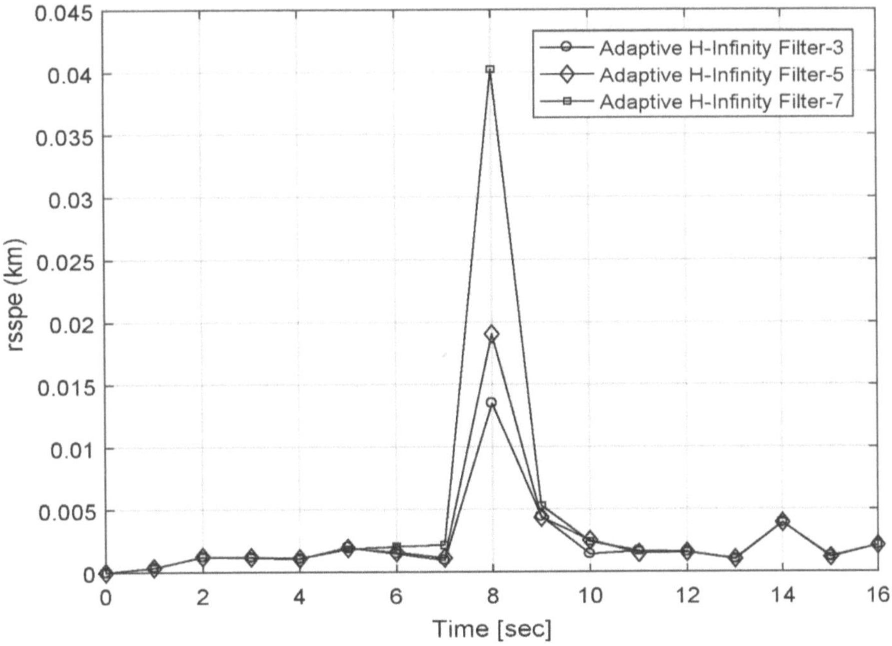

FIGURE 10.14 RSSPE for evasive maneuvering target tracking for AHIF with windom lengths – 3, 5 and 7.

TABLE 10.5
PFE for HIF, FKF, AHIF for Evasive Maneuvering

Algorithm	PFE – X	% Reduction in PFE, w.r.t. (HIF/FKF)	PFE – Y	% Reduction in PFE, wrt (HIF/FKF)
HIF	2.028645072	–	2.224569761	–
FKF	0.564246727	–	0.61066931	–
AHIF – 3	0.043088829	97.87/92.36	0.049574268	97.77/91.88
AHIF – 5	0.05855819	97.1/97.1	0.064330919	89.26/89.46
AHIF – 7	0.117524236	94.2/94.28	0.127203565	79.17/79.16

effects in MSDF. An initial discussion of FLHIFis presented. In fuzzy Kalman filter is discussed. On similar lines FLHIF is developed. Target tracking for nonmaneuvering objects using HIF and FLHIF

For nonmaneuvering objects sampling time $S = 1$ sec is considered. The total number of scans j considered for the simulation is 25. The state-transition matrix M and process noise matrix N considered is

$$M = \begin{bmatrix} 1 & S \\ 0 & 1 \end{bmatrix}$$

$$N = \begin{bmatrix} S^2/2 \\ S \end{bmatrix}$$

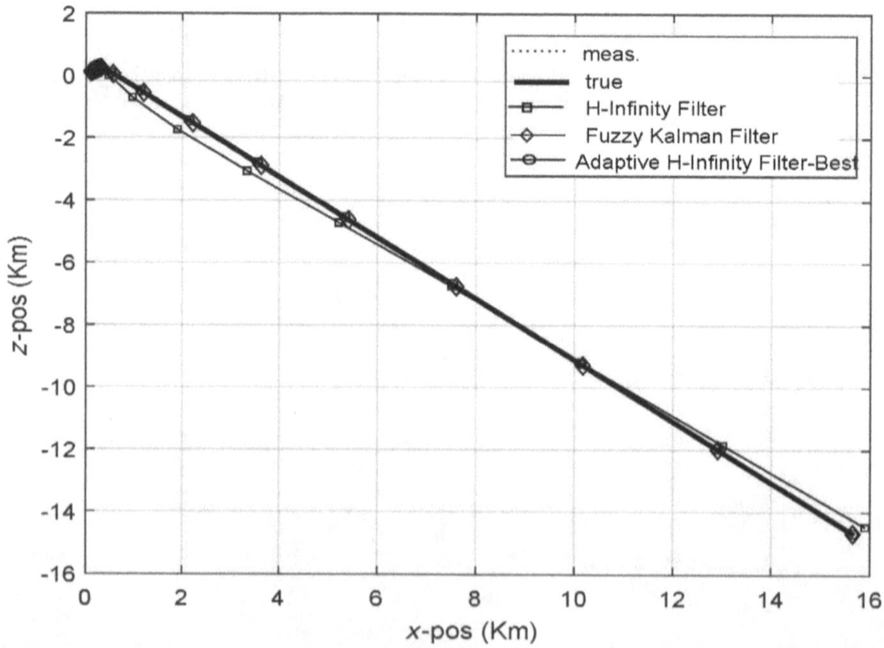

FIGURE 10.15 Evasive maneuvering tracking results for true and estimated positions using HIF, FKF, AHIF for window length = 3.

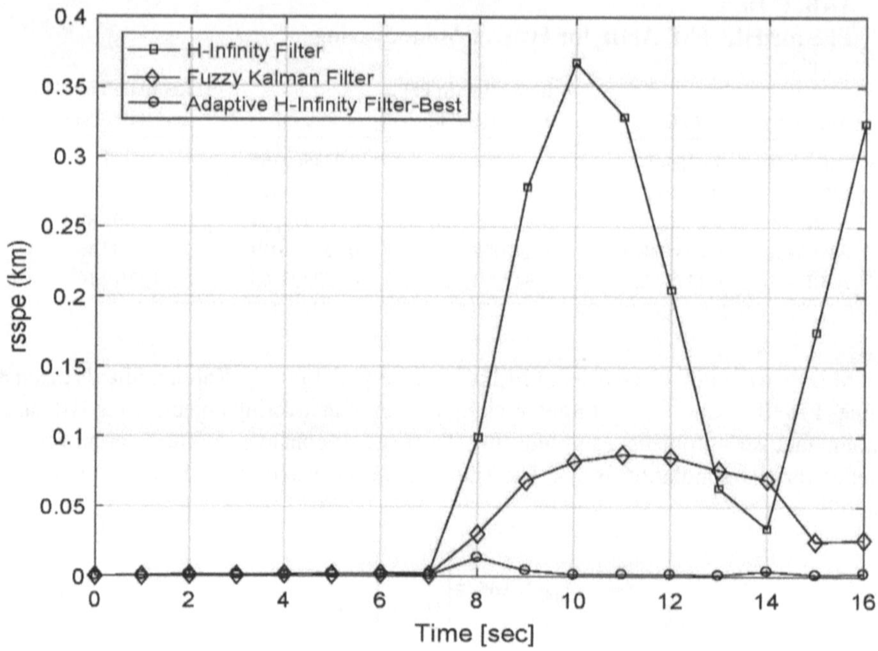

FIGURE 10.16 RSSPE by HIF, FKF, AHIF for evasive maneuvering target tracking for window length = 3.

TABLE 10.6
PFE for H-Infinity and Fuzzy Logic H-Infinity Filter when No Missing Data

Percentage Fit Error	PFE – X	PFE – Y
H-infinity filter	0.64405	1.4483
Fuzzy logic H-infinity filter	0.60628	1.2912

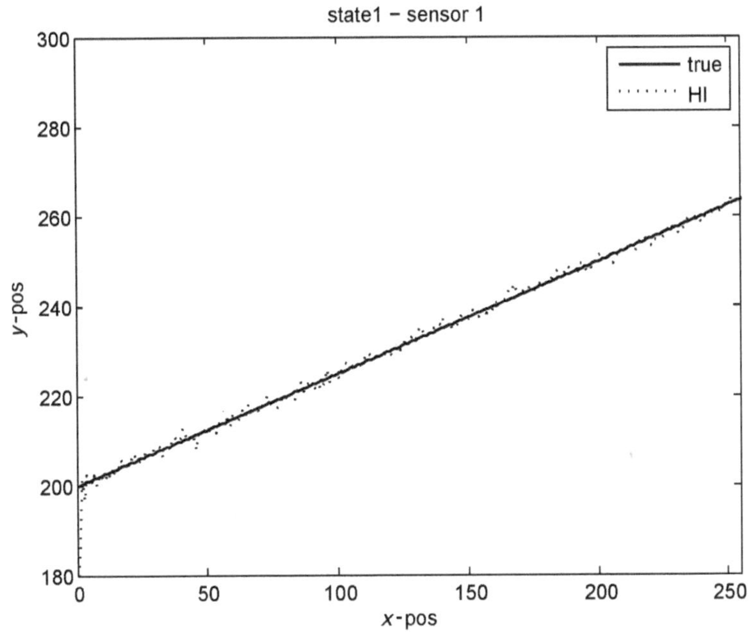

FIGURE 10.17 State 1 of sensor 1 in nonmaneuvering target tracking for true position and estimated position using the H-infinity filter for no missing data.

Here, Table 10.6 represents the PFE for H-infinity and fuzzy-logic H-infinity filter when there is no missing data. It is clearly demonstrated that the fuzzy-logic H-infinity filter performs better compare to the HIF in terms of PFE.

Figures 10.17 and 10.18 demonstrate x and y positions of the true and estimated position state (state 1) of sensor 1 using the H-infinity filter, fuzzy H-infinity filter and fuzzified fuzzy H-infinity, respectively, for nonmaneuvering targets when no information is missing. Simulated results shows the superiority of the fuzzified fuzzy H-infinity filter over the H-infinity filter and the fuzzy H-infinity filter in terms of the true and estimated position. During nonmaneuvering phase minimal percentage fit errors are observed in FLHIF compare to HIF and FLHIF in x and y direction.

Figures 10.19 and 10.20 demonstrate the true and estimated state (state 1, state 2 and fused state) of sensor 1 and sensor 2 using the H-infinity filter and the fuzzy H-infinity filter, respectively, for nonmaneuvering targets when information is collected efficiently from all the sensors. Simulated results show the superiority of the fuzzified fuzzy H-infinity filter over the H-infinity filter and the fuzzy H-infinity filter in terms of true and estimated position.

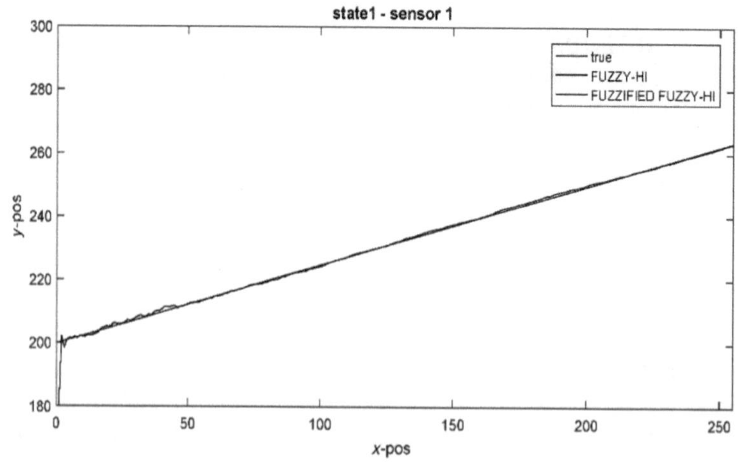

FIGURE 10.18 State 1 of sensor 1 in nonmaneuvering target tracking for true position and estimated position using the fuzzy H-infinity filter for no missing data.

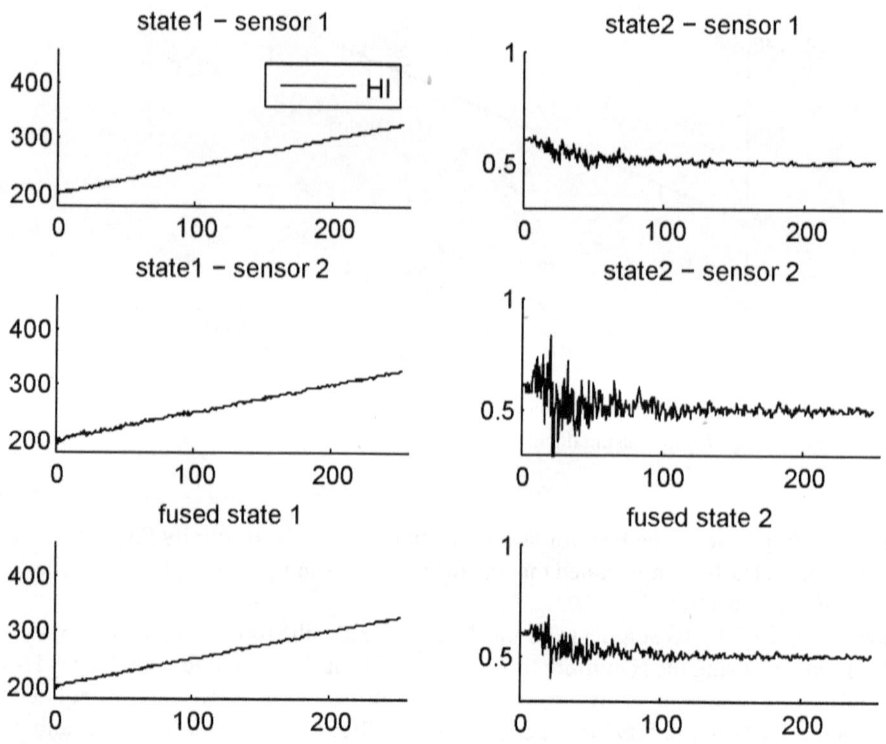

FIGURE 10.19 State 1 and 2 of all sensor scenarios (1, 2 and fused) in nonmaneuvering target tracking for estimated position using the H-infinity filter.

Table 10.7 represents PFE for H-infinity and fuzzy logic H-infinity filter when there is data missing. There are a total of 500 samples used in the testing model and 100 samples from 300 to 400 samples are missing. And it is clearly demonstrated that the fuzzy logic H-infinity filter performs better compared to the H-infinity filter (HIF) in terms of PFE.

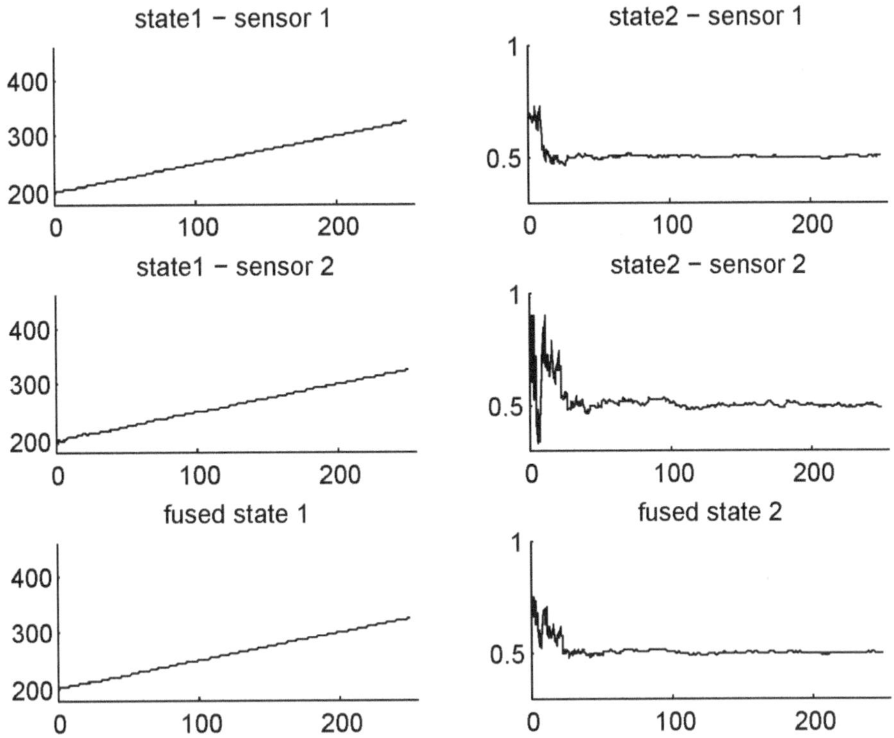

FIGURE 10.20 State 1 and 2 of all sensor scenarios (1, 2 and fused) in nonmaneuvering target tracking for estimated position using the fuzzy H-infinity filter.

Figures 10.21 and 10.22 demonstrate the x and y position of the true and estimated position state (state 1) of sensor 1 using the H-infinity filter, fuzzy H-infinity filter and fuzzified fuzzy H-infinity, respectively, for nonmaneuvering targets when data is missing. Simulated results show the superiority of the fuzzified fuzzy H-infinity filter over the H-infinity filter and the fuzzy H-infinity filter in terms of the true and estimated position. During nonmaneuvering phase, minimal percentage fit errors are observed in FFHIF compare to HIF and FLHIF in x and y directions.

Figures 10.23 and 10.24 demonstrate the true and estimated state (state 1, state 2 and fused state) of sensor 1 and sensor 2 using the H-infinity filter and the fuzzy H-infinity filter, respectively, for nonmaneuvering targets when sensor data are missing. However, still it can sufficiently provide exact information of nonmaneuvering targets. Simulated results shows the superiority of the fuzzy H-infinity filter over the H-infinity filter in terms of the true and estimated position.

To achieve an established performance in nonmaneuvering target tracking systems, FLHIF is offered and detailed modelling is presented. The FLHIF discussed encompasses FLHIF for every sensor to remove local estimation errors at filter level. A fuzzy logic-based system is considered in HIF to minimize estimation errors and ill effects of outliers at fusion level.

TABLE 10.7
PFE for H-Infinity and Fuzzy Logic H-infinity Filter when Missing Data

Percentage Fit Error	PFE – *Y*	PFE – *Y*
H-infinity filter	1.1385	1.4483
Fuzzy logic H-infinity filter	0.62382	1.2912

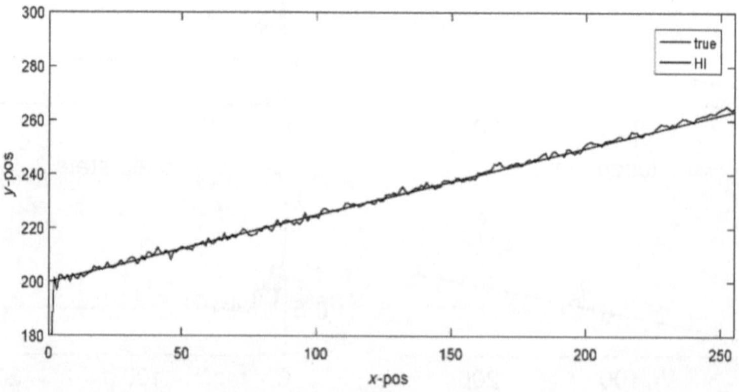

FIGURE 10.21 State 1 of sensor 1 in nonmaneuvering target tracking for true position and estimated position using the H-infinity filter for missing data.

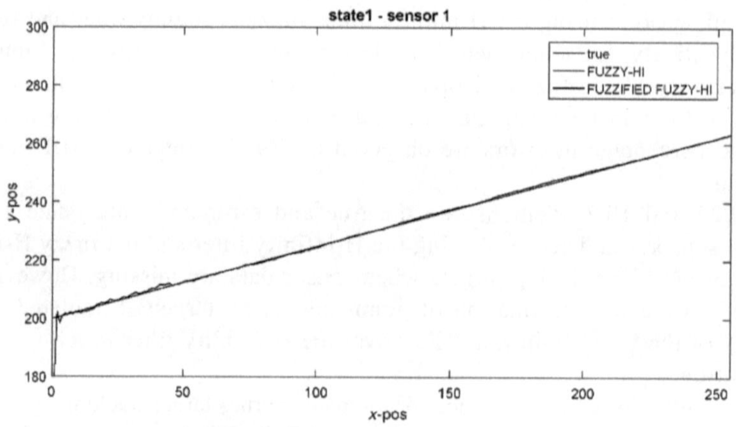

FIGURE 10.22 State 1 of sensor 1 in nonmaneuvering target tracking for true position and estimated position using the fuzzy H-infinity filter for missing data.

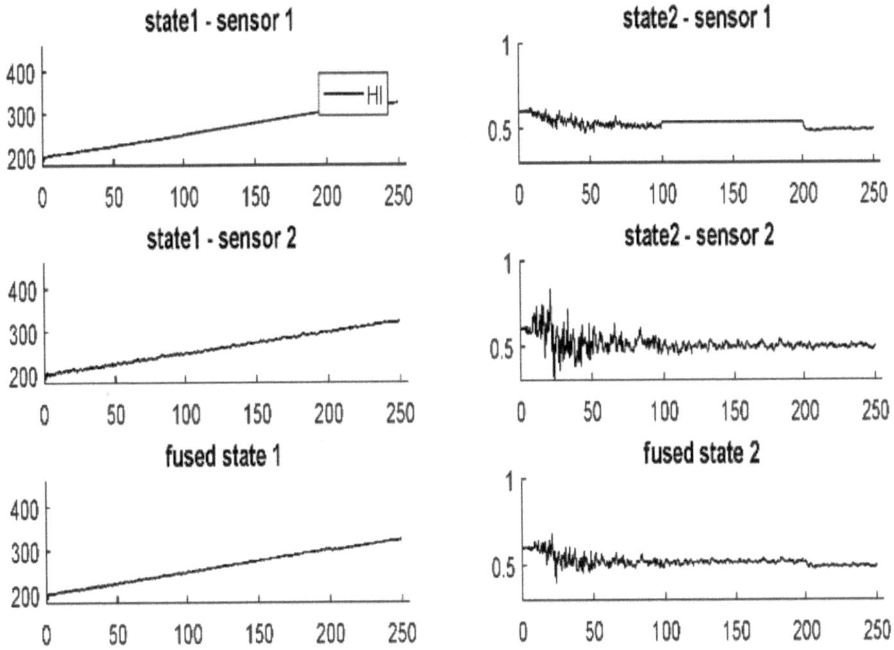

FIGURE 10.23 State 1 and 2 of all sensor scenarios (1, 2 and fused) in nonmaneuvering target tracking for estimated position using the H-infinity filter.

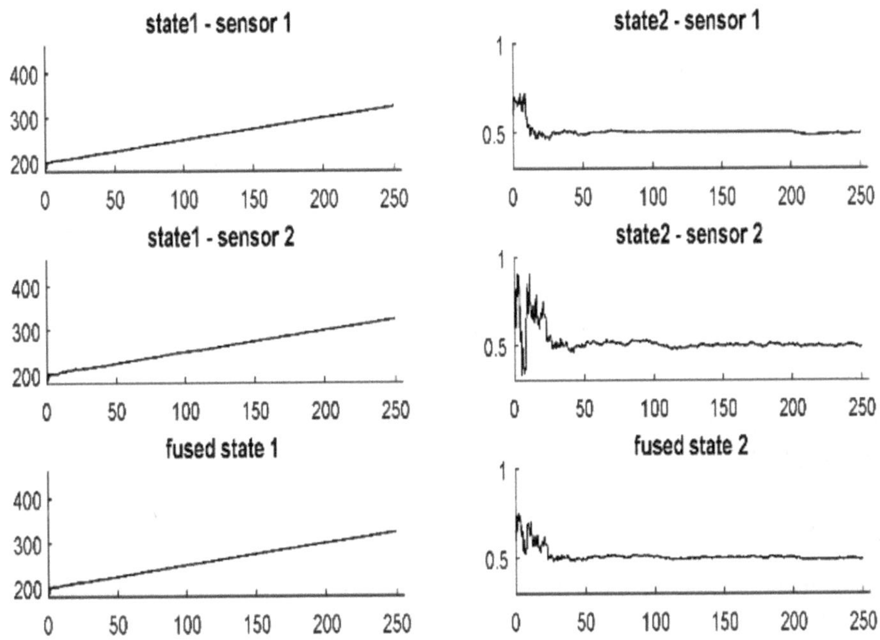

FIGURE 10.24 State 1 and 2 of all sensor scenarios (1, 2 and fused) in nonmaneuvering target tracking for estimated position using the fuzzy H-infinity filter.

EXERCISES

EXERCISE 10.1

How does the integration of fuzzy logic with H-infinity control theory enhance the robustness and performance of the filtering system?

EXERCISE 10.2

Are there any specific tuning techniques or strategies for optimizing the performance of fuzzification in the fuzzy H-infinity filter?

EXERCISE 10.3

What are the advantages and limitations of H-infinity fuzzification filtering compared to other filtering methods?

EXERCISE 10.4

How do you evaluate the performance of an H-infinity filter in target tracking, and what metrics are commonly used for assessment?

EXERCISE 10.5

Why is it necessary to implement H-infinity fuzzification filtering for target tracking?

EXERCISE 10.6

How does DOM contribute to the control strategy of an adaptive H-infinity filter fuzzification scheme?

EXERCISE 10.7

In what scenario might a low DOM value indicate the need for parameter adjustment in an adaptive H-infinity filter?

EXERCISE 10.8

How does the sliding window concept enhance the robustness and performance of adaptive H-infinity filtering in dynamic or nonstationary environments?

EXERCISE 10.9

What are the trade-offs involved in selecting the size and characteristics of the sliding window in adaptive H-infinity filtering?

REFERENCES

1. Kashyap, S.K., and Raol, J.R., *Fuzzy logic applications in filtering and fusion for target tracking*, Defence Science Journal, National Aerospace Laboratories, Bangalore–560 017, Vol. 58, No. 1, pp. 120–135, Jan. 2008.

2. Reshma, V., *Fuzzy Logic Augmented H-Infinity Filter for Target Tracking*. Ph.D. Thesis, Electronics and Communications Engineering, Jain University, Bangalore, 2017.
3. Girija, G., and Raol, J.R., Evaluation of adaptive Kalman filtering methods for target tracking applications, Paper AIAA 2001-4106; doi:10.2514/6.2001-4106, AIAA Guidance, Navigation, and Control Conference and Exhibit, Montreal, Canada, 2001.
4. Simon, D.J., H-infnity estimation for fuzzy membership function optimization, *International Journal of Approximate Reasoning*, Vol. 40, No. 3, pp. 224–242, 2005.
5. Escamilla, P.J., and Mort, N., Development of a fuzzy logic-based adaptive Kalman filter. In *Proc. of the European Control Conference ECC'01*, Porto, Portugal, pp. 1768–1773. Sept. 4–7, 2001.
6. Naidu, V.P.S., Girija, G., and Raol, J.R., Data association and fusion algorithms for tracking in presence of measurement loss, *Journal of the Institution of Engineers (I)*, Vol. 86, pp. 17–28, May 2005.
7. Naidu, V.P.S., Girija, G., and Santhakumar, N., *Three model IMM-EKF for tracking targets executing evasive maneuvers*. AIAA-66928, 45th AIAA conference on Aerospace Sciences, Reno, USA, 2007.

APPENDIX 10A: FL-BASED ADAPTIVE TUNING

Three methods of FL-based methods for adaptive tuning for KF/HI are discussed here.

10A.1 FOR *R* WITH KNOWN *Q*

In this case, the adaptation/tuning of R is carried out assuming Q is known; large values of R imply inaccurate measurement data. In such a case, we give less weightage to the measurements and higher weightage (of course relatively) to the prediction in the KF. We have the theoretical covariance of the innovation given by

$$S(k) = H(k)\tilde{P}(k)H(k)^T + R(k) \tag{10A.1}$$

The actual/sample covariance of the innovations is computed by using a moving average across a window, the size of which is chosen empirically to give some good smoothing, on average. The difference between the theoretical and the actual covariances is used to derive adjustments for R: a variable called the degree of matching is defined to get the size of the difference

$$\text{DOM}(k) = S(k) - \hat{S}(k) \tag{10A.2}$$

The DOM is used by fuzzy inference system (FIS) to derive new/required R values. From eqn. (10A.2) it is clear that increasing R increases S. Hence, R can be varied to reduce the DOM value; the following rule set is used:

Rule 1: If $\text{DOM}(k) \cong 0$, $S(k)$ and $\hat{S}(k)$ are nearly equal, then maintain R at the same value.
Rule 2: If $\text{DOM}(k) > 0$, means that $S(k) > \hat{S}(k)$, then decrease R.
Rule 3: If $\text{DOM}(k) < 0$, means that $S(k) < \hat{S}(k)$, then increase R.

The adaptation of $R(i,i)$ is equivalent to the adaptation in $\text{DOM}(i,i)$. The FIS generates tuning of R by creating a correction factor $\Delta R(k)$ to be added or subtracted to all the diagonal elements of the R matrix at each instant of time using

$$R(k) = R(k-1) + \Delta R(k) \tag{10A.3}$$

So, it is clear that $\text{DOM}(k)$ is the input to the FIS and $\Delta R(k)$ is the output which is generated sequentially.

10A.2 FOR *Q* WITH KNOWN *R*

If the large values of Q occur, these imply large uncertainties in the process model/disturbances, hence, less weightage should be given to the predicted value of the state and more weightage to the measurement data. By incorporating time-propagation covariance equation of the KF into eqn. (10A.2) we get

$$S(k) = H(k)\left(\varnothing(k)\hat{P}(k-1)\varnothing(k)^T + Q(k)\right)H(k)^T + R(k) \tag{10A.4}$$

It is assumed that $R(k)$ is known, then if Q increases, S increases. This mismatch between $S(k)$ and $\hat{S}(k)$ can be used to create a correction term using a similar procedure as for estimating $R(k)$; update the value of Q using the following rules:

Rule 1: If $\text{DOM}(k) \cong 0$, $S(k)$ and $\hat{S}(k)$ are nearly equal, then maintain Q at the same value.
Rule 2: If $\text{DOM}(k) > 0$, means that $S(k) > \hat{S}(k)$, then decrease Q.
Rule 3: If $\text{DOM}(k) < 0$, means that $S(k) < \hat{S}(k)$, then increase Q.

For each element in the diagonal matrix Q, an FIS can generate the tuning factor, i.e. $\Delta Q(k)$ is obtained and added/subtracted to correct the $Q(k)$ value for each element in the main diagonal of the matrix Q as

$$Q(k) = Q(k-1) + \Delta Q(k) \tag{10A.5}$$

If there is no direct correspondence between the dimensions of S, Q and DOM, empirical considerations can be used in the FIS to overcome this problem; using the preceding three rules, a typical fuzzy system for the input DOM and output ΔR could be formulated as follows:

If DOM is negative medium, then ΔR = increase large.
If DOM is negative small, then ΔI = increase.
If DOM is zero, then ΔR = maintain.
If DOM is positive small, then ΔR = decrease.
If DOM is positive medium, then ΔR = decrease large

In FL, the input variable DOM defines the universe of discourse U_{DOM} and the output variables ΔR define the universe of discourse $U_{\Delta R}$. Commonly used MFs are trapezoidal, triangular, Gaussian or their combination. A suitable defuzzification procedure is used to get the crisp values at each step.

10A.3 A SPECIFIC FL APPROACH

Another specific approach is obtained from eqn. (10A.5):

$$S(k+1) = H\widetilde{P}H^T + R(k+1) = H\{\Phi\,\hat{P}(k)\,\Phi^T + \hat{Q}(k)\}\,H^T + R(k+1) \tag{10A.6}$$

$$= H(\Phi\hat{P}(k)\$^T + \sigma^2(k)\bar{Q}(k))\,H^T + R(k+1) \tag{10A.7}$$

In eqn. (10A.9), $\hat{Q}(k) = \sigma^2(k)\bar{Q}(k)$, where $\bar{Q}(k)$ is a fixed at an *a priori* known covariance matrix. Then, current $Q(.)$ is altered (in fact $\sigma^2(\cdot)$) at each sampling instant based on, if the innovation,

(i) is neither too near nor too far from zero, then leave the estimate of $Q(k)$ almost unchanged,
(ii) is very near to zero, then reduce the estimate of $Q(k)$, and
(iii) is very far from zero, then increase the estimate of $Q(k)$.

This is intuitively appealing since, it achieves the covariance matching. In fact, this mechanism would adjust the value of $Q(.)$ in such a proportion so as to adjust the value of $S(.)$ in eqn. (10A.9), and hence, in turn match with the actual covariance of the residual, thereby achieving the covariance matching. This adjustment mechanism can be implemented using FL; at each sampling instant, the input variable (membership function, MF) is given by the parameter

$$r_s(k+1) = \frac{r(k+1)}{\sqrt{s(k+1)}} \tag{10A.8}$$

In eqn. (10A.11), $r(k+1)$ is the actual innovation component and $s(k+1)$ is the $(k+1)$th value of S, then $r_s(k+1)$ gives the measure of relative amplitude of innovation compared to its theoretical assumed value. The following If … Then … rules can be used to generate output variables; the fuzzy rule-based system has r_s as input variables and Ψ as output variables, with MFs of r_s and Ψ denoted as m_r and m_Ψ

If r_s is near zero, then Ψ is near zero.
If r_s is small, then Ψ is near one.
If r_s is medium zero, then Ψ is a little larger than one.
If r_s is moderately large, then Ψ is moderately larger than one.
If r_s is large, then Ψ is large.

11 H-Infinity-Based Observer

11.1 INTRODUCTION

The objective of an observer is to reconstruct the states of a dynamic system using the knowledge of the system IO data. In the case of a linear system, its states are estimated using the Luenberger observer (LO) or Kalman filter (KF); whereas for a nonlinear system one uses an extended Luenberger observer (ELO) or an extended Kalman filter (EKF) [1]. In this chapter, we present an HI-based observer and derive asymptotic results.

11.2 HI-BASED NONLINEAR OBSERVER

The H-infinity (norm-based) theory is by now well established in the control and estimation literature [2]; and HI filtering algorithms are available in the deterministic setting; and these do not require any statistical assumptions on state/process noise and measurement noise. In fact, the state variables and unknown disturbance signals in the HI framework are considered as generalized 'random variables' rather than the stochastic processes; and the related Kalman filter 'covariance'-type matrices are called Gramians [2]. The HI-based estimators are alternatives to the well-known stochastic counterpart like KFs and EKFs. Interestingly enough, the deterministic observers [2] can also be considered as some other type of alternative to stochastic estimators. In general, for such deterministic observers, including the HI-based estimators, one can use the Lyapunov method for establishing the asymptotic stability results for the observer error dynamics. In [3], extensive theoretical treatment of HI-based control and nonlinear filtering algorithms is given. An HI-based control algorithm is studied in [4] that considers the use of weighting functions for performance and robustness improvement.

Fuzzy logic (FL) is being increasing used in science and engineering problems [5,6], and is one of the most important components of the so-called soft computing paradigms, which can be used in designing artificial intelligence (AI)-based systems for: (i) control, (ii) robotics, (iii) data-fusion applications and (iv) estimation. FL provides better parameterization for modelling of linear and nonlinear dynamic systems, and helps in handling a type of uncertainty, called vagueness. Once, the event's probability, meaning uncertainty ceases, that is when it is ascertained that the event has occurred, the vagueness would remain. Fuzzy logic is a rule-based expert decision-making support system that becomes an important ingredient of AI-based systems.

The work presented here is an extension of [7] and [8], however, it is based on the concept of HI theory and the use of fuzzy logic, and hence, is a novel step. A FMF is used as a nonlinear operator on the observer residuals. The observer gain is dependent on state-error Gramian that is based on HI theory. The asymptotic stability condition for the observer error dynamics is derived using the

DOI: 10.1201/9781003620211-11

Lyapunov energy functional, and it is felt that a Lyapunov–Krasovskii functional is not required for the present study as was used in [7]. Also, the approaches of [8] and [9], which add an unstable term and use a fixed-point linearization, respectively, are avoided in the present work, without any loss of theoretical rigor; and the linearization is carried out at every sampling instant.

The approach of the Lyapunov energy functional to study the asymptotic convergence is a standard process in control theory, and it does not need any specific theorem from control and estimation theories, and from any other design method. The condition arrived as a result of the present analysis is required to be satisfied for the local stability of the observer thus designed. The combination of the use of a FMF to modulate observer residuals, utilize the gain (and associated matrix differential equation) from the HI theory, and the derivation of the asymptotic result for this combination constitute a new application and contribution of this chapter in the domain of nonlinear observer theory. The performance of the observer is illustrated by implementing the algorithm in MATLAB. The FL- and HI-based extensions to the existing deterministic observers would enhance their capability of robustness and handling of uncertainty in the empirical data when used for state and parameter estimation.

Let the nonlinear system be given as

$$\begin{aligned}\dot{x}(t) &= f\left(x(t),u(t)\right)+g(x)w(t)\\ z &= h_1(x)\\ y(t) &= h_2(x)+q(x)w(t)\end{aligned} \tag{11.1}$$

For eqn. set (11.1), the initial conditions for the states are assumed to be appropriately specified; $x(t)$ is a state vector, $u(t)$ is a control input, $y(t)$ is a measurement vector, $w(t)$ is an unknown disturbance process that might be present in the state and/or measurements; and the variables have appropriate dimensions. Also, it is assumed that the system's nonlinear functions f, g and h are continuously differentiable, and that the system of eqn. (11.1) has a unique solution. Then, a conventional nonlinear observer for the system of eqn. (11.1) is considered as

$$\begin{aligned}\dot{\hat{x}}(t) &= f\left(\hat{x}(t),u(t)\right)+L(t)\left(y(t)-\hat{y}(t)\right)\\ \hat{z}(t) &= h_1(\hat{x})\\ \hat{y}(t) &= h_2(\hat{x})\end{aligned} \tag{11.2}$$

The observer gain matrix, $L(t)$ is of appropriate dimension, and it is determined by using the cost function based on the HI concept, which in turn is based on the HI norm:

$$\text{HI}(\text{norm}) = \frac{\sum_{k=0}^{N} (\hat{x}(k)-x(k)^T(\hat{x}(k)-x(k))}{(\hat{x}_0-x)^T P_0(\hat{x}_0-x)+\sum_{k=0}^{N} w^T(k)w(k)+\sum_{k=0}^{N} v^T(k)v(k)}$$

The HI norm is the ratio of the output error (state-estimation error) energy (or variance) to the total input energies of the disturbances; the latter includes the error in the initial state conditions, $x(0)$, and the noise variances: $w(.)$, and $v(.)$. In general, this HI norm is to be kept bounded below a certain level. The HI-based filtering algorithms [2] are all based on this concept. So, one does not get an optimal solution, but a robust solution, since the input disturbances/uncertainties to output-estimation error (maximum) gain is kept bounded (minimized).

An HI norm as applicable to the dynamics of eqn. (11.1) is chosen; the main point is to seek that L_2 gain from the input disturbance energy to the output estimation error energy that obeys the inequality

$$\int_{t_0}^{T}\left\|\tilde{z}(\tau)\right\|^2 d\tau \le \int_{t_0}^{T}\left\|w(\tau)\right\|^2 d\tau;\quad T > t_0 \tag{11.3}$$

In eqn. (11.3), the estimation error is given as

$$\tilde{z}(t) = z(t) - \hat{z}(t) = h_1(x) - h_1(\hat{x}) \tag{11.4}$$

The cost function, eqn. (11.3), is reformulated as

$$J(w) \triangleq \frac{1}{2}\int_{t_0}^{T}\left(\left\|\tilde{z}(t)\right\|^2 - \gamma^2\left\|w(t)\right\|^2\right)dt \tag{11.5}$$

In eqn. (11.5), γ (>0) specifies the bound on error gain from the input energy to the output energy and can be considered as a tuning parameter for the observer as well as an upper bound for the HI norm. This also means that the actual error energy (uncertainty-variance) gain is bounded from above by the square of this factor, so the observer might not be optimal but it would be suboptimal, and yet robust. This would signify that the performance of the observer is robust to the level of γ. In many cases, though an optimal filter might have been used, due to use of the real data (that might have been affected with noise processes), one might not obtain the optimal practical solution, and in that case one accepts the performance of the filter at a certain level; this then is tantamount to using and accepting some kind of robust performance. Following eqn. (11.5), the observer gain can be chosen as

$$L(t) = P(t)H^T(t) \tag{11.6}$$

For use in eqn. (11.6), the Gramian $P(t)$ is obtained as the solution of the following ordinary matrix differential equation:

$$\begin{aligned}&\dot{P}(t) = P(t)A^T(t) + A(t)P(t) + P(t)\left[\frac{1}{\gamma^2}C^T(t)C(t) - H^T(t)H(t)\right]P(t) + B(t)B^T(t);\\ &\text{with } P(0) = P_0; P(T) = \text{steady state value}\end{aligned} \tag{11.7}$$

An equation of type of eqn. (11.7) is known as a matrix Riccati differential equation. However, for this chapter it can be called s RTOMDE (Riccati-type ordinary matrix differential equation). Various Jacobians needed in eqn/ (11.6), and eqn. (11.7) are obtained as

$$A(t) = \frac{\partial f(\cdot)}{\partial \hat{x}(t)}; C(t) = \frac{\partial h_1}{\partial \hat{x}(t)}; H(t) = \frac{\partial h_2(\cdot)}{\hat{x}(t)} \tag{11.8}$$

$$B(t) = \frac{\partial g(\cdot)}{\partial \hat{x}(t)}; D(t) = \frac{\partial q(\cdot)}{\partial \hat{x}(t)} \tag{11.9}$$

In eqn. (11.9), the matrix R that is found in KF gain formula $\{K(t) = P(t)H^T(t)R^{-1}\}$ is not present, however it can be easily incorporated for the sake of tuning the observer, and can be regarded as an additional weighting factor. The observer gain, eqn. (11.6), is based on eqn. (11.7) and hence, in turn on the HI theory. In eqns. (11.6)–(11.9), although various quantities are function of $(\hat{x}, t)$, the dependence only on 't' is mentioned for the sake of simplicity.

First, the basic observer structure given in eqn. (11.2) is expanded as

$$\begin{aligned}
\dot{\hat{x}}(t) &= f(\hat{x}(t),u(t)) + L(t)(y(t) - \hat{y}(t)) \\
\dot{\hat{x}}(t) &= f(\hat{x}(t),u(t)) + L(t)(Hx(t) + Dxw(t) - H\hat{x}(t)) \\
\dot{\hat{x}}(t) &= f(\hat{x}(t),u(t)) + L(t)(H(t)e(t) + Dx(t)w(t)) \\
\hat{z}(t) &= h_1(\hat{x}); \quad \hat{z}(t) \cong C(t)\hat{x} \\
\hat{y}(t) &= h_2(\hat{x}); \quad \hat{y}(t) \cong H(t)\hat{x}
\end{aligned} \tag{11.10}$$

In eqn. (11.10), the residuals and the state errors are given as

$$r(t) = y(t) - \hat{y}(t); \quad e(t) = x(t) - \hat{x}(t) \tag{11.11}$$

In eqn. (11.10), the original system's nonlinear functions are retained, however, only for the error part are the corresponding Jacobians used, since these are also required in eqn. (11.6), and eqn. (11.7), and then are subsequently used in the stability analysis for simplicity.

11.3 NONLINEAR FMF-BASED OBSERVER

Now, it is proposed to use FL in the observer design by way of using FMF on the residuals as follows:

$$\begin{aligned}
\dot{\hat{x}}(t) &= f(\hat{x}(t),u(t)) + L(t) f_m\{(y(t) - \hat{y}(t))\} \\
\dot{\hat{x}}(t) &= f(\hat{x}(t),u(t)) + L(t) f_m(Hx(t) + Dxw(t) - H\hat{x}(t)) \\
\dot{\hat{x}}(t) &= f(\hat{x}(t),u(t)) + L(t) f_m(H(t)e(t) + Dx(t)w(t))
\end{aligned} \tag{11.12}$$

In eqn/ (11.12), f_m is an FMF that operates on the observer residuals $r(t)$, (in turn on $e(t)$) and obtains the output that is used in the observer structure, thus providing a nonlinear function of the residuals effect. The idea of using FMF is to provide the multivalued (unquantized) effect on the observer residuals; the membership function varies and takes values between 0 and 1. If FMF is not used, then the basic system is equivalent to the use of crisp logic (like yes or no response; the MF here has either a '0' or '1' value), and it is a nonadaptive approach. With the use of FMF, one gets an effect of adaptive tuning, and the scope of this is enlarged because of the availability of several types of such FMFs. Also, FL allows to incorporate If…Then…, rules, thereby the designer can use their heuristic knowledge and experience in FL-based system and control; this also extends to their use as an adaptive tuning mechanism in estimation [25], and MSDF. It is noted that in eqn. (11.12), the nonlinear function, i.e. FMF f_m is a function of not only the state errors, but also of the function of $x(t)w(t)$. Now, by subtracting the corresponding expressions of eqn. (11.12) from eqn. (11.1), one obtains the observer error dynamics as

$$\begin{aligned}
\dot{e}(t) = A(t)e(t) - L(t) f_m\{He(t) + D(t)xw(t)\} \\
+\phi(.) + B(t)xw(t)
\end{aligned} \tag{11.13}$$

In eqn. (11.13), the term in {.} is a composite residual term and it arises due to the form of the measurement equation in eqn. (11.1). In eqn. (11.13), the nonlinear function 'phi' is expressed as

$$\phi(.) = -A(t)e(t) + f(x,t) - f(\hat{x},t) \tag{11.14}$$

The form of eqn. (11.13) is obtained by adding and subtracting the term related to the Jacobian, $A(t)$, in the original equations of the error dynamics. In eqn. (11.14), the short form of $\phi(.)$ is given as

$$\phi(.) = \phi\left(x(t), \hat{x}(t)\right) \tag{11.15}$$

Mostly, for simplicity compact forms of $f(t)$, and $\phi(t)$, are used by avoiding the arguments, since, these would be implied any way from their defining forms; also $u(t)$ can be omitted.

Due to the combined use of FMF and the HI-based gain (and RTOMD equation) in the observer design, it is necessary to establish the stability result analytically for the observer error dynamics.

11.4 ASYMPTOTIC RESULT FOR OBSERVER ERROR DYNAMICS

Certain conceptually logical conditions are necessary to be considered for establishing the local asymptotic behavior of the observer error dynamics of eqn. (11.13):

1. The solution of the RTOMD equation. (11.7) should be bounded as

$$p_l I \le P(t) \le p_u I \tag{11.16}$$

In eqn. (11.16), $p_1, p_u > 0$ are positive constants (which are the lower and upper bound, respectively); the matrix $P(t)$ is theoretically positive definite and symmetric matrix; and I is the identity matrix.

2. The nonlinearity in eqns. (11.14), (11.15) is bounded

$$\|\phi(.)\| \le \rho_1 \|x(t) - \hat{x}(t)\|^2 \tag{11.17}$$

In eqn. (11.17), the bounding constants are equal to or greater than zero.

3. FMF nonlinearity $f_m(.)$ is bounded

$$\|H^T f_m(.)\| \le \rho \|x(t) - \hat{x}(t)\|^2 \tag{11.18}$$

In eqn. (11.18), the bounding constant is equal to or greater (but finite) than zero. However, since in the present case the nonlinear function is a FMF, the overall bound on the nonlinearity could also be $0 \le \|f_m(.)\| \le 1$. Thus, if the conditions 1, 2 and 3 are satisfied, and the time derivative of the LE functional is negative definite, then the nonlinear observer error dynamics as in eqn. (11.13) are locally asymptotically stable; this result is established next.

The normalized LE functional of the error dynamics, eqn. (11.13), is considered as

$$V(t) = e^T(t) Y(t) e(t) \tag{11.19}$$

In eqn. (11.19), $Y(t)$ is the normalizing (weighting) matrix and can be recognized as an information matrix, $Y(t) = P^{-1}(t)$. Since, the observer gain is from the H-infinity theory, the matrix $P(t)$ is regarded as the 'covariance' Gramian matrix; and hence, the inverse of matrix $P(t)$ can be called either information matrix or the 'information Gramian'. In such a case the variables $x(.)$, $z(.)$, $y(.)$, and $e(t)$ are treated as the generalized random variables.

LE functional, eqn. (11.19), is positive definite due to condition 1, and the inequality of eqn. (11.16) is

$$\frac{1}{p_u}\|e(t)\|^2 \leq e^T(t)Y(t)e(t) \leq \frac{1}{p_l}\|e(t)\|^2 \tag{11.20}$$

The time derivative of the LE functional, eqn. (11.19) (under the constraints of the error dynamics, eqn. (11.13), of gain in eqn. (11.6) and of Gramian $P(t)$ in eq. (11.7)), should be negative definite for all time, for the stability of the presented observer (error dynamics). For simplicity, the error dynamics (keeping in mind that the quantities A, L, H, D, x, w and B are dependent on time t) is rewritten as

$$\dot{e}(t) = Ae(t) - Lf_m\{He(t) + Dxw\} + \phi(.) + Bxw \tag{11.21}$$

Then the time derivative of eqn. (11.19), utilizing eqn. (11.21) is obtained as

$$\begin{aligned}\dot{V}(t) = {} & e^T(t)\dot{Y}(t)e(t) + e^T(t)Y(t)\big[Ae(t) - Lf_m\{He(t) + Dxw\} \\ & + \phi(.) + Bxw] + \big[Ae(t) - Lf_m\{He(t) + Dxw\} \\ & + \phi(.) + Bxw]^T Y(t)e(t)\end{aligned} \tag{11.22}$$

Expressing in terms of the composite residuals, and regrouping the various terms, eqn. (11.22) is written as

$$\begin{aligned}\dot{V}(t) = {} & e^T(t)\dot{Y}(t)e(t) + e^T(t)\{Y(t)A + A^TY(t)]e(t) \\ & + e^T(t)Y(t)\big[-Lf_m\{r_c\} + \phi(\cdot) + Bxw\big] \\ & + \big[-f_m^T\{r_c\}L^T + \phi^T(\cdot) + w^Tx^TB^T\big]Y(t)e(t)\end{aligned} \tag{11.23}$$

Since, the matrix $Y(t)$ is symmetrical, and also, using eqn. (11.6) for the gain $L(t)$, eqn. (11.23) can be further simplified as

$$\begin{aligned}\dot{V}(t) = {} & e^T(t)\dot{Y}(t)e(t) + e^T(t)\{Y(t)A + A^TY(t)]e(t) \\ & - e^T(t)H^Tf_m\{r_c\} - f_m^T(r_c)He(t) \\ & + e^T(t)Y(t)Bxw + w^Tx^TB^TY(t)e(t) \\ & + 2e^T(t)Y(t)\phi(\cdot)\end{aligned} \tag{11.24}$$

Next, substitute for $\dot{Y}(t) = -Y(t)\dot{P}(t)Y(t)$; this can be easily verified since, $Y(t)P(t) = I$; and using eqn. (11.7) in eqn. (11.24), obtain the following:

$$\begin{aligned}\dot{V}(t) = {} & -e^T(t)Y(t)\{P(t)A^T(t) + A(t)P(t) \\ & + P(t)\left[\frac{1}{\gamma^2}C^T(t)C(t) - H(t)^T H(t)\right]P(t) \\ & + B(t)B^T(t)\}Y(t)e(t) + e^T(t)\{Y(t)A + A^TY(t)]e(t) \\ & - e^T(t)H^Tf_m\{r_c\} - f_m^T(r_c)He(t) + e^T(t)Y(t)Bxw \\ & + w^Tx^TB^TY(t)e(t) + 2e^T(t)Y(t)\phi(.)\end{aligned} \tag{11.25}$$

In eqn. (11.25), there are terms of equal magnitudes and with +/- signs, and hence, these terms cancel out, and one obtains

$$\begin{aligned}\dot{V}(t) &= -e^T(t)Y(t)BB^TYe(t) - e^T(t)H^T f_m\{r_c\} - f_m^T(r_c)He(t) \\ &\quad -\frac{1}{\gamma^2}e^T(t)C^TCe(t) + e^T(t)H^THe(t) \\ &\quad + e^T(t)Y(t)Bxw + w^Tx^TB^TY(t)e(t) + 2e^T(t)Y(t)\phi(.)\end{aligned} \tag{11.26}$$

Next, using the conditions from eqns. (11.16)–(11.18), and substituting the following values in eqn. (11.26):

(i) $\|C^TC\| \le c^2; \|H^TH\| \le h^2$ (c and h are positive constants)
(ii) $\|e(t)\|^2 \le \varepsilon^2$
(iii) $S = BB^T$ (positive definite matrix)
(iv) the jointly considered bound as

$$\|Bx(t)w(t)\| \le \rho_2 \|e(t)\|^2; \rho_2 > 0; \|e(t)\| \le \varepsilon = d$$

(where 'd' is the small positive constant; +ve square root); one obtains, just by regrouping only:

$$\dot{V}(t) \le -\left\{\frac{s_1}{p_u^2} - \frac{2\rho_1}{p_u}\varepsilon\right\}\|e(t)\|^2 - \left\{-h^2 + \frac{1}{\gamma^2}c^2\|e(t)\|^2 + \left\{\frac{2\rho_2}{p_u} - 2\rho\right\}\varepsilon\right\}\|e(t)\|^2 \tag{11.27}$$

In eqn. (11.27), $s(.)$ is the smallest (positive) eigenvalue of the matrix S. Further, simple rearrangement of eqn. (11.27) gives

$$\dot{V}(t) \le -\left\{\frac{s_1}{p_u^2} - \frac{2\rho_1}{p_u}\varepsilon\right\}\|e(t)\|^2 - \left\{-h^2 + \frac{1}{\gamma^2}c^2 - \left[\frac{2\rho_2}{p_u} - 2\rho\right]\varepsilon\right\}\|e(t)\|^2 \tag{11.28}$$

In order that the time derivative of the LE functional is -ve definite, the terms in the '{.}' should be +ve, which can be seen from the following simple rearrangement of eqn. (11.28)

$$\dot{V}(t) \le -\left\{\frac{s_1}{2\rho_1 p_u} > d; k_c = \frac{(c^2 - h^2\gamma^2)p_u}{2\gamma^2(\rho_2 - \rho p_u)} < d\right\}\|e(t)\|^2 \tag{11.29}$$

From eqn. (11.29), the following condition is obtained:

$$\frac{(c^2 - h^2\gamma^2)p_u}{\gamma^2(\rho_2 - \rho p_u)} < d < \frac{s_1}{\rho_1 p_u} \tag{11.30}$$

Since, all the constants and bounds appearing in the {.;.} of eqn. (11.29) are individually positive, and hence, if it is ensured that the second composite constant, k_c, in eqn. (11.29) is positive, then, the time derivative of the Lyapunov energy functional is bounded from above as in eqn. (11.29). As the

Lyapunov energy functional is positive definite because of eqns. (11.16) and (11.20), and its time derivative is locally negative definite as in eqn. (11.29), the observer error dynamics of the newly proposed FMF-based H-infinity observer for nonlinear systems are locally asymptotically stable.

11.5 EVALUATION OF THE NONLINEAR OBSERVER

The proposed joint FL- and HI-based observer for nonlinear dynamic system is validated using numerical simulations carried out in MATLAB. The simulations are done for a period of 20 sec with a sampling interval of 0.01 sec. The following nonlinear dynamic system is used:

$$\begin{aligned} \dot{x}_1(t) &= -0.1x_1(t) + 0.3x_1(t)x_2(t) \\ \dot{x}_2(t) &= 10x_2(t) - 0.7x_1(t)x_2(t) - 0.1x_2^2(t) \end{aligned} \tag{11.31}$$

$$y(t) = x_2(t)$$

The dynamic model in eqn. (11.31) is the Lotka–Volterra (L–V) system that is an interactive model of nonlinear systems, and there has been good interest in the application of the mathematical system theory to population biology systems. It is a two species (predator–prey) L–V dynamic system. The proposed nonlinear observer is given as

$$\begin{aligned} \dot{\hat{x}}_1(t) &= -0.1\hat{x}_1(t) + 0.3\hat{x}_1(t)\hat{x}_2(t) + L_1 f_m\left(y(t) - \hat{x}_2(t)\right) \\ \dot{\hat{x}}_2(t) &= 10\hat{x}_2(t) - 0.7\hat{x}_1(t)\hat{x}_2(t) - 0.1\hat{x}_2^2 + L_2 f_m\left(y(t) - \hat{x}_2(t)\right) \end{aligned} \tag{11.32}$$

In eqn. (11.32), $f_m(r(t))$ is an output of the FL-based process (FMF) and is, in general, a nonlinear function of inputs which are the residuals 'r' (and often derivative of 'r') of the observer. The f vector consists of the modified residuals' sequence, expressed in the discrete-time domain, for the sake of clarity:

$$f_m(k+1) = \left[r(k+1) \quad \dot{r}(k+1) \right] \tag{11.33}$$

The f vector consists of two inputs, r_x and $\dot{r}_x$, and single output $f_m(k+1)$, where $\dot{r}_x$ is computed by

$$\dot{r}_x = \frac{\left\{ r_x(k+1) - r_x(k) \right\}}{T} \tag{11.34}$$

Here, T is the sampling interval in seconds. Appropriate FL-based MF (FMF/FIS; fuzzy inference system) has been used and the technique is evaluated using MATLAB-based FL tool box,where necessary:(evalfis:Type=sugeno;NumInputs=2;NumOutputs=1;NumRules=4;AndMethod= 'prod';OrMethod='max';ImpMethod='prod';AggMethod='max'). It is affirmed here, that the asymptotic result derived using FMF applied only to residual term $r(t)$, eqn. (11.29) is equally applicable when it is also applied to the derivative of $r(t)$ eqn. (11.34). The states' initial conditions used for the simulation and the observer are $x(0) = [15\ 2]'$; and $\hat{x}(0) = \left[10 \quad 5\right]'$. In order to implement the observer algorithm, the matrix RTOMD eqn. (11.7) is required to be solved, and for which, the following transformation [1] is used:

$$a = P(t)b \tag{11.35}$$

$$\dot{b} = A^T b + \left(H^T H - \frac{1}{\gamma^2} C^T C \right) a \tag{11.36}$$

$$\dot{a} = \left(BB^T\right)b + Aa \tag{11.37}$$

Eqns. (11.36) and (11.37) are solved by the transition matrix method, and then using eqn. (11.35) one gets $P(t)$. For data simulation and observer states, the dynamic eqns. (11.31) and (11.32) are solved using the Euler integration method. The required Jacobian matrix to be used in RTOMD eqn. (11.7) is represented in the discrete form as

$$A(k) = \begin{bmatrix} -0.1 + 0.3x_2(k-1) & 0.3x_1(k-1) \\ -0.7x_2(k-1) & 10 - 0.7x_1(k-1) - 0.2x_2(k-1) \end{bmatrix} \tag{11.38}$$

The performance of the FL- and HI-based nonlinear observer is illustrated in Figure 11.1 for two states x_1 and x_2. Figure 11.2 depicts the convergence of the eigenvalues of the matrix P (left graph), and the true and predicted measurements. The performance of the observer with a certain gamma factor and the use of FMF that modulates the residuals in some nonlinear way has been thus found to be very satisfactory, as also can be seen from the % fit errors (PFE) obtained by using the formula; PFE = 100 × norm (errors)/norm (true signal): (i) x_1 state errors PFE: 1.8525; (ii) x_2 state errors PFE: 1.2937; and (iii) residuals PFE: 1.2906.

Figure 11.2 shows that the proposed FMF and HI-based nonlinear observer is asymptotically stable. It would be straightforward to extend the presented results to multi-sensory data-fusion requirements as a new application.

Figure 11.3 shows the phase portrait obtained from the proposed nonlinear observer. The present plot compares well, qualitatively with the one obtained in [8]; however, the observer design approaches are somewhat different.

Thus, a new nonlinear observer for a nonlinear system is presented that is based on the use of FMF as a nonlinear function of the residuals and their time derivatives, and the concept of H-infinity filter

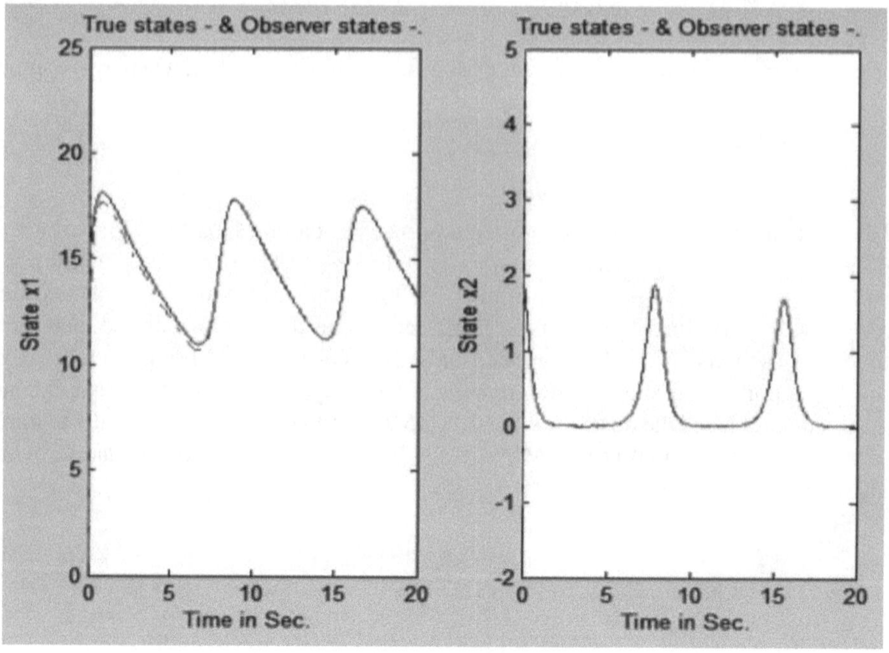

FIGURE 11.1 Time history match of the true (-) and observer states (-.).

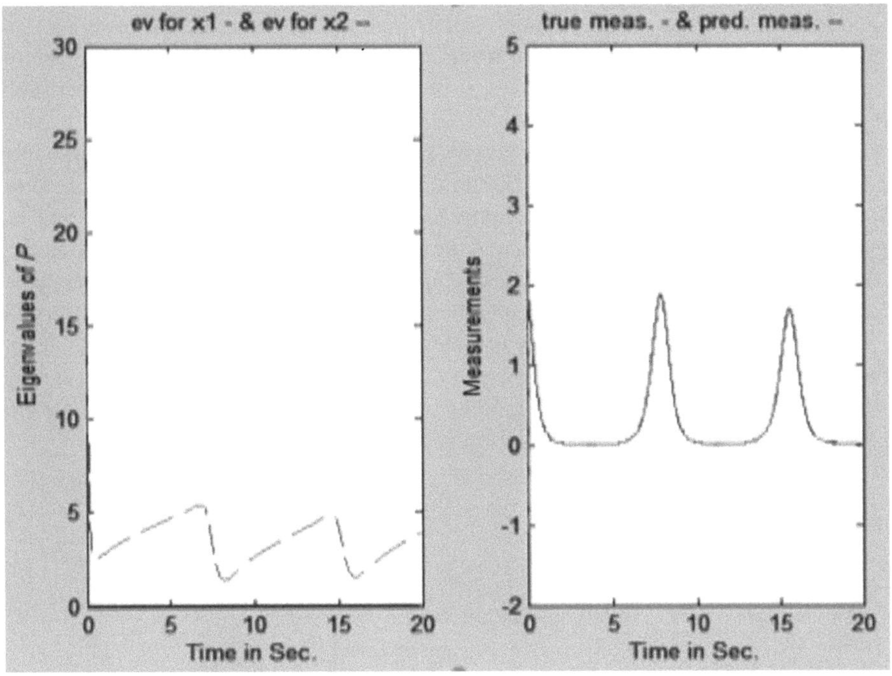

FIGURE 11.2 Eigenvalues of the Gramian P showing the satisfaction of the condition of asymptotic stability, related to state $x1$ (-), and state $x2$ (--); left graph; and the true (-), and predicted measurements (--).

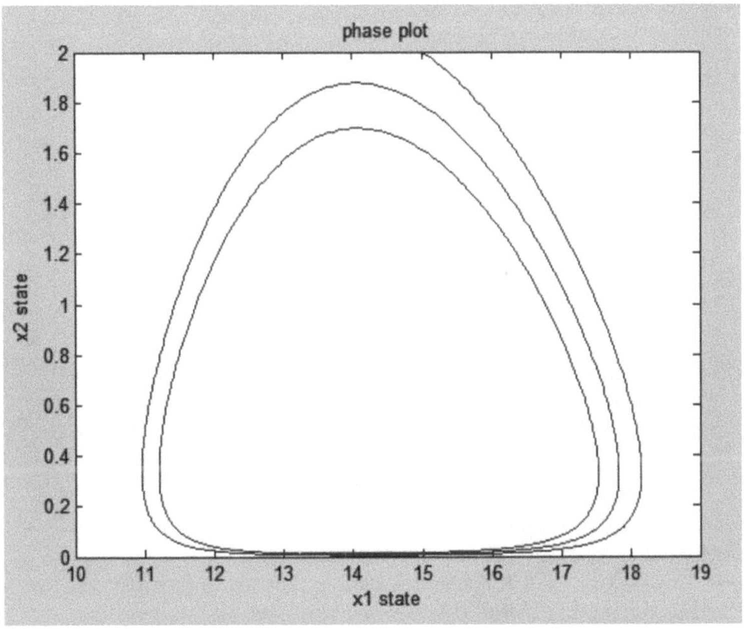

FIGURE 11.3 The state-space trajectory-phase plot from the nonlinear observer.

theory. For such an observer, the asymptotic stability result has been derived using the Lyapunov energy functional, and no specific theorem from control and estimation theory was required to be used. The performance of the observer has been validated by using the L–V two species (population dynamics) nonlinear model simulated in MATLAB. The observer's design is straightforward, and it requires a selection of a proper FMF and recursive solution of the RTOMD equation; the former can be done using the FL toolbox of MATLAB, and the latter can be done using the transition matrix method as described. The application of the approach to the general problem of estimation and target tracking is straightforward. The performance of the proposed FMF and HI-based nonlinear observer is very satisfactory and also corroborates the asymptotic result derived using the Lyapunov energy (LE) functional, as demonstrated and validated by the behaviour of the eigenvalues of the Gramian matrix, $P(t)$. The major aspect of this study is that with the theoretical basis provided, the extension of the deterministic nonlinear observer based on the fuzzy logic and the H-infinity theory, paves the way for further research towards obtaining more robust observers, which can handle a range of uncertainty in practical application for target tracking.

EXERCISES

Exercise 11.1

What is the main goal of using the Lyapunov method in observer design?

Exercise 11.2

How does the derivative of the Lyapunov function contribute to stability analysis?

Exercise 11.3

How does the incorporation of fuzzy membership functions (FMFs) enhance the adaptability in observer design compared to nonadaptive approaches?

Exercise 11.4

What three conditions outlined establishing the local asymptotic behaviour of the observer error dynamics?

Exercise 11.5

Why is establishing the stability result analytically for the observer error dynamics necessary, and what are the key reasons for its importance?

REFERENCES

1. Reshma, V., Fuzzy Logic Augmented H-Infinity Filter for Target Tracking. Ph.D. Thesis, Electronics and Communications Engineering, Jain University, Bangalore, 2017.
2. Hassibi, B., Sayed, A.H., and Kailath, T., Linear estimation in Krein spaces-Part I: Theory, *IEEE Transactions on Automatic Control, USA*, Vol. 41, No. 1, pp. 18–33, Jan. 1996.
3. Aliyu, M.D.S., *Nonlinear H-Infinity Control Hamiltonian Systems and Hamilton-Jacobi Equation*, CRC Press, Boca Raton, 2011. https://books.google.co.in/books?isbn=1439854858, accessed January 2017.
4. Ferreira, H.C., Rocha, P.H., and Sales, R.M., Nonlinear H∞-control and the Hamilton-Jacobi-Isaacs equation. In Proc. of the 17th World Congress, the International Federation of Automatic Control (IFAC), Seoul, Korea, July 6–11, 2008.

5. Kashyap, S.K., and Raol, J.R., Fuzzy logic applications in filtering and fusion for target tracking, Defence Science Journal, National Aerospace Laboratories, Bangalore–560 017, Vol. 58, No. 1, pp. 120–135, Jan. 2008.
6. Escamilla, P.J., and Mort, N., Development of a fuzzy logic-based adaptive Kalman filter. In *Proc. of the European Control Conference ECC'01, Porto, Portugal*, pp. 1768–1773, Sept. 4–7, 2001.
7. Raff, T., and Allgower, F., An EKF-based observer for nonlinear time-delay systems. In Proc. of the American Control Conference, Minneapolis, MN, USA, 2006, Vol. 4, pp. 31303133, July 2006, http://bibbase.org/network/publication/raff-allgwer-anekfbasedobserver
8. Vaidyanathan, S., Nonlinear observer design for L-V system. In IEEE International conference on Computational Intelligence and Computing Research (ICCIC), 2010. Coimbatore, India, Dec. 28–29, 2010, pp. 1–5.
9. Reif, K., Sonnemann, F., and Unbehauen, R., An EKF-based nonlinear observer with prescribed degree of stability, *Automatica*, Vol. 34, pp. 1119–1123, Sept. 1998, www.researchgate.net/publication/220157426, accssed January 2017.

12 Deterministic Nonlinear Estimators–Observers and Stability Results*

12.1 INTRODUCTION

A practical approach of solving accurately the nonlinear systems in modelling of a dynamic system is of great demand and importance in real-life engineering problems. The KF/EKF does not determine the discrepancy in the model of a nonlinear system, which is used in the filter. The reason is that it assumes the requirement of a complete state-space model, EKF is used and the prior knowledge of nonlinear functions 'f' and 'h' are presupposed. Any error or the so-called discrepancy in the model will cause a mismatch between the estimated and the true states of the dynamic system. This mismatch could be due to a few reasons; a Gaussian assumption made might degrade the performance of the algorithm and an assumption on the covariance matrix 'P' of the state estimate error is also made, which would be required for calculating the Kalman gain 'K'. These are the limitations of the EKF, which can be overcome by the method based on the 'principle of model error'. This method of model error computes not only the states of the system but also the model discrepancy with respect to time. The method of model error starts with using the known (deficient) model for the state estimation and computes the deterministic discrepancy. Once it is calculated, we can fit another model to this discrepancy and estimate the parameters related to states using a regression LS method. Then, combining the previously used deficient model, for the estimation of states and the new additional model identified will yield accurate model of the nonlinear system considered. This method of evaluating the model error and the states of the dynamic system are very useful in modelling of many complex systems like large flexible structure, aerospace systems and in robotics which are nonlinear systems. Also, an attempt is made to use H-infinity (HI/H∞) theory to obtain robust nonlinear system identification, with predictable performance in the face of unknown modelling errors and disturbances.

This chapter presents the theory of model error, or the so-called deterministic discrepancy, and its robustification using the H-infinity (HI) norm. Also, the continuous-time HI filter is derived from the discrete-time HI filter. Certain important aspects of DF in the context of using model error-based estimators are studied and evaluated with simulated data in MATLAB [1].

12.2 CONCEPT OF MODEL ERROR AND COST-FUNCTION FORMULATION

Consider a nonlinear system

$$\dot{x} = f\left(x(t), u(t), t\right) + d(t) \tag{12.1}$$

The unmodelled disturbance is constituted by $d(t)$ and is assumed to be piecewise continuous. This disturbance is not the process noise term as in the KF theory. Therefore, it cannot manage the true

* This chapter is contributed by Jitendra, R. Raol and Swetha Amit.

DOI: 10.1201/9781003620211-12

process noise as the output error method. In control theory, the term $d(t)$ constitutes control force or input, $u(t)$. This is calculated using the optimization method by minimizing the cost function [2]:

$$J = \sum_{k=1}^{N}\left[z(k)-h(\hat{x}(k),k)\right]^{T} R^{-1}\left[z(k)-h(\hat{x}(k),k)\right]+\int_{t_0}^{t_f} d^{T}(t)Qd(t)dt \tag{12.2}$$

We make assumption $E\{v(k)\}=0$; $E\{v(k)v^{T}(k)\}=R(k)$, 'h' is the measurement model. The Q is the covariance matrix of the process noise and is a very important parameter as it acts as a tuning device for the estimator. The value of Q can be chosen in such a way that [1] the equality is satisfied.

$$R(k)=\left[z(k)-h(\hat{x}(k),k)\right]\left[z(k)-h(\hat{x}(k),k)\right]^{T} \tag{12.3}$$

In eqn. (12.3), $R(.)$ is the presumed covariance matrix of the measurement noise. The right side of the equation depicts the difference between the actual and predicted measurements from the estimator. This type of equality is known as a covariance constraint. The virtue of this viewpoint is that it obtains accurate estimates even in the presence of unmodelled effects. Only R is assumed and no other statistical assumptions are made, this is unlike KF theory. The estimation method used is based on the LS method. With this procedure, the model gets improved, because initially unmodelled effects are considered in '$d(t)$', and a new model is arrived at by fitting a parametric model to '$d(t)$'; then we have an accurate model = deficient model + model fitted to the discrepancy. This involves the two-point boundary value problem (TPBVP), which is solved by using the method of invariant embedding, which in turn is the result of Pontryagin's conditions, and this is developed next. The dynamic system is given,

$$\dot{x}=f(x(t),u(t),t);x(t_0)=x_0 \tag{12.4}$$

Then, the cost function is defined as

$$J=\phi(x(t_f),t_f)+\int_{t_0}^{t_f}\psi(x(\tau),u(\tau),\tau)d\tau \tag{12.5}$$

In eqn. (12.5), among the two terms, the first one postulates the cost penalty of the state $x(t_f)$ and second term supposes the deviation of $x(t)$ and $u(t)$ with respect to time; $u(t)$ is in the interval $t_0 \le t \le t_f$, which is done so that the cost function J is minimized and eqn. (12.4) is not violated.

12.3 PONTRYAGIN'S CONDITIONS

A concept of Lagrange's multiplier is used to hold the restrain for the cost functional J:

$$J_a=\phi(x(t_f,t_f)+\int_{t_0}^{t_f}\left[\psi(x(\tau),u(\tau),\tau)+\lambda^{T}\left(-f(x(\tau),u(\tau),\tau)+\dot{x}\right)\right]d\tau \tag{12.6}$$

In eqn. (12.6), λ is Lagrange's multiplier, which gives a way to incorporate the condition, eqn. (12.4), which is the restrain on the states of the dynamic system. In the absence of the constraint, the 'u' might be such that the system would not follow the dynamics, which otherwise it is supposed to follow/obey. The Lagrange multiplier can also be called adjoint variables or co-states; since in the process of solving the equation we estimate the state equation too. The circumstances of eqn. (12.4) need to be strictly met, only then are eqns. (12.5) and (12.6) identical and the same, but now the dynamic constraint is taken care. Eqn. (12.6) can be rewritten as [3]

$$J_a = \phi\left(x(t_f), t_f\right) + \int_{t_0}^{t_f} \left[H_a\left(x(\tau), u(\tau), \tau\right) - \lambda^T(\tau) x(\tau) \right] d\tau + (\lambda^T x)_{t_f} - (\lambda^T x)_{t_0} \tag{12.7}$$

Here,

$$H_a = \psi\left(x(\tau), u(\tau), \tau\right) - \lambda^T(\tau) f\left(x(\tau), u(\tau), \tau\right) \tag{12.8}$$

In eqn. (12.7), the last three terms related to λ are obtained from the term $\int_{t_0}^{t_f} \lambda^T \dot{x}\, d\tau$ of eqn. (12.6) by 'integration by parts', and the remaining two terms of eqn. (12.6) are clubbed and we get eqn. (12.8) to define it as Hamiltonian, H. From eqn. (12.7), by the concept of differentials, we get

$$\delta J_a = \left(\left(\frac{\delta \psi}{\delta x} \right)^T \delta x \right)_{t_f} + \lambda^T \delta x\Big|_{t_f} - \lambda^T \delta x\Big|_{t_0} + \int_{t_0}^{t_f} \left[\left(\left(\frac{\delta H}{\delta x} \right)^T - \lambda^T \right) \delta x + \left(\frac{\delta H}{\delta u} \right)^T \delta u \right] d\tau \tag{12.9}$$

From eqn. (12.9), Pontryagin's conditions are obtained as

$$\lambda\left(t_f\right) = -\frac{\delta \psi}{\delta x}\bigg|_{t_f} \tag{12.10}$$

$$\frac{\delta H}{\delta x} = \lambda \tag{12.11}$$

$$\frac{\delta H}{\delta u} = 0 \tag{12.12}$$

$\delta x\left(t_0\right) = 0$, presuming that initial conditions $x\left(t_0\right)$ are independent of $u(t)$. Eqn. (12.10) is called the transversality condition. In the TPBVP, boundary conditions are set for states at t_0 and for the co-state, λ at t_f eqn. (12.10), then from eqns. (12.8) and (12.11), we obtain,

$$\lambda = \left(\frac{\delta H}{\delta x} \right) = -\left(\frac{\delta f}{\delta x} \right)^T \lambda + \left(\frac{\delta \psi}{\delta x} \right) \tag{12.13}$$

$$\frac{\delta H}{\delta u} = 0 = -\left(\frac{\delta f}{\delta u} \right)^T \lambda + \left(\frac{\delta \psi}{\delta u} \right) \tag{12.14}$$

Because of the split-boundary conditions, the problem as such is relatively difficult to solve. A method to solve TPBVP is to guess and estimate on $\lambda(t_0)$ and use $x(t_0)$ to configure at the final time t_f. After this, the boundary condition $\lambda(t_f) = -\frac{\delta \phi}{\delta x}\Big|_{t_f}^T$ is checked. If the condition is not met, it is recapitulated once again with new $\lambda(t_0)$ (or start from the end and traverse backward and verify other conditions) and so on, until the convergence of the algorithm is obtained; this procedure is called the multiple shooting method. The method of invariant embedding that obtains the recursive solution to the TPBV problem is discussed in the next section.

12.4 METHOD OF INVARIANT EMBEDDING

It is always feasible to analyse the system with a general solution and then evaluate the particular case which is basically our original problem. The method of invariant embedding (IE) belongs to this category of analysis. The meaning of this is as follows: first we find the general solution to the problem in which our original problem is embedded. Once the general solution is found, the particular solution that is the original problem is solved using some special conditions, which were kept invariant till then. Considering the equations from the TPBVP eqns. (12.1) and (12.13), in general,

$$\dot{x} = \phi(x(t), \lambda(t), t) \tag{12.15}$$

$$\dot{\lambda} = \psi(x(t), \lambda(t), t) \tag{12.16}$$

The protectorate for $\phi$and ψ on $x(t)$ and $\lambda(t)$ emerge from eqns. (12.1), (12.15) and (12.16), hence, here, we have a general TPBVP with boundary conditions $\lambda(t_0) = a$ and $\lambda(t_f) = b$. Now, $\lambda(t_f) = b$ are considered to be free variables, making the problem more general, which has our original problem embedded in it. There is dependency of $x(t_f)$ on t_f and $\lambda(t_f)$. Therefore, the dependency can be regarded as

$$x(t_f) = r(c, t_f) = r(\lambda(t_f), t_f) \tag{12.17}$$

with $t_f \to t_f + \Delta t$, neglecting higher-order terms,

$$\lambda(t_f + \Delta t) = \lambda(t_f) + \dot{\lambda}(t_f)\Delta t = c + \Delta c \tag{12.18}$$

We also get, using eqn. (12.16) in eqn. (12.18)

$$c + \Delta c = c + \psi(x(t_f), \lambda(t_f), t_f)\Delta t \tag{12.19}$$

and therefore, we get

$$\Delta c = \psi(r, c, t_f)\Delta t \tag{12.20}$$

Additionally, we obtain, like eqn. (12.18),

$$x(t_f + \Delta t) = x(t_f) + \dot{x}(t_f)\Delta t = r(c + \Delta c, t_f + \Delta t) \tag{12.21}$$

and therefore, using eqn. (12.15) in eqn. (12.21), we get

$$r(c + \Delta c, t_f + \Delta t) = r(c, t_f) + \phi(x(t_f), \lambda(t_f), t_f)\Delta t = r(c, t_f) + \phi(r, c, t_f)\Delta t \tag{12.22}$$

Using Taylor's series, we obtain

$$r\left(c+\Delta c, t_f+\Delta t\right)=r\left(c, t_f\right)+\frac{\partial r}{\partial c}\Delta c+\frac{\partial r}{\partial t_f}\Delta t \tag{12.23}$$

Comparing eqn. (12.22) and eqn. (12.23), we procure,

$$\frac{\partial r}{\partial t_f}\Delta t+\frac{\partial r}{\partial c}\Delta c=\phi\left(r, c, t_f\right)\Delta t \tag{12.24}$$

or, using eqn. (12.20) in eqn. (12.24), we get

$$\frac{\partial r}{\partial t_f}\Delta t+\frac{\partial r}{\partial c}\psi\left(r, c, t_f\right)\Delta t=\phi\left(r, c, t_f\right)\Delta t \tag{12.25}$$

Eqn. (12.25) untangles to

$$\frac{\partial r}{\partial t_f}+\frac{\partial r}{\partial c}\psi\left(r, c, t_f\right)=\phi\left(r, c, t_f\right) \tag{12.26}$$

We see that eqn. (12.26) connects dissimilarity of the terminal condition $x\left(t_f\right)=r\left(c, t_f\right)$ to the state and co-state differential functions, eqns. (12.15) and (12.16), and in order to estimate an optimal estimate $\hat{x}\left(t_f\right)$, $r\left(b, t_f\right)$ is evaluated as

$$\hat{x}\left(t_f\right)=r\left(b, t_f\right) \tag{12.27}$$

Eqn. (12.27) can be transfigured to an initial value problem by using approximation

$$r\left(c, t_f\right)=S\left(t_f\right)+\hat{x}\left(t_f\right) \tag{12.28}$$

Substituting eqn. (12.28) in eqn. (12.26), we get

$$\frac{\partial S\left(t_f\right)}{\partial t_f}c+\frac{\partial \hat{x}\left(t_f\right)}{\partial t_f}+S\left(t_f\right)\psi\left(r, c, t_f\right)=\phi\left(r, c, t_f\right) \tag{12.29}$$

Next, expanding ϕ and ψ about $\phi\left(\hat{x}, b, t_f\right)$ and $\psi\left(\hat{x}, b, t_f\right)$, we obtain

$$\begin{aligned}\phi\left(r, c, t_f\right)&=\phi\left(\hat{x}, b, t_f\right)+\phi_{\hat{x}}\left(\hat{x}, b, t_f\right)\left(r\left(c, t_f\right)-\hat{x}\left(t_f\right)\right)=\phi\left(\hat{x}, b, t_f\right)\\&\quad+\phi_{\hat{x}}\left(\hat{x}, b, t_f\right)S\left(t_f\right)c\end{aligned} \tag{12.30}$$

and

$$\psi\left(r, c, t_f\right)=\psi\left(\hat{x}, b, t_f\right)+\psi_{\hat{x}}\left(\hat{x}, b, t_f\right)S\left(t_f\right)c \tag{12.31}$$

Utilizing expressions of eqns. (12.30) and (12.31), in eqn. (12.29), we obtain a composite state estimation equation,

$$\frac{\partial S(t_f)}{\partial t_f}c+\frac{\partial \hat{x}(t_f)}{\partial t_f}+S(t_f)\left[\psi(\hat{x},b,t_f)+\psi_{\hat{x}}(\hat{x},b,t_f)S(t_f)c\right]$$
$$=\phi(\hat{x},b,t_f)+\phi_{\hat{x}}(\hat{x},b,t_f)S(t_f)c \quad (12.32)$$

Eqn. (12.32) is in abstraction a sequential state-estimation algorithm but a composite one involving $\hat{x}$, $S(t_f)$ and c, hence it is necessary to separate by substituting the specific expressions for ϕ and ψ in eqn. (12.32). We can do this when we specify the dynamic systems for which we need to obtain the estimators. This is done in the next section after arriving at TPBVP and then using eqn. (12.32).

12.4.1 Continuous-Time Estimator

Considering a dynamic system

$$\dot{x}=f(x(t),t)+d(t) \quad (12.33)$$

$$z(t)=Hx(t)+v(t) \quad (12.34)$$

The cost-function equation can be written as

$$J=\int_{t_0}^{t_f}\left[(z(t)-Hx(t))^T R^{-1}(z(t)-Hx(t))+(d^T(t)Qd(t))\right]dt \quad (12.35)$$

Here, $d(t)$ is the model discrepancy with $x(t)$. $R(t)$ is the spectral density matrix of noise covariance.

Formulating J by using Lagrange multipliers:

$$J_a=\int_{t_0}^{t_f}\left[(z(t)-Hx(t))^T R^{-1}(z(t)-Hx(t))+(d^T(t)Qd(t))\right.$$
$$\left.+\lambda^T(\dot{x}(t)-f(x(t),t)-d(t))\right]dt \quad (12.36)$$

Comparing with eqn. (12.7) and eqn. (12.8), we get

$$H=(z(t)-Hx(t))^T R^{-1}(z(t)-Hx(t))+d^T(t)Qd(t)-\lambda^T[(f(x(t),t)+d(t)]$$

$$H=\psi-\lambda^T f_m(x(t),d(t),t) \quad (12.37)$$

With straightforward development paralleling eqn. (12.9), we get

$$\dot{\lambda}^T=\left(\frac{\delta H}{\delta x}\right)=\left(\frac{\delta \psi}{\delta x}\right)-\lambda^T\left(\frac{\delta f_m}{\delta x}\right) \quad (12.38)$$

$$\dot{\lambda} = \left(\frac{\delta\psi}{\delta x}\right)^T - \left(\frac{\delta f_m}{\delta x}\right)^T \lambda = -f_{\hat{x}}^T \lambda - 2H^T R^{-1}\left(z(t) - Hx(t)\right) \tag{12.39}$$

$$\text{and } \frac{\partial H}{\partial d} = 0 = 2Qd - \lambda; \text{ leading to} \Rightarrow d = \frac{1}{2}Q^{-1}\lambda \tag{12.40}$$

Thus the TPBVP is given by

$$\dot{x} = f\left(x(t), t\right) + d(t)$$

$$\dot{\lambda} = -f_{\hat{x}}^T \lambda - 2H^T R^{-1}\left(z(t) - Hx(t)\right)$$

$$d = \frac{1}{2}Q^{-1}\lambda \tag{12.41}$$

On comparing with eqn. (12.15) and eqn. (12.16), we get

$$\phi\left(x(t), \lambda(t), t\right) = f\left(x(t), t\right) + d(t) \tag{12.42}$$

and

$$\psi\left(x(t), \lambda(t), t\right) = -f_{\hat{x}}^T \lambda - 2H^T R^{-1}\left(z(t) - Hx(t)\right) \tag{12.43}$$

We also have

$$\psi_{\hat{x}} = 2H^T R^{-1} H - \left[\frac{\partial}{\partial \hat{x}}\left(f_{\hat{x}}^T\right)\lambda\right] \tag{12.44}$$

$$\phi_{\hat{x}} = f_{\hat{x}} \tag{12.45}$$

Substituting eqns. (12.42) to (12.45) in eqn. (12.32) and observing t_f as a function of 't', we get

$$\begin{aligned}\dot{S}(t)\lambda + \dot{\hat{x}}(t) + S(t)\Big[-f_{\hat{x}}^T \lambda - 2H^T R^{-1}\left(z(t) - Hx(t)\right) + 2H^T R^{-1} H S(t)\lambda \\ - \frac{\partial}{\partial \hat{x}}\left(f_{\hat{x}}^T \lambda\right) S(t)\lambda\Big] = f\left(x(t), t\right) + \frac{1}{2}Q^{-1}\lambda + f_{\hat{x}} S(t)\lambda\end{aligned} \tag{12.46}$$

Now, separating the terms relating to λ from eqn. (12.46) to obtain

$$\dot{\hat{x}} = f\left(x(t), t\right) + 2S(t) H^T R^{-1}\left(z(t) - Hx(t)\right) \tag{12.47}$$

$$\begin{aligned}\dot{S}(t)\lambda = f_{\hat{x}}^T S(t)\lambda + f_{\hat{x}} S(t)\lambda - 2S(t) H^T R^{-1} H S(t)\lambda + \frac{1}{2}Q^{-1}\lambda \\ + S(t)\frac{\partial}{\partial \hat{x}}\left(f_{\hat{x}}^T \lambda\right) S(t)\lambda\end{aligned} \tag{12.48}$$

We divide eqn. (12.48) by λ and for $\lambda \to 0$, we obtain

$$\dot{S}(t) = f_{\hat{x}}^T S(t) + f_{\hat{x}} S(t) - 2S(t) H^T R^{-1} H S(t) + \frac{1}{2} Q^{-1} \quad (12.49)$$

Solving the above equations, gives a straightforward expression for model error (discrepancy), comparing eqn. (12.47) to eqn. (12.33):

$$\hat{d}(t) = 2S(t) H^T R^{-1} \left(z(t) - Hx(t)\right) \quad (12.50)$$

Eqns. (12.47), (12.49) and (12.50) depict the model error-estimation algorithm for the continuous-time system using the IE method from eqns. (12.33) and (12.34), in a recursive form. Eqn. (12.49) is known as the matrix Riccati equation.

In the process of implementing the algorithm, the matrix differential eqn. (12.49) needs to be solved. Consider the following transformation:

$$a = Sb \quad (12.51)$$

and using eqn. (12.49),

$$\dot{S}b = f_{\hat{x}}^T Sb + f_{\hat{x}} Sb - 2S(t) H^T R^{-1} H Sb + \frac{1}{2} Q^{-1} b \quad (12.52)$$

$$\dot{S}b + 2SH^T R^{-1} H Sb - f_{\hat{x}}^T Sb = f_{\hat{x}} a + \frac{1}{2} \alpha^{-1} Q^{-1} b \quad (12.53)$$

Also, $\dot{a} = \dot{S}b + S\dot{b}$ and $\dot{S}b = \dot{a} - S\dot{b}$. Applying, $\dot{S}b$ in eqn. (12.54) and describing $\dot{b}$ as in eqn. (12.32), we obtain

$$\dot{b} = -f_{\hat{x}}^T b + 2H^T R^{-1} H a \quad (12.54)$$

$$a = \frac{1}{2} \alpha^{-1} Q^{-1} b + f_{\hat{x}} a \quad (12.55)$$

Eqns. (12.54) and (12.55) can be solved by the transition matrix method. Q acts as the weighting matrix in the model error. It imparts normalization of the second term in the cost function eqn. (12.36).

12.4.2 Discrete-Time Estimator

Consider the nonlinear system

$$x(k+1) = g\left(x(k), k\right) \quad (12.56)$$

$$z(k) = h\left(x(k), k\right) \quad (12.57)$$

where 'g' is the vector function and z is the vector of observables in the interval $t_0 < t_j < t_N$. Writing above equations explicitly with the model error, and measurement noise, we obtain

$$x(k+1) = f(x(k),k) + d(k) \tag{12.58}$$

$$z(k) = h(x(k),k) + v(k) \tag{12.59}$$

where

f = deficient model, v = measurement noise with zero mean for the covariance matrix R and d = model error. The 'd' is resolved by minimizing the following criterion:

$$J = \sum_{k=0}^{N}\left[\left(z(k) - h(x(k),k)\right)^T R^{-1}\left(z(k) - h(x(k),k)\right)\right] + d^T(k)Qd(k) \tag{12.60}$$

With co-state/Lagrange multiplier, we obtain

$$\begin{aligned} J_a = \sum_{k=0}^{N}&\left[\left(z(k) - h(x(k),k)\right)^T R^{-1}\left(z(k) - h(x(k),k)\right)\right] + d^T(k)Qd(k) \\ &+ \lambda^T\left[x(k+1) - f(x(k),k) - d(k)\right] \end{aligned} \tag{12.61}$$

$$H = \left[z(k) - h(x(k),k)\right]^T R^{-1}\left[z(k) - h(x(k),k)\right] + d^T(k)Qd(k) - \lambda^T\left(f(x(k),k) + d(k)\right)$$

$$H = \psi - \lambda^T f_m(x(k),d(k),k) \tag{12.62}$$

Here, H is called the Hamiltonian. In eqn. (12.62), there is term ψ which is given by

$$\psi = \left[z(k) - h(x(k),k)\right]^T R^{-1}\left[z(k) - h(x(k),k)\right] + d^T(k)Qd(k) \tag{12.63}$$

By using Pontryagin's conditions we get the following equations:

$$\lambda(k-1) = \left(\frac{\delta H}{\delta x}\right)^T = -\left(\frac{\delta f}{\delta x}\right)^T \lambda(k) + \left(\frac{\delta \psi}{\delta x}\right)^T \tag{12.64}$$

$$\lambda(k-1) = \left(\frac{\delta \psi}{\delta x}\right)^T - \left(\frac{\delta f_m}{\delta x}\right)^T \lambda(k) = -f_{\hat{x}}^T \lambda - 2H^T R^{-1}\left(z(k) - Hx(k)\right) \tag{12.65}$$

$$\frac{\delta H}{\delta u} = 0 = -\left(\frac{\delta f}{\delta u}\right)^T \lambda(k) + \left(\frac{\delta \psi}{\delta u}\right) \tag{12.66}$$

leading to

$$d(k) = -\frac{1}{2}Q^{-1}\lambda(k) \tag{12.67}$$

Therefore, the TPBV Problem is stated as follows:

$$x(k+1) = f(x(k),k) + d(k)$$

$$\lambda(k-1)=-f_{\hat{x}}^{T}\lambda-2H^{T}R^{-1}\left(z(k)-Hx(k)\right)$$

$$d(k)=\frac{1}{2}Q^{-1}\lambda(k) \tag{12.68}$$

We know the discrete-time nonlinear system can be represented as

$$g\left(x(k),k\right)=f\left(x(k),k\right)+d(k) \tag{12.69}$$

$$h\left(x(k),k\right)=-f_{\hat{x}}^{T}\lambda-2H^{T}R^{-1}\left(z(k)-Hx(k)\right) \tag{12.70}$$

$$\psi_{\hat{x}}=2H^{T}R^{-1}H-\left[\frac{\partial}{\partial\hat{x}}\left(f_{\hat{x}}^{T}\right)\lambda(k)\right] \tag{12.71}$$

$$\phi_{\hat{x}}=f_{\hat{x}} \tag{12.72}$$

Substituting the above equations in the last equation of invariant embedding eqns. (12.32), we obtain the following equations:

$$\begin{aligned}S(k+1)\lambda(k)+x(k+1)+S(k)\Big[-f_{\hat{x}}^{T}\lambda(k)-2H^{T}R^{-1}\left(z(k)-Hx(k)\right)\\+2H^{T}R^{-1}HS(k+1)\lambda(k)-\frac{\partial}{\partial\hat{x}}\left(f_{\hat{x}}^{T}\lambda(k)\right)S(k+1)\lambda(k)\Big]\\=f\left(x(k),k\right)+\frac{1}{2}Q^{-1}\lambda+f_{\hat{x}}S(k)\lambda(k)\end{aligned} \tag{12.73}$$

By separating the terms related to λ, we obtain

$$x(k+1)=f\left(x(k),k\right)+2S(k)H^{T}R^{-1}\left(z(k)-Hx(k)\right) \tag{12.74}$$

$$\begin{aligned}S(k+1)\lambda(k)=f_{\hat{x}}^{T}S(k)\lambda(k)+f_{\hat{x}}S(k)\lambda(k)-2S(k)H^{T}R^{-1}HS(k)\lambda(k)\\+\frac{1}{2}Q^{-1}\lambda(k)+S(k)\frac{\partial}{\partial\hat{x}}\left(f_{\hat{x}}^{T}\lambda(k)\right)S(k)\lambda(k)\end{aligned} \tag{12.75}$$

Dividing the above equation by λ and for λ→0,

$$\begin{aligned}S(k+1)=f_{\hat{x}}^{T}S(k)+f_{\hat{x}}S(k)-2S(k)H^{T}R^{-1}HS(k)\\+\frac{1}{2}Q^{-1}+S(k)\frac{\partial}{\partial\hat{x}}\left(f_{\hat{x}}^{T}\lambda(k)\right)S(k)\lambda(k)\end{aligned} \tag{12.76}$$

Resolving the equations gives a clear-cut expression for the model error;

$$d(k)=2S(k)H^{T}R^{-1}\left(z(k)-Hx(k)\right) \tag{12.77}$$

Eqns. (12.74), (12.76), (12.77) present the IE solution for the DTS estimation algorithm for model error determination.

12.5 NONLINEAR OBSERVERS

In the previous sections we have studied so-called deterministic state estimators that are based on the theory of IE, wherein a general (TPBV) is resolved, of which the required solution of the test problem is a special case. These are called deterministic estimators, because they handle only the measurement noise and not the process or state noise.

In this section, an observer for continuous-time nonlinear dynamic system is studied. This observer utilizes the gain from the IE theory. A Lyapunov energy (LE) functional is utilized for deriving the condition for the local asymptotic stability for the observer's state errors; such a theoretical result for the invariant-embedding-based estimator/observer is novel. The performance of the observer is demonstrated by implementing the observer algorithm in MATLAB. The simulation results support the asymptotic result of the IE-based observer error dynamics.

12.5.1 Observer Structure with IE Gain

The nonlinear system considered is

$$\dot{x} = f(x(t),t) + d(t) \tag{12.78}$$

$$y(t) = Hx(t) + v(t)$$

In eqn. (12.78), $d(.)$ is a deterministic signal or an unknown deterministic disturbance, and $y(.)$ is a measurement vector; $v(.)$ can be regarded as a noise and covariance matrix as R; or as an unknown disturbance with known intensity R. A nonlinear observer [2] for the system of eqn. (12.78) is given as

$$\dot{\hat{x}}(t) = f(\hat{x}(t),t) + \hat{d}(t) + L(t)(y(t) - \hat{y}(t))$$

$$\hat{y}(t) = H\hat{x}(t) \tag{12.79}$$

The observer gain is chosen from eqn. (12.47) as

$$L(t) = 2S(t)H^T R^{-1} \tag{12.80}$$

For eqn. (12.80), $S(t)$ is obtained as the solution of the observer Riccati differential (ORD) eqn. (12.48),

$$\dot{S}(t) = S(t)A^T(t) + A(t)S(t) - 2SH^T R^{-1} HS(t) - \frac{1}{2}Q^{-1}\alpha^{-1}$$

$$\text{with } S(0) = S_0; S(T) = S_{ss}; \text{ a steady state value} \tag{12.81}$$

The Jacobian matrix $A(t)$ needed in eqn. (12.80) is obtained as

$$A(t) = \frac{\delta f(.)}{\delta \hat{x}(t)} \tag{12.82}$$

Eqns. (12.80) and (12.81) are from the IE estimator for the continuous-time nonlinear dynamic system. In the domain of deterministic IE estimators and observers, the matrices Q and R, are called weighting matrices or the Gramians. In that case, the corresponding variables are termed as the 'generalized random' variables. The observer residuals and the state errors are given as

$$r(t) = y(t) - \hat{y}(t); e(t) = x(t) - \hat{x}(t) \tag{12.83}$$

The state error dynamics are obtained by subtracting eqn. (12.79) from eqn. (12.78):

$$\dot{e}(t) = A(t)e(t) + d(t) - \hat{d}(t) + \psi(t) - L(t)He(t) - L(t)v(t) \tag{12.84}$$

In eqn. (12.84), the nonlinear function is

$$\psi(.) = -A(t)e(t) + f(x,t) - f(\hat{x},t)$$

$$\psi(.) = \psi(x(t), \hat{x}(t)) \tag{12.85}$$

12.5.2 Asymptotic Stability with Lyapunov Method

Certain important conditions need to be considered for establishing the local asymptotic behaviour of the error dynamics of eqn. (12.84) [4,5]:

(a) The solution of the RTD eqn. (12.48) should be considered as bounded

$$s_l I \leq S(t) \leq s_u I \tag{12.86}$$

With, $s_l, s_u > 0$ are positive constants; and are the lower and upper bounds on $S(t)$, respectively; the matrix $S(t)$ is supposed to be positive definite and symmetric; I is the identity matrix. This condition is natural, because we need a bounded solution to the observer error dynamics.

(b) The nonlinearity in eqn. (12.85) of the error dynamics is bounded

$$\|\psi(.)\| \leq \rho_1 \|x(t) - \hat{x}(t)\|^2 \tag{12.87}$$

With the bounding constants equal to or greater than zero. This is again a natural requirement for the same reason as for condition (a).

(c) The deterministic error is bounded

$$\|d(t) - \hat{d}(t)\| = \|e_d(.)\| \leq \rho_2 \|x(t) - \hat{x}(t)\|^2 \tag{12.88}$$

If all three conditions (a) to (c) are satisfied, and the time derivative of the LE functional is negative definite, then, the observer error dynamics in eqn. (12.84) are locally asymptotically stable. Now, consider the normalized LE functional as

$$V(t) = e^T Y(t) e(t) \tag{12.89}$$

In eqn. (12.89), $Y(t)$ is the normalizing matrix regarded as an information Gramian, for $Y(t) = S^{-1}(t)$. Since, for the deterministic observers and IE-based estimators, one deals with unknown, but deterministic disturbance processes; $S(t)$ is considered as the state-error Gramian matrix; and hence, the variables $x(.), y(.)$ and $e(t)$ are called the generalized 'random' variables. The LE functional is positive definite as seen from the following condition, the inequality:

$$\frac{1}{s_u}\|e(t)\|^2 \le e^T(t)Y(t)e(t) \le \frac{1}{s_l}\|e(t)\|^2 \tag{12.90}$$

Then, in order to establish the asymptotic convergence result for the considered nonlinear IE-based observer, the time derivative of the LE functional, eqn. (12.89), (after the use of the constraints of the error dynamics, IE gain, and the Gramian $S(t)$ in the time-derivative formula), should be negative definite for all time t. For simplicity, the error dynamics are written (with the knowledge that the quantities A, L, H, x, v are dependent on time t) as

$$\dot{e}(t) = A(t)e(t) + e_d(t) + \psi(t) - LHe(t) - Lv \tag{12.91}$$

Then, the time derivative is obtained as

$$\begin{aligned}\dot{V}(t) = e^T(t)\dot{Y}(t)e(t) + e^T(t)\left[A^TY(t) + Y(t)A\right]e(t) + 2e^T(t)Y(t)\\ e_d(t) - 2e^T(t)Y(t)LHe(t) + 2e^T(t)Y(t)\psi(.) - 2e^T(t)Y(t)Lv(t)\end{aligned} \tag{12.92}$$

Here, $\dot{Y}(t) = -Y(t)\dot{S}(t)Y(t)$. After substituting the gain and RTD eqns. In eqn. (12.92) and after cancelling out several common terms and simplifying without any approximations, one gets

$$\begin{aligned}\dot{V}(t) = -\frac{1}{2}e^T(t)Y(t)\alpha^{-1}2Q^{-1}Y(t)e(t) + 2e^T(t)H^TR^{-1}He(t) + 2e^T(t)Ye_d(t)\\ + 2e^T(t)Y(t)\psi(.) - 4e^T(t)H^TR^{-1}He(t) - 4e^T(t)H^TR^{-1}v(t)\end{aligned} \tag{12.93}$$

We assume

(i) $\|H^TH\| \le h^2; \|R^{-1}\| \le r$, (here, h, and r, are positive constants);

(ii) $\frac{1}{q_u}I \le Q^{-1}\gamma^{-2} \le \frac{1}{q_l}I; \|e(t)\|^2 \le \varepsilon^2$; with $k = \varepsilon$

(iii)
$$\|v(t)\| \le \rho_3\|e(t)\|^2 \tag{12.94}$$

The constants in all these bounds assumed known and/or pre-specified quantities. These are assumed finite. Thus, one obtains

$$\begin{aligned}\dot{V}(t) \le -\frac{1}{2s_u^2q_u}\|e(t)\|^2 + \frac{2h^2}{r}\|e(t)\|^2 + \frac{2\rho_2}{s_u}\varepsilon e(t)^2 + \frac{2\rho_1}{s_u}\varepsilon\|e(t)\|^2\\ -\frac{4h^2}{r}e(t)^2 - \frac{4h\rho_3}{r}\varepsilon\|e(t)\|^2\end{aligned} \tag{12.95}$$

After rearranging one obtains

$$\dot{V}(t) \le -\frac{1}{2s_u^2q_u}\|e(t)\|^2 - \frac{4h^2}{r}\|e(t)\|^2 - \left\{\frac{2h^2}{r} + \frac{2\rho_2}{s_u}\varepsilon + \frac{2\rho_1}{s_u}\varepsilon - \frac{4h\rho_3}{r}\varepsilon\right\}\|e(t)\|^2 \tag{12.96}$$

$$\dot{V}(t) \le -\left(\frac{1}{2s_u^2 q_u} + \frac{4h^2}{r}\right)\|e(t)\|^2 \le -\left\{-\frac{2h^2}{r} + \left(-\frac{2\rho_2}{s_u} - \frac{2\rho_1}{s_u} + \frac{4h\rho_3}{r}\right)\varepsilon\right\}\|e(t)\|^2 \quad (12.97)$$

$$\dot{V}(t) \le -\left\{\frac{8s_u^2 q_u h^2 + r}{2s_u^2 q_u r};\frac{2s_u h^2}{\left(4s_u\rho_3 h - 2r(\rho_1 + \rho_2)\right)}\right\}\|e(t)\|^2 \quad (12.98)$$

Since, all the constants and bounds appearing in the bracket {.;.} of eqn. (12.88) are individually positive, and hence, if it is ensured that the denominator of the second composite constant is positive, then, the time derivative of the LE functional is bounded from above as in eqn. (12.88). Because, initially the LE functional is positive definite as defined earlier, and its time derivative is locally negative definite, the IE gain-based observer error dynamics are locally asymptotically stable.

12.5.3 Performance Evaluation

The IE gain-based observer for nonlinear dynamic system is validated using numerical simulation carried out in MATLAB. The data are generated for a period of 20 sec with a sampling interval of 0.01 sec. The nonlinear dynamic system used for the purpose is given by

$$\dot{x}_2(t) = -0.1x_1(t) + 0.3x_1(t)x_2(t)$$

$$\dot{x}_2(t) = 10x_2(t) - 0.7x_1(t)x_2(t) - 0.1x_2^2(t)$$

$$y(t) = x_2(t) \quad (12.99)$$

The model, eqn. (12.99), is the Lotka–Volterra (L–V) dynamic system that is often used as an interactive model for nonlinear systems [6]. It has become very popular for the study of the mathematical system theory in relation to the population in biology systems. The IE-based nonlinear observer is obtained as

$$\dot{\hat{x}}_1(t) = -0.1\hat{x}_1(t) + 0.3\hat{x}_1(t)\hat{x}_2(t) + L_1\left(y(t) - \hat{x}_2(t)\right)$$

$$\dot{\hat{x}}_2(t) = 10\hat{x}_2(t) - 0.7\hat{x}_1(t)\hat{x}_2(t) - 0.1\hat{x}_2^2(t)L_2\left(y(t) - \hat{x}_2(t)\right) \quad (12.100)$$

The observer gain is taken as in eqn. (12.80), and the initial conditions of the states used for the simulation and the observer are, respectively, $x(0) = [15 \quad 2]'$ and $\hat{x}(0) = [10 \quad 5]'$. The ORDE eqn. (12.81) is solved by using the transition matrix method. First, the Gramian matrix $S(t)$ is transformed, and then the ORDE is partitioned and reformulated as

$$a = S(t)b \quad (12.101)$$

$$\dot{b} = -A^T b + \left(\frac{2}{r}H^T H\right)a \quad (12.102)$$

$$\dot{a} = -\frac{1}{2}\left(\gamma^{-2}Q^{-1}\right)b + Aa \quad (12.103)$$

Eqns. (12.102) and (12.103) are like any state-space differential equations, and are solved by the transition matrix method. Then, using eqn. (12.101), the solution $S(t)$ of the ORDE is obtained. For the generation of the simulated data, the dynamics are solved by the Euler integration method with a small step size. The required Jacobian matrix, $A(t)$ to be used in ORDE is represented in the discrete form as

$$A(k)=\begin{bmatrix} -0.1+0.3x_2(k-1) & 0.3x_1(k-1) \\ -0.7x_2(k-1) & 10-0.7x_1(k-1)-0.2x_2(k-1) \end{bmatrix} \tag{12.104}$$

In eqn. (12.104), $A(.)$, being time-dependent is evaluated at every time step. The performance of the IE-based nonlinear observer is illustrated in Figure 12.1 for two states x_1 and x_2.

Figure 12.2 shows the convergence of the eigenvalues of the matrix $S(t)$ (left plot), and the true and predicted measurements (right plot). The time history match, Figure 12.1, is found to be very good. The performance of the IE-based observer has been found to be very satisfactory, as is also supported by the % fit errors (PFEs) obtained by using the formula: PFE=100×norm (errors)/norm (true signal), Table 12.1.

Figure 12.2 clearly shows that the IE-based nonlinear observer is asymptotically stable, because the eigenvalues of the Gramian matrix asymptotically converge. The theoretical result established is corroborated, at least qualitatively, by the actual behaviour of the IE-based observer's error dynamics. It should be straightforward to extend the present results to MSDF that would use the IE-based estimators.

A LEF (Lyapunov energy functional)-based asymptotic result for a nonlinear observer that is based on the theory of IE has been established; and its working has been validated using the L–V two species nonlinear model simulated in MATLAB. The performance of this IE-based nonlinear observer has been found to be very satisfactory and the results of the numerical evaluation also

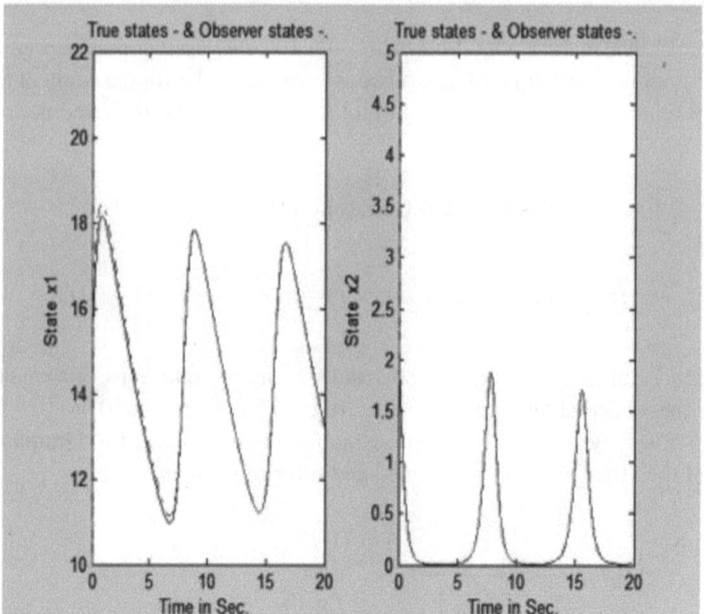

Time history match (the true (-) & observer states (-.)) for the IE-based nonlinear observer.

FIGURE 12.1 Time history match (the true (-) and observer states (-.)) for the IE-based nonlinear observer.

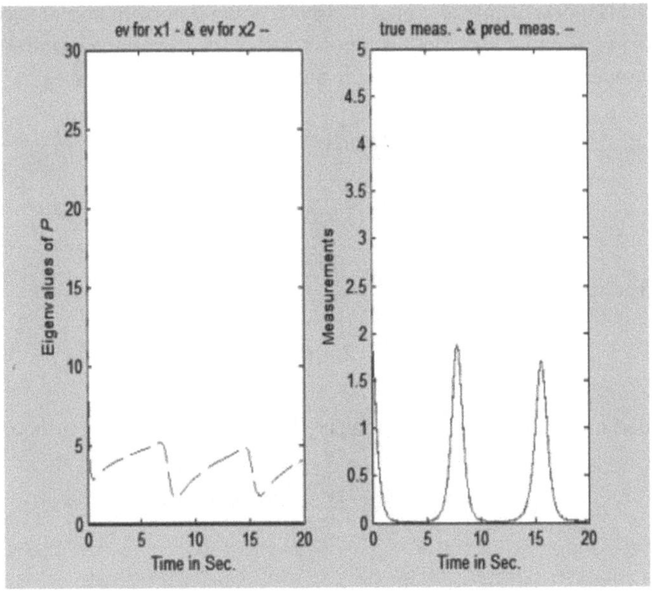

IE-based nonlinear observer: Eigenvalues of the Gramian S matrix depict the satisfaction of the condition of asymptotic stability {for states: x1 (-), & x2 (--); the left graph}. The true (-), and predicted measurements (--) are shown in the right graph.

FIGURE 12.2 IE-based nonlinear observer: eigenvalues of the Gramian S matrix depict the satisfaction of the condition of asymptotic stability (for states: $x1$ (-), and $x2$ (--); the left graph). The true (-), and predicted measurements (--) are shown in the right graph.

TABLE 12.1
PFEs for the IE-Based Observer

PFEs: for $r = 0.001$; $Q = [0.9\ 0;0\ 0.85]$	$\alpha = 0.2$	$\alpha = 0.5$	$\alpha = 1.0$	$\alpha = 1.8$
State x_1	1.103	1.005	0.86	0.673
State x_2	1.461	1.534	1.575	1.6
Measurements	1.46	1.534	1.577	1.605

support the asymptotic result derived using the LE functional; the latter result is validated by the behaviour of the eigenvalues of the Gramian matrix, $S(t)$.

12.5.4 Observer with IEHI (IE and HI)-Based Gain

A nonlinear observer for the continuous-time system using the gain from the theory of IE and HI is presented [7]; and it also uses the matrix Riccati-type differential (RTD) equation, which is needed in the computation of the gain. The asymptotic stability result for the observer is derived using the LEF. The nonlinear system considered is

$$\dot{x} = f\left(x(t),t\right) + d(t)$$

$$y(t) = Hx(t) + v(t) \tag{12.105}$$

In eqn. (12.105), $d(.)$ is a deterministic signal or an unknown deterministic disturbance, and $y(.)$ is a measurement vector; $v(.)$ can be regarded as a noise and R the covariance matrix; or as an unknown disturbance with known intensity R. A nonlinear observer for this is given as

$$\dot{\hat{x}}(t) = f(\hat{x}(t),t) + \hat{d}(t) + L(t)(y(t) - \hat{y}(t))$$

$$\hat{y}(t) = H\hat{x}(t) \tag{12.106}$$

The observer gain is chosen as

$$L(t) = 2S(t)H^T R^{-1} \tag{12.107}$$

Here, $S(t)$ is obtained as the solution of the matrix RTD equation (from HI theory)

$$\dot{S}(t) = S(t)A^T(t) + A(t)S(t) - 2SH^T R^{-1} HS(t) - \frac{1}{2}Q^{-1}\gamma^{-2};$$

$$\text{with } S(0) = S_0; S(T) = \text{a steady state value} \tag{12.108}$$

The Jacobian matrix $A(t)$ needed in eqn. (5.31) is obtained as

$$A(t) = \frac{\delta f(.)}{\delta \hat{x}(t)} \tag{12.109}$$

Eqns. (12.107), and (12.108) are from the combined theories of IE and HI (IEHI) for continuous-time nonlinear dynamic system estimator. In the domain of deterministic IEHI estimators and observers, the matrices Q and R, are called weighting matrices or the Gramians. The residuals and the state errors from the observer are given as

$$r(t) = y(t) - \hat{y}(t); e(t) = x(t) - \hat{x}(t) \tag{12.110}$$

The state-error dynamics are obtained as usual

$$\dot{e}(t) = A(t)e(t) + d(t) - \hat{d}(t) + \psi(t) - L(t)He(t) - L(t)v(t) \tag{12.111}$$

As usual the nonlinear function in eqn. (12.111) is

$$\psi(.) = -A(t)e(t) + f(x,t) - f(\hat{x},t)$$

$$\psi(.) = \psi(x(t),\hat{x}(t)) \tag{12.112}$$

12.5.5 Asymptotic Stability Result

Certain important conditions need to be considered for establishing the local asymptotic behaviour of the error dynamics:

(a) The solution of the RTD eqn. (12.108) should be considered as bounded

$$s_l I \le S(t) \le s_u I \tag{12.113}$$

(b) The nonlinearity in eqn. (12.112) of the error dynamics is bounded [8]

$$\|\psi(.)\| \le \rho_1 \|x(t) - \hat{x}(t)\|^2 \tag{12.114}$$

(c) The deterministic error is bounded

$$\|d(t) - \hat{d}(t)\| = (e_d(.)] \le \rho_2 \|x(t) - \hat{x}(t)\|^2 \tag{12.115}$$

In a basic observer theory, such a deterministic error, or model deficiency is generally not considered. Now, if all three conditions (a) to (c) are satisfied, and the time derivative of the LEF is negative definite, then, the observer error dynamics are locally asymptotically stable; of course some more feasible conditions are needed that will be evolved, when the stability result is derived. Now, consider the normalized LE functional,

$$V(t) = e^T Y(t) e(t) \tag{12.116}$$

The LEF is positive definite as seen from the following condition:

$$\frac{1}{s_u}\|e(t)\|^2 \le e^T(t) Y(t) e(t) \le \frac{1}{s_l}\|e(t)\|^2 \tag{12.117}$$

Then, in order to establish the asymptotic convergence for the considered nonlinear IEHI-based observer, the time derivative of the LEF; (after the use of the constraints of the error dynamics, of gain, and of the Gramian $S(t)$;in the time-derivative formula) should be negative definite for all time t. The error dynamics are written as

$$e(t) = A(t)e(t) + e_d(t) + \psi(t) - LHe(t) - Lv \tag{12.118}$$

Then, the time derivative of the LEF (TDLF) is obtained as

$$\begin{aligned}\dot{V}(t) = e^T(t)Y(t)e(t) + e^T(t)Y(t)\left[A(t)e(t) + e_d(t) + \psi(t) - LHe(t) - Lv\right] \\ +\left[e^T(t)A^T(t) + e_d^T(t) + \psi^T(.) - e^T(t)H^T L^T - v^T L^T\right]Y(t)e(t)\end{aligned} \tag{12.119}$$

Next, substitute

(i) for $Y(t) = -Y(t)S(t)Y(t)$, gain formula, and the RTDE in eqn. (12.119);

(ii) $\|H^T H\| \le h^2; \|R^{-1}\| \le r$, (here, h, and r, are positive constants);

(iii) $\frac{1}{q_u} I \le Q^{-1}\gamma^{-2} \le \frac{1}{q_l} I; \|e(t)\|^2 \le \varepsilon^2$; with $k = \varepsilon$;

(iv) $\|v(t)\| \le \rho_3 \|e(t)\|^2$ (12.120)

and the basic bounds, and then after cancelling out several common terms and simplifying (without any approximations) one gets, in terms the norms of the individual terms:

$$\text{TDLF} = \frac{1}{2s_u^2 q_u}\left\|e(t)\right\|^2 - \frac{2h^2}{r}e(t)^2 + \frac{2\rho_2}{s_u}\varepsilon\left\|e(t)\right\|^2 + \frac{2\rho_1}{s_u}\varepsilon e(t)^2 - \frac{4h\rho_3}{r}\varepsilon\left\|e(t)\right\|^2 \quad (12.121)$$

Just rearranging one obtains

$$\text{TDLF} = -\left[\frac{4h\rho_3}{r}\varepsilon - \frac{1}{2s_u^2 q_u}\right]\left\|e(t)\right\|^2 - \left[\frac{2h^2}{r} - \frac{2(\rho_1\rho_2)}{s_u}\varepsilon\right]\left\|e(t)\right\|^2 \quad (12.122)$$

$$\text{TDLF} = -\left\{k = \varepsilon > \frac{r}{8s_u^2 q_u h\rho_3}; k = \varepsilon < \frac{h^2 s_u}{r\rho_{12}}\right\}\left\|e(t)\right\|^2 \quad (12.123)$$

If the condition in eqn. (12.123) is satisfied for the chosen constants, then the observer's error dynamics would converge and the proposed IEHI-gain based nonlinear observer would be asymptotically stable.

12.5.6 Performance Evaluation

The IEHI-based observer for nonlinear dynamic system is evaluated using numerical simulation carried out in MATLAB. The data are generated for a period of 20 sec with a sampling interval of 0.01 sec. The nonlinear dynamic system used for the purpose is given by

$$x_1(t) = -0.1x_1(t) + 0.3x_1(t)x_2(t)$$

$$x_2(t) = 10x_2(t) - 0.7x_1(t)x_2(t) - 0.1x_2^2(t)$$

$$y(t) = x_2(t) \quad (12.124)$$

The IEHI-based nonlinear observer is given as

$$\hat{x}_1(t) = -0.1\hat{x}_1(t) + 0.3\hat{x}_1(t)\hat{x}_2(t) + L_1(y(t) - \hat{x}_2(t))$$

$$\hat{x}_2(t) = 10\hat{x}_2'(t) - 0.7\hat{x}_1(t)\hat{x}_2(t) - 0.1\hat{x}_2^2(t)L_2(y(t) - \hat{x}_2(t)) \quad (12.125)$$

The initial conditions of the states used for the simulation and the observer are, respectively, $x(0) = [15 \quad 2]'$ and $\hat{x}(0) = [10 \quad 5]'$. The RTDE is solved by using the transition matrix method. First, the Gramian matrix $S(t)$ is transformed, and then the RTDE is partitioned and reformulated as

$$a = S(t)b \quad (12.126)$$

$$b = -A^T b + \left(\frac{2}{r}H^T H\right)a \quad (12.127)$$

$$a = -\frac{1}{2}\left(\gamma^{-2}Q^{-1}\right)b + Aa \quad (12.128)$$

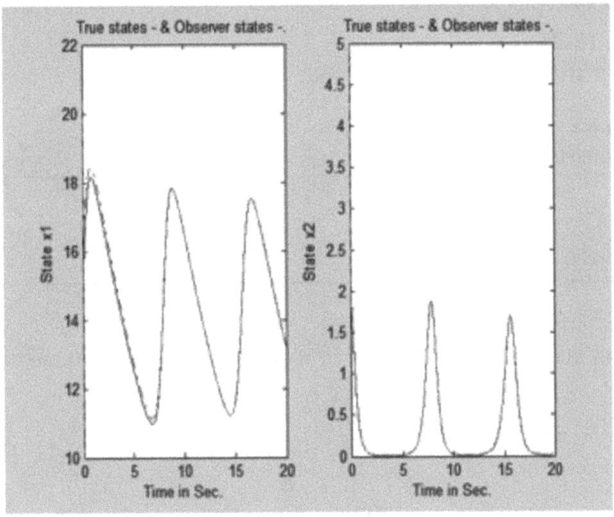

Time history (the true (-) and observer states (-.)) for the IEHI-based nonlinear observer. {left graph};

FIGURE 12.3 Time history (the true (-) and observer states (-.)) for the IEHI-based nonlinear observer (left graph).

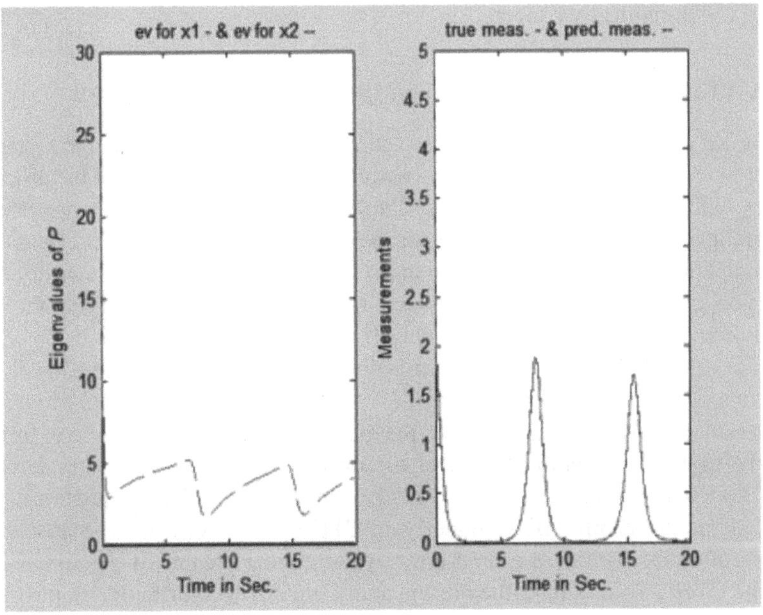

Eigenvalues of S(t) for states: x1 (-), & x2 (--);True (-), predicted meas. (--) {right graph}.

FIGURE 12.4 Eigenvalues of $S(t)$ for states: $x1$ (-), and $x2$ (--); true (-), predicted meas. (--) (right graph).

TABLE 12.2
PFEs for the IEHI-Based Observer

PFEs: for r = 0.001; Q = [0.9 0;0 0.85]	γ^2 = 0.04	γ^2 = 0.09	γ^2 = 0.25	γ^2 = 0.49
State x_1	2.2517	0.1891	1.1738	1.1065
State x_2	1.6233	1.3955	1.4803	1.5329
Measurements	1.6016	1.3816	1.4764	1.5328

The required Jacobian matrix, $A(t)$ to be used in RTDE is represented in the discrete form as

$$A(k)=\begin{bmatrix} -0.1+0.3x_2(k-1) & 0.3x_1(k-1) \\ -0.7x_2(k-1) & 10-0.7x_1(k-1)-0.2x_2(k-1) \end{bmatrix} \tag{12.129}$$

The performance of the IEHI-based nonlinear observer is illustrated in Figure 12.3 for two states x_1 and $x_{2.}$ Figure 12.4 shows the convergence of the eigenvalues of the matrix $S(t)$ (left plot), and the true and predicted measurements (right plot). The time history match, Figure 12.3, is found to be very good.

The performance of the IEHI-based observer has been found to be very satisfactory, as is also supported by the PFEs 00×norm (errors)/norm (true signal), Table 12.2. Figure 12.4 clearly shows that the IEHI-based nonlinear observer is asymptotically stable, because the eigenvalues of the Gramian matrix asymptotically converge.

12.6 NEW CONTINUOUS-TIME HI NORM-BASED ESTIMATOR

Conventionally KF and its variants are used for estimation of states of a dynamic system from its IO data using measurements. This process requires assumptions on the statistics of the process and measurement noises. Often such information is not available. H-infinity filtering theory has provided an alternative paradigm in a deterministic framework for the estimator of the states of a dynamic system with minimum assumptions [7]. Another alternative in the deterministic domain is the theory of observers.

In this section, first the HI-based *a posteriori* filter is briefly described. Then, this discrete-time filter (HIPF) is converted into a continuous-time HI-based filter (CTHF). Interestingly, the resulting CTHF is similar to the HIPF with some additional terms. The continuous-time development offers good flexibility to study analytical results.

Then, an observer is presented that utilizes the gain and the state error Gramian from this CTHF. Using LEF, the asymptotic stability result of the observer's error dynamics is derived. The performance of the CTHF is evaluated using MATLAB-based numerical simulations. These results establish the asymptotic stability of the underlying CTHF. The latter is true because the asymptotic convergence result is based on the observer error dynamics that use the CTHF gain and the Gramian matrix from the CTHF. The result of the present study is a novel contribution in the literature of HI theory of filters and the observers. These results establish the local stability of the underlying CTHF.

12.6.1 H-Infinity *A Posteriori* Filter

Considering a linear discrete-time system

$$x(k+1)=\phi x(k)+Gw(k) \tag{12.130}$$

$$z(k)=Hx(k)+v(k) \tag{12.131}$$

The HIPF discrete-time algorithm is given here. First, the (covariance) state error Gramian (SEG) propagation is given as

$$P(k+1) = \phi P(k)\phi^T + \phi P(k-1)\phi^T + Q - \phi P(k)\begin{bmatrix} H^T & L^T \end{bmatrix} R_e^{-1} \begin{bmatrix} H \\ L \end{bmatrix} P(k)\phi^T \quad (12.132)$$

The composite measurement (covariance) Gramian matrix is given as

$$R_e = \begin{bmatrix} I & 0 \\ 0 & -\gamma^2 I \end{bmatrix} + \begin{bmatrix} H \\ L \end{bmatrix} P(k)\begin{bmatrix} H^T & L^T \end{bmatrix} \quad (12.133)$$

Here, γ is a factor that specifies bound on the gain: from the input energies which arise because of the disturbances ($w(.)$, and $v(.)$), and the input error in the state initial condition to the output state error energy (this gain is HI norm). The HIPF filter gain is given as

$$K = P(k+1)H^T\left(HP(k+1)H^T + I\right)^{-1} \quad (12.134)$$

The measurements-data update of the state estimate is given as

$$\hat{x}(k+1) = \tilde{x}(k) + K\left(z(k+1) - H\tilde{x}(k)\right) \quad (12.135)$$

The previous state estimate is obtained as $\tilde{x}(k) = \phi\hat{x}(k)$ (12.136)

12.6.2 Derivation of the Continuous-Time HI Filter

A continuous-time linear dynamic system is given as

$$\dot{x}(t) = Ax(t) + w(t) \quad (12.137)$$

$$z(t) = Hx(t) + v(t) \quad (12.138)$$

The main idea here is to convert the HIPF discrete-time filter to a continuous-time filter, and hence, we utilize the following substitutions, as a first-order approximation:

$$\phi(k) = I + A\Delta t \quad (12.139)$$

$$Q(k) = GQG^T\Delta t; \quad \text{and } R(k) = \frac{R}{\Delta t} \quad (12.140)$$

The approach is to use the difference equations of the HIPF and then observe their behaviour as the discrete-time step $\Delta t \to 0$. First, the HIPF gain is handled as follows:

$$\frac{1}{\Delta t}K(k) = \frac{1}{\Delta t}P(k+1)H^T\left(HP(k+1)H^T + I\right)^{-1} \quad (12.141)$$

$$\frac{1}{\Delta t}K(k) = P(k+1)H^T\left(\Delta t HP(k+1)H^T + I\Delta t\right)^{-1}$$

$$\frac{1}{\Delta t}K(k)=P(k+1)H^T\left(\Delta tHP(k+1)H^T+R\right)^{-1};R=I\Delta t \tag{12.142}$$

In the limit as the $\Delta t \to 0$, the continuous-time HI gain is obtained as

$$\lim_{\Delta t\to 0}\frac{1}{\Delta t}K(k)=\lim_{\Delta t\to 0}P(k+1)H^T\left(\Delta tHP(k+1)H^T+R\right)^{-1} \tag{12.143}$$

$$K(t)=\lim_{\Delta t\to 0}\frac{1}{\Delta t}K(k)=P(t)H^TR^{-1};\text{wherein } R(t)\text{is given as } R=I\Delta t; \tag{12.144}$$

Next, eqn. (12.132) is handled:

$$\begin{aligned}P(k+1)&=(I+A\Delta t)P(k)(I+A\Delta t)^T+GQG^T\Delta t\\&\quad-(I+A\Delta t)P(k)\begin{bmatrix}H^T & L^T\end{bmatrix}R_e^{-1}\begin{bmatrix}H\\L\end{bmatrix}P(k)(I+A\Delta t)^T\end{aligned} \tag{12.145}$$

Simplifying eqn. (12.145) at the first stage one gets, neglecting the higher product term,

$$\begin{aligned}P(k+1)&=P(k)+AP(k)\Delta t+P(k)A^T\Delta t+GQG^T\Delta t\\&\quad-(I+A\Delta t)\begin{bmatrix}P(k)H^T & P(k)L^T\end{bmatrix}R_e^{-1}\begin{bmatrix}HP(k)\\LP(k)\end{bmatrix}(I+A\Delta t)^T\end{aligned} \tag{12.146}$$

Now, the gains are specified as

$$K=PH^TR^{-1};HP=R^TK^T \tag{12.147}$$

$$K_l=PL^T;LP=K_l^T \tag{12.148}$$

Substituting the gain formulas in the SEG equation one obtains

$$\begin{aligned}P(k+1)&=P(k)+AP(k)\Delta t+P(k)A^T\Delta t+GQG^T\Delta t-(I+A\Delta t)\\&\quad\begin{bmatrix}KR & K_l\end{bmatrix}R_e^{-1}\begin{bmatrix}R^TK^T\\K_l^T\end{bmatrix}(I+A\Delta t)^T\end{aligned} \tag{12.149}$$

Next, the central part of the last term is written as follows for simplicity:

$$P_{cd}=\begin{bmatrix}KR & K_l\end{bmatrix}R_e^{-1}\begin{bmatrix}R^TK^T\\K_l^T\end{bmatrix} \tag{12.150}$$

Further simplifying one obtains, neglecting the higher-product term,

$$P(k+1)=P(k)+AP(k)\Delta t+P(k)A^T\Delta t+GQG^T\Delta t-(I+A\Delta t)P_{cd}(I+A\Delta t)^T \tag{12.151}$$

$$P(k+1)=P(k)+AP(k)\Delta t+P(k)A^T\Delta t+GQG^T\Delta t-P_{cd}-AP_{cd}\Delta t-P_{cd}A^T\Delta t \tag{12.152}$$

Now, forming the differential from eqn. (12.152), and dividing both the sides by Δt one obtains

$$\frac{P(k+1)-P(k)}{\Delta t}=\frac{1}{\Delta t}\Big[\{AP(k)+P(k)A^T\}\Delta t+GQG^T\Delta t-P_{cd} \\ -\{AP_{cd}+P_{cd}A^T\}\Delta t\Big] \tag{12.153}$$

Then, applying the limit $\Delta t\to 0$, one obtains,

$$\lim_{\Delta t\to 0}\frac{P(k+1)-P(k)}{\Delta t}=\lim_{\Delta t\to 0}\frac{1}{\Delta t}\Big[\{AP(k)+P(k)A^T\}\Delta t+GQG^T\Delta t-P_{cd} \\ -\{AP_{cd}+P_{cd}A^T\}\Delta t\Big] \tag{12.154}$$

$$\dot{P}(t)=AP(k)+P(k)A^T+GQG^T-\lim_{\Delta t\to 0}\frac{1}{\Delta t}P_{cd}-AP_{cd}-P_{cd}A^T \tag{12.155}$$

Next, the limit of the fourth term is evaluated as follows:

$$\lim_{\Delta t\to 0}\frac{1}{\Delta t}P_{cd}=\lim_{\Delta t\to 0}\frac{1}{\Delta t}\left\{\begin{bmatrix}KR & K_l\end{bmatrix}R_e^{-1}\begin{bmatrix}R^TK^T\\ K_l^T\end{bmatrix}^T\right\} \tag{12.156}$$

$$\lim_{\Delta t\to 0}\frac{1}{\Delta t}P_{cd}=\begin{bmatrix}KR & K_l\end{bmatrix}\left\{\lim_{\Delta t\to 0}\frac{1}{\Delta t}(R_e)^{-1}\right\}\begin{bmatrix}R^TK^T\\ K_l^T\end{bmatrix}^T \tag{12.157}$$

The limit of the bracketed term is evaluated as follows:

$$\lim_{\Delta t\to 0}\frac{1}{\Delta t}(R_e)^{-1}=\lim_{\Delta t\to 0}\frac{1}{\Delta t}\left\{\begin{bmatrix}I & 0\\ 0 & -\gamma^2 I\end{bmatrix}+\begin{bmatrix}H\\ L\end{bmatrix}P(k)\begin{bmatrix}H^T & L^T\end{bmatrix}\right\}^{-1} \tag{12.158}$$

$$=\lim_{\Delta t\to 0}\left\{\begin{bmatrix}I\Delta t & 0\\ 0 & -\gamma^2 I\Delta t\end{bmatrix}+\begin{bmatrix}H\\ L\end{bmatrix}P(k)\begin{bmatrix}H^T & L^T\end{bmatrix}\Delta t\right\}^{-1} \tag{12.159}$$

Evaluating the limit in eqn. (12.159), one obtains the following:

$$R_{ec}^{-1}=\begin{bmatrix}I\Delta t & 0\\ 0 & -\gamma^2 I\Delta t\end{bmatrix}^{-1} \tag{12.160}$$

Substituting eqn. (12.160) in eqn. (12.157), one gets

$$P_c=\begin{bmatrix}KR & K_l\end{bmatrix}(R_{ec})^{-1}\begin{bmatrix}R^TK^T\\ K_l^T\end{bmatrix}^T \tag{12.161}$$

Substituting eqn. (12.161) in eqn. (12.155) the final equation for the time propagation of the SEG matrix is obtained as

$$\dot{P}(t) = AP(t) + P(t)A^T + GQG^T - P_c - AP_{cd} - P_{cd}A^T \tag{12.162}$$

In a similar manner, one can obtain the continuous-time equations for estimation of the states as follows:

$$\hat{x}(k+1) = \phi\hat{x}(k) + K\left(z(k+1) - H\phi\hat{x}(k)\right) \tag{12.163}$$

$$\dot{\hat{x}} = A\hat{x} + PH^T R^{-1}\left(z - H\hat{x}\right) \tag{12.164}$$

Considering appropriate initial conditions: $x(0), P(0)$, etc; the CTHF gain is given as

$$K(t) = P(t)H^T R^{-1}; R = I\Delta t; K_l(t) = P(t)L^T \tag{12.165}$$

The continuous-time composite Gramian matrix is given as

$$P_c = \begin{bmatrix} KR & K_l \end{bmatrix} (R_{ec})^{-1} \begin{bmatrix} R^T K^T \\ K_l^T \end{bmatrix};$$

$$R_{ec}^{-1} = \begin{bmatrix} I\Delta t & 0 \\ 0 & -\gamma^2 I\Delta t \end{bmatrix}^{-1} \tag{12.166}$$

The continuous-time composite Gramian matrix with discrete measurement variance Gramian is given as

$$P_{cd} = \begin{bmatrix} KR & K_l \end{bmatrix} R_e^{-1} \begin{bmatrix} R^T K^T \\ K_l^T \end{bmatrix}^T;$$

$$R_e = \begin{bmatrix} I & 0 \\ 0 & -\gamma^2 I \end{bmatrix} + \begin{bmatrix} H \\ L \end{bmatrix} P(k) \begin{bmatrix} H^T & L^T \end{bmatrix} \tag{12.167}$$

It is to be noted that in the equation of $P_{cd}(.)$,eqn. (12.167), the discrete measurement 'variance' Gramian matrix continues (as was in the discrete-time HIPF); however, in matrix $P_c(.)$, eqn. (12.166), the continuous-time Gramian is used. The SEG can be obtained by solving the following RTDE:

$$\dot{P}(t) = AP(t) + P(t)A^T + GQG^T - P_c - AP_{cd} - P_{cd}A^T \tag{12.168}$$

$$\dot{\hat{x}} = A\hat{x} + PH^T R^{-1}\left(z - H\hat{x}\right);$$

$$\dot{\hat{x}} = A\hat{x} + K(t)\left(z - H\hat{x}\right) \tag{12.169}$$

12.6.3 Observer Based on the CTHF

A continuous-time observer that uses the gain from the newly derived CTHF is proposed, and the asymptotic stability result is derived. A nonlinear continuous-time system is taken as

$$\dot{x} = f(x,t) \tag{12.170}$$

$$z = Hx \tag{12.171}$$

A nonlinear observer is specified as

$$\dot{\hat{x}}(t) = f(\hat{x}(t),t) + L_o(t)(y(t) - \hat{y}(t))$$

$$\hat{y}(t) = H\hat{x}(t) \tag{12.172}$$

The $L_o(t)$ is the observer gain matrix of appropriate dimension, and is taken from the CTHF as

$$L_o(t) = P(t)H^T R^{-1} \tag{12.173}$$

The matrix $P(t)$ is obtained as the solution of the observer RTDE which is also directly based on the CTHF:

$$\dot{P}(t) = AP + PA^T + GQG^T - P_c - AP_{cd} - P_{cd}A^T \tag{12.174}$$

The required Jacobian is obtained as

$$A(t) = \frac{\delta f(.)}{\delta \hat{x}(t)} \tag{12.175}$$

The following observer error dynamics are obtained by the usual process

$$\dot{e}(t) = A(t)e(t) - L_o(t)H(x(t) - \hat{x}(t)) + \phi(.)$$

$$= A(t)e(t) - L_o(t)He(t) + \phi(.) \tag{12.176}$$

The new nonlinear function is

$$\phi(.) = -A(t)e(t) + f(x,t) - f(\hat{x},t) \tag{12.177}$$

The short forms for f and $\phi(.)$ can be used for simplicity,

$$\phi(.) = \phi(x(t),\hat{x}(t),t);\, f(x,t) = f(x(t),t) \tag{12.178}$$

The state errors are given as

$$e(t) = x(t) - \hat{x}(t) \tag{12.179}$$

12.6.4 Asymptotic Analysis of the Observer

The following conditions need to be specified:

1. The solution of the matrix RTDE should be bounded

$$p_l I \leq P(t) \leq p_u I \tag{12.180}$$

Here, $p_l, p_u > 0$ are constants (for, $P(t)$ is theoretically positive definite and symmetrical matrix), and are the lower and upper bounds, respectively.

2. The nonlinearity in the error dynamics should be bounded, because the nonlinear function should be bounded,

$$\|\phi(.)\| \leq \rho \|x(t) - \hat{x}(t)\|^2 \tag{12.181}$$

Then, the nonlinear observer error dynamics would be asymptotically stable, if basically conditions 1 and 2 are satisfied; of course some more conditions are required, that would evolve as the derivation proceeds further. First, the LE functional is considered to establish the asymptotic stability of the error dynamics

$$V(t) = e^T(t)Y(t)e(t) \tag{12.181}$$

One can easily see that the LE functional is positive definite because of the condition 1:

$$\frac{1}{p_u}\|e(t)\|^2 \leq e^T(t)Y(t)e(t) \leq \frac{1}{p_l}\|e(t)\|^2 \tag{12.182}$$

The time derivative of the LE functional, eqn. (12.181), under the constraints governed by the state dynamics, gain formula of CTHF, and the SEG equation should be negative definite for the asymptotic stability of the observer error dynamics. This time derivative is obtained as

$$\dot{V}(t) = e^T(t)\dot{Y}(t)e(t) + e^T(t)Y(t)\dot{e}(t) + \dot{e}^T(t)Y(t)e(t) \tag{12.183}$$

Since, $\dot{Y} = -Y\dot{P}Y$, substituting error expression in eqn. (12.183), the following expressions are obtained

$$\begin{aligned}\dot{V}(t) = &-e^T(t)Y(t)\left[AP + PA^T + GQG^T - P_c - AP_{cd} - P_{cd}A^T\right]Y(t)e(t)\\ &+ e^T(t)Y(t)\dot{e}(t) + \dot{e}^T(t)Y(t)e(t)\end{aligned} \tag{12.184}$$

$$\dot{e}(t) = A(t)e(t) - L_o(t)He(t) + \phi(.) \tag{12.185}$$

$$\begin{aligned}\dot{V}(t) = &-e^TY(t)\left[AP + PA^T + GQG^T - P_c - AP_{cd} - P_{cd}A^T\right]Y(t)e(t) + e^TY(t)\\ &\left\{Ae(t) - L_o(t)He(t) + \phi(.)\right\} + \left\{Ae(t) - L_o(t)He(t) + \phi(.)\right\}^T Y(t)e(t)\end{aligned}$$

$$\begin{aligned}\dot{V}(t) = &-e^TY(t)\left[AP + PA^T + GQG^T - P_c - AP_{cd} - P_{cd}A^T\right]Y(t)e(t) + e^TY(t)\\ &\left\{Ae(t) - L_oHe(t) + \phi(.)\right\} + \left\{e^T(t)A^T - e^T(t)H^TL_o^T + \phi^T(.)\right\}Y(t)e(t)\end{aligned} \tag{12.186}$$

Simplifying further and cancelling some common terms (without any approximations), one obtains

$$\begin{aligned}\dot{V}(t) &= -e^T(t)YGQG^TYe(t) + e^T(t)YP_cYe(t) + e^T(t)YAP_{cd}Ye(t)\\ &+ e^T(t)YP_{cd}A^TYe(t) + e^T(t)Y\{Ae(t) - L_oHe(t) + \phi(.)\}\\ &+ \{e^T(t)A^T - e^T(t)H^TL_o^T + \phi^T(.)\}Ye(t)\end{aligned} \tag{12.187}$$

$$\begin{aligned}\dot{V}(t) &= -e^T(t)YGQG^TYe(t) + e^T(t)YP_cYe(t) + e^T(t)YAP_{cd}Ye(t)\\ &+ e^T(t)YP_{cd}A^TYe(t) + e^T(t)YAe(t) - e^T(t)YL_oHe(t)\\ &+ e^T(t)Y\phi(.) + e^T(t)A^TYe(t) - e^T(t)H^TL_o^TYe(t) + \phi^T(.)Ye(t)\end{aligned} \tag{12.188}$$

$$\begin{aligned}\dot{V}(t) &= -e^T(t)YGQG^TYe(t) + e^T(t)YP_cYe(t) + e^T(t)YAP_{cd}Ye(t)\\ &+ e^T(t)YP_{cd}A^TYe(t) + e^T(t)YAe(t) - e^T(t)H^TR^{-1}He(t)\\ &+ e^T(t)Y\phi(.) + e^T(t)A^TYe(t) - e^T(t)H^TR^{-1}He(t) + \phi^T(.)Ye(t)\end{aligned} \tag{12.189}$$

Next, the following additional bounds are specified:

(a) It is assumed that,

$$\|R^{-1}\| \le 1/r; \|H^TH\| \le h^2,$$

(r and h are positive constants);

$$\|e(t)\|^2 \le \varepsilon^2 \tag{12.190}$$

(b) $\|P_c\| \le b; \|AP_{cd}\| = \|P_{cd}A^T\| \le c; \|A\| \le a$ (12.191)

After substituting these bounds one gets the following:

$$\dot{V}(t) = -\frac{q_l}{p_u^2}\|e(t)\|^2 + \frac{b}{p_u^2}\|e(t)\|^2 + \frac{2c}{p_u^2}e(t)^2 + \frac{2a}{p_u}\|e(t)\|^2 + \frac{2\rho\varepsilon}{p_u}\|e(t)\|^2 - \frac{2h^2}{r}\|e(t)\|^2 \tag{12.192}$$

Here, q_l is the smallest (positive) eigenvalue of the matrix GQG^T, that is positive definite and symmetric. Combining certain terms in eqn. (12.192), one gets

$$\dot{V}(t) = -\left\{\frac{q_l}{p_u^2} - \frac{b}{p_u^2} - \frac{2c}{p_u^2}\right\}\|e(t)\|^2 - \left\{\frac{2h^2}{r} - \frac{2a}{p_u} - \frac{2\rho\varepsilon}{p_u}\right\}\|e(t)\|^2 \tag{12.193}$$

$$\dot{V}(t) = -\left\{\left[\frac{q_l}{p_u^2} - \frac{b}{p_u^2} - \frac{2c}{p_u^2}\right];\left[\frac{2h^2}{r} - \frac{2a}{p_u} - \frac{2\rho\varepsilon}{p_u}\right]\right\}\|e(t)\|^2 \tag{12.194}$$

For $\|e(t)\| \le \varepsilon = k$, the following condition is obtained:

$$\dot{V}(t) = -\left\{q_l > (b+2c); k < \frac{h^2p_u - ar}{r\rho}\right\}\|e(t)\|^2 \tag{12.195}$$

Since, various constants and bounds (all defined earlier) appearing in the {.;.} of eqn. (12.195) are positive, then for the specified conditions on q_l and k it is evident that the time derivative of the LEF is negative definite. Hence, the error dynamics of the nonlinear observer with the gain and SEG matrix from the continuous-time HI filter, CTHF are asymptotically stable; and interestingly this result also establishes that the continuous-time HI filtering algorithm would be asymptotically stable. The latter is true because the asymptotic convergence result is based on the observer error dynamics that use the CTHF gain and the SEG matrix from the CTHF, which is a novel concept.

12.6.5 Performance Evaluation of the CTHF

The CTHF algorithm is implemented in MATLAB and validated using simulated data. The state-space model considered is given as

$$\dot{x}(t) = Ax(t) + Bu + w \tag{12.196}$$

The system matrices are

$$A = \begin{bmatrix} 0.06 & -2.0 \\ 0.8 & -0.8 \end{bmatrix}; B = \begin{bmatrix} -0.6 \\ 1.5 \end{bmatrix} \tag{12.197}$$

The measurement model is given as

$$z(t) = Hx(t) + v \tag{12.198}$$

The simulated data for 10 sec are generated using the models, eqn. (12.197), and eqn. (12.198), with a sampling interval of 0.05 sec. Appropriate additive process noises in states and measurement are used. Control input signal $u(.)$ is a doublet that excites the modes of the system, eqn. (12.196). The % fit errors (PFE metrics) of the measurement residuals, and the states erroes are computed as: PFE (PSE) = 100×cov(of measurements or *state errors*)/cov(of true or actual meas. or *states*). If the covariance measure is found to be ill-conditioned, then 'norm' can be used instead. The H-infinity norm is evalauted as

$$\text{HI}(\text{norm}) = \frac{\sum_{K=0}^{N}\left(\hat{x}(k) - x(k)^T\right)\left(\hat{x}(k) - x(k)\right)}{\left(\hat{x}_0 - x^T\right)P_0\left(\hat{x}_0 - x\right) + \sum_{k=0}^{N} w^T(k)w(k) + \sum_{k=0}^{N} v^T(k)v(k)} \tag{12.199}$$

The dynamic equations for simulation as well as for the filtering algorithm are solved by the RK4 method of integration. The RTDE is solved by using the transition matrix method. The performance metrics are given in Table 12.3; from which one can see that the CTHF algorithm performs very satisfactorily.

TABLE 12.3
Performance Metrics for the CTHF Algorithm

HI norm	PFE of residuals
0.0269	4.8968
PSE of state x^1	**PSE of state x^2**
0.2038	0.1213

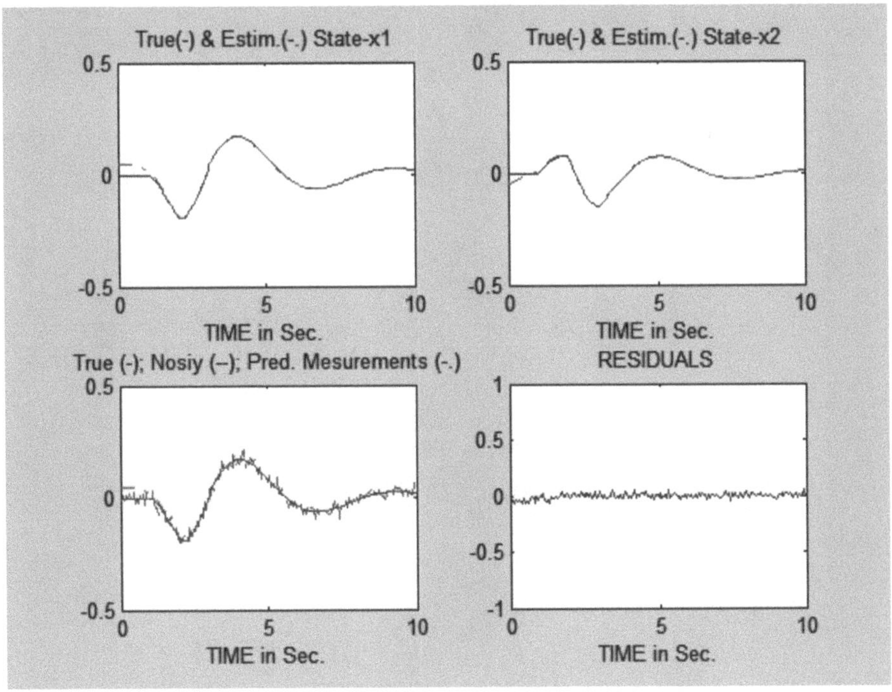

FIGURE 12.5 Time histories match for CTHF: states, measurements; residuals.

Figures 12.5 and 12.6 show satisfactory trajectory matching; and the state errors are within their theoretical bounds.

Figure 12.7 depicts the time history of the norm of $P(t)$, which corroborates the asymptotic convergence of the observer as well as the continuous-time H-infinity filtering algorithm.

H-infinity *a posteriori* filter has been converted to a continuous-time H-infinity filter. Then, an observer is presented which uses the gain and the SEG from the continuous-time HI filter. Asymptotic stability result of the observer's error dynamics is derived using the LEF. The performance of the CTHF is evaluated using numerical simulations carried out in MATLAB. These results establish the stability of the underlying CTHF. This type of result is novel in the literature on HI theory of filters and the observers. Such studies have utilization in applications of the CTHF algorithm and observers in communication systems, wireless sensor networks, mechanical/aerospace engineering (aircraft trajectory and parameter estimation and target tracking), and robotics.

12.7 DATA FUSION WITH IE- AND HI-BASED ESTIMATORS

In this section, sensor data fusion [8], which utilizes model error estimators are studied. These estimators are based on the principle of invariant embedding and the H-infinity concept; for both continuous- and discrete-time nonlinear systems. The model error (ME) estimators are used in state-vector fusion process *as if*, it combines the processed information from two individual 'fictitious' sensor channels. Actually, two deficient models are used in the first place, and the ME estimators are used which give corrected states from the measurements. The concept of DF is employed to enhance the predictive capability of the estimated state trajectory. The presented approach is novel in utilization of IE/HI-based model error estimators in the DF process.

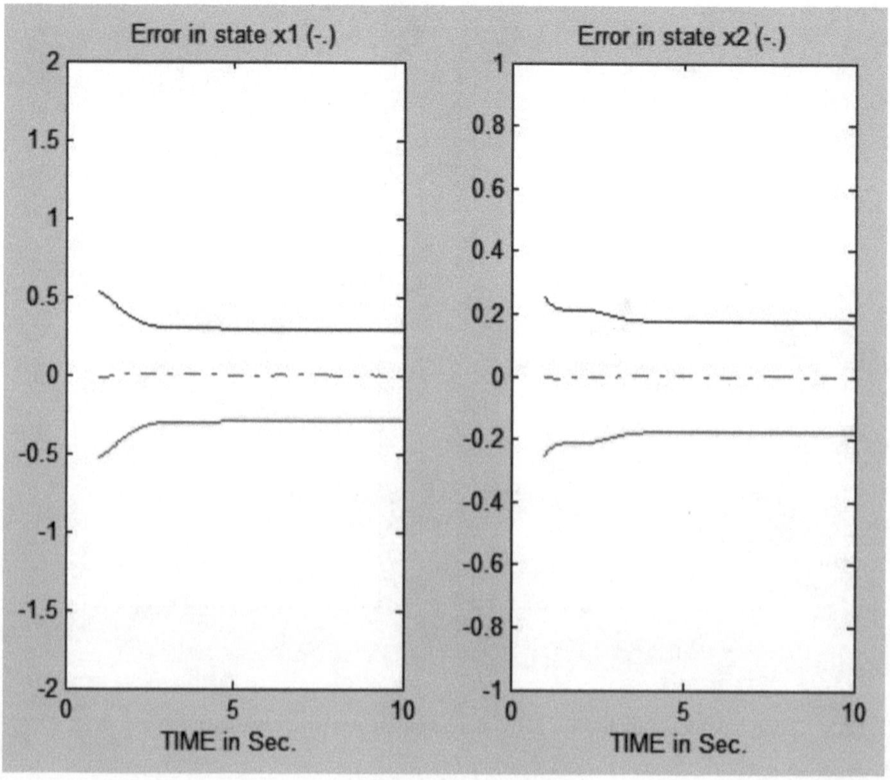

FIGURE 12.6 State errors with their bounds for CTHF.

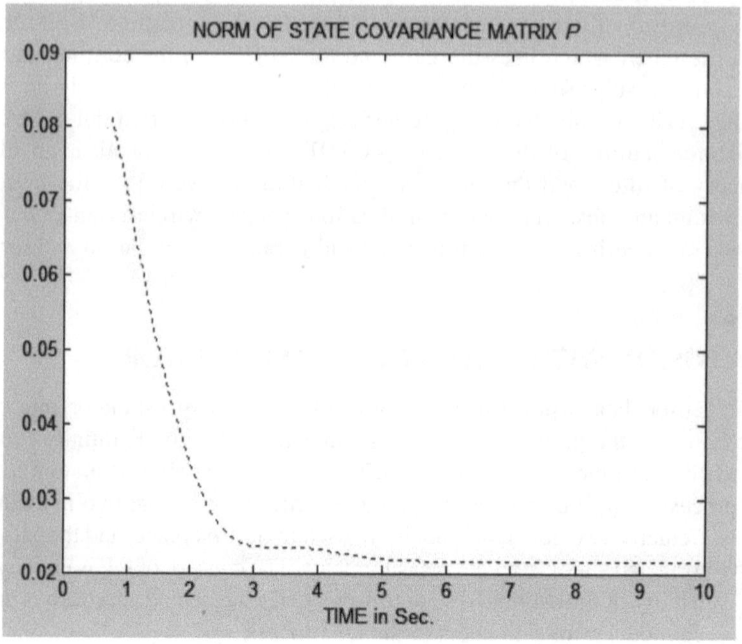

FIGURE 12.7 Norm of $P(.)$ showing the convergence of CTHF.

12.7.1 Model Error Estimators

This section deals with the SDF scheme which utilizes ME estimators. The estimators used are based on the method of IE and HI frameworks. The DF is evaluated for both continuous- and discrete-time systems. The performances of the ME estimator-based DF scheme and of the continuous observer are evaluated by simulation carried out in MATLAB. The presented results validate the theoretical asymptotic behaviour of the nonlinear observer, as well as the convergence properties of the data-fusion scheme, since both are based on IE/HI theory. This type of validation study is a novel feature.

First, the main aim is to calculate the ME from the noisy measurements which are available from a given nonlinear dynamic system; it is presumed that the empirical data are obtained from an actual nonlinear system [4]. Then, one fits only a primarily known model that might be deficient, because the true model is not available. Let us consider expression for nonlinear system as

$$\dot{x} = f\left(x(t), u(t), t\right) + d(t) \tag{12.200}$$

Here, $d(.)$ is the 'unmodelled' deterministic disturbance. In control theory, term $d(.)$ represents an input, $u(t)$, which is determined using an optimization technique by minimizing the cost function

$$J = \sum_{k=1}^{N}\left[z(k) - h\left(\hat{x}(k), k\right)\right]^T R^{-1}\left[z(k) - h\left(\hat{x}(k), k\right)\right] + \int_{t_0}^{t_f} d^T(t) Q d(t)\, dt \tag{12.201}$$

From the theory developed in Sections 12.3 and 12.4, we have the following TPBVP:

$$\dot{x} = \phi\left(x(t), \lambda(t), t\right) \tag{12.202}$$

$$\dot{\lambda} = \psi\left(x(t), \lambda(t), t\right) \tag{12.203}$$

It is seen that the dependencies for ϕ and ψ on $x(t)$ and $\lambda(t)$ arise from the form of eqn. (12.200), and eqn. (12.201): a general TPBVP with boundary conditions as $\lambda(0) = a$ and $\lambda(t_f) = b$.

(a) A continuous-time nonlinear dynamic system is represented by

$$\dot{x} = f\left(x(t), t\right) + d(t) \tag{12.204}$$

$$z(t) = Hx(t) + v(t) \tag{12.205}$$

The ME estimator based on the IE method is given as

$$\dot{\hat{x}} = f\left(x(t), t\right) + 2S(t) H^T R^{-1}\left(z(t) - Hx(t)\right) \tag{12.206}$$

$$S(t) = S(t) f_{\hat{x}}^T + f_{\hat{x}} S(t) - 2S(t) H^T R^{-1} H S(t) + \frac{1}{2} Q^{-1} \tag{12.207}$$

Explicitly the model error is given as

$$\hat{d}(t) = 2S(t) H^T R^{-1}(z(t) - Hx(t) \tag{12.208}$$

(b) A discrete-time nonlinear dynamic system is represented by

$$x(k+1) = g(x(k),k) \tag{12.209}$$

$$z(k) = h(x(k),k) \tag{12.210}$$

Here, 'g' and 'h' are the vector-valued function. These equations can also be written as

$$X(k+1) = f(x(k),k) + d(k) \tag{12.211}$$

$$Z(k) = h(x(k),k) + v(k) \tag{12.212}$$

Here, 'f' is deficient model. The discrete-time IE recursive algorithm is given as

$$\hat{x}(k+1) = f_{\hat{x}}(\hat{x}(k),k) + 2S(k+1)H^T(k+1)R^{-1}\left[z(k+1) - h(\hat{x}(k+1),k+1)\right] \tag{12.213}$$

$$S(k+1) = \left[I + 2P(k+1)H^T(k+1)R^{-1}H(k+1)\right]^{-1} P(k+1) \tag{12.214}$$

$$P(k+1) = f_{\hat{x}}(\hat{x}(k),k)S(k)f_{\hat{x}}^T(\hat{x}(k),k) + \frac{1}{2}Q^{-1} \tag{12.215}$$

$$\hat{d}(k) = 2S(k)H^T(k)R^{-1}\left[(z(k) - h(\hat{x}(k),k)\right] \tag{12.216}$$

(c) Next, the H∞ (HI) norm provides a measure of a worst-case system's transfer function gain given as

$$\|G\|_\infty = \sup_w |G(j\omega)| \tag{12.217}$$

The HI norms of error system matrix are given as
for a continuous system

$$\|\tilde{z}\|_2^2 = \frac{1}{2\pi}\int_{-\infty}^{\infty} \tilde{z}(j\omega)^H \tilde{z}(j\omega)d\omega = \int_{-\infty}^{\infty} \tilde{z}(t)^H \tilde{z}(t)dt \tag{12.218}$$

for a discrete system

$$\|\tilde{z}\|_2^2 = \sum_{n=-\infty}^{\infty}\int_{-\infty}^{\infty} \tilde{z}(n)^H \tilde{z}(n) = \frac{1}{2\pi j}\oint\int_{-\infty}^{\infty} \tilde{z}(e^{j\omega T})^H \tilde{z}(e^{j\omega T})\frac{de^{j\omega T}}{e^{j\omega T}} \tag{12.219}$$

The error is given as $\tilde{z} = z - \hat{z}$, and $d = \begin{bmatrix} w \\ n \end{bmatrix}$ as input disturbance. For an HI norm criterion, the transfer function, from the input disturbances to the estimate error, is $G_{\tilde{z}d}$, and it has an upper bound;

$$\left\| G_{\hat{z}d} \right\|_{\infty}^{2} < \gamma^2 \quad (12.220)$$

This criterion is also written as

$$\sup_{w} \frac{\left\| z - \hat{z} \right\|_2^2}{\left\| \begin{matrix} w \\ n \end{matrix} \right\|_2^2} < \gamma^2 \quad (12.221)$$

The model error estimator equations for a continuous-time system are

$$\dot{\hat{x}}(t) = 2H^T R^{-1} S(t)\big(z(t) - Hx(t)\big) + f\big(x(t), t\big) \quad (12.222)$$

$$\dot{S}(t) = S(t) f_{\hat{x}}^T - 2S(t) H^T R^{-1} H S(t) - \frac{1}{2}\gamma^2 Q^{-1} + f_{\hat{x}} S(t) \quad (12.223)$$

$$\hat{d}(t) = 2S(t) H^T R^{-1}\big(z(t) - Hx(t)\big) \quad (12.224)$$

Eqn. (12.223) is matrix RTDE.

The model error estimator equations for a discrete-time system are

$$\hat{x}(k+1) = f_{\hat{x}}\big(\hat{x}(k), k\big) + 2S(k+1) H^T(k+1) R^{-1}\big[z(k+1) - h(\hat{x}(k+1), k+1\big] \quad (12.225)$$

$$S(k+1) = \big[I + 2P(k+1) H^T(k+1) R^{-1} H(k+1)\big]^{-1} P(k+1) \quad (12.226)$$

$$P(k+1) = f_{\hat{x}}\big(\hat{x}(k), k\big) S(k) f_{\hat{x}}^T\big(\hat{x}(k), k\big) - \frac{1}{2}\gamma^2 Q^{-1} \quad (12.227)$$

$$\hat{d}(k) = 2S(k) H^T(k) R^{-1}\big[z(k) - h\big(\hat{x}(k), k\big)\big] \quad (12.228)$$

12.7.2 DF Scheme Utilizing ME Estimators

These ME estimators are evaluated in the SVF scheme, Figure 12.8.

The data from a system are received from/via some information channels S_1 and S_2; then, a model is postulated (that is primarily not known accurately) and hence, is a deficient model (DFM) of the system: DFM1, DFM2 (and DFM3). The IE and IEHI-based ME estimators are used to estimate the states of the system as well as the model errors. The estimated states are fused using the state-vector fusion formulae:

$$\text{Fused State}: x_f = x_1 + S_1 \times \text{inv}(S_1 + S_2) \times (x_2 - x_1) \quad (12.229)$$

$$\text{Covariance(Gramian)} = S_f = S_1 - S_1 \times \text{inv}(S_1 + S_2) \times (S_1)' \quad (12.230)$$

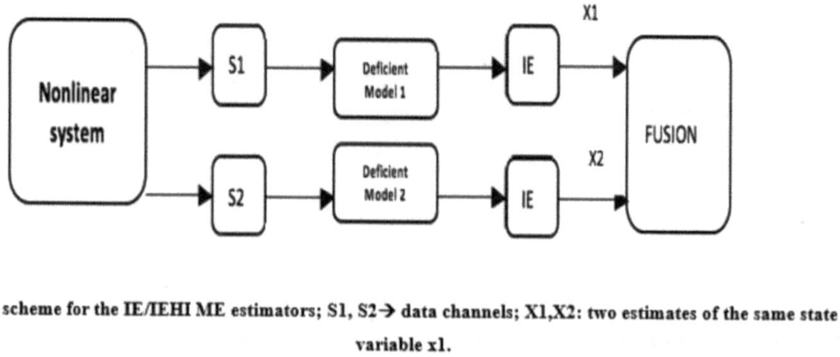

FIGURE 12.8 DF scheme for the IE/IEHI ME estimators; $S1$, $S2\rightarrow$ data channels; $X1$, $X2$: two estimates of the same state variable $x1$.

12.7.3 Performance Evaluation of the DF Scheme

(a) A nonlinear discrete-time system is considered for analysis:

$$X_1(k+1) = 0.8X_1(k) + 0.223X_2(k) + 2.5\cos(0.3k) + 0.8\sin(0.2k) - 0.05X_1^3(k) \tag{12.231}$$

$$X_2(k+1) = 0.5X_2(k) + 0.1\cos(0.4k) \tag{12.232}$$

The ME is determined by removing the terms from eqn. (12.231) in turn: (i) X_1^3, (ii) X_1, X_1^3, and (iii) X_1, X_2, X_1^3. To the ME estimated, a regression model is fitted: $d(k) = a_1X_1(k) + a_2X_1^2(k) + a_3X_1^3(k) + a_4X_2(k)$ to evaluate the parameters of DTS. 100 samples are used from eqns. (12.231) and (12.232). Even though the term X_1^2 is not present in the model, it is only added to check the robustness of the algorithm.

Table 12.4 depicts the parameter estimation compared with the true values for the three cases by using estimators in the IE/HI framework with suitable gamma values (in HI cases). The results are appealing; also, the term a_2 is estimated to be zero. Tables 12.5–12.7 show the percentage measurement residuals' fit errors, state errors and HI norm, which are reasonably low. Figures 12.9 to 12.12 depict performance graphs for the discrete-time case for the DF, from which it can be seen that the DF scheme that uses (at a time any) two DFMs (deficient models), gives very satisfactory SVF results.

The norm of the 'covariance' Gramian (SEG, $S(t)$, studied in the context of DF scheme using IE/HI estimators) is shown in a typical graph in Figure 12.11, which again corroborates the asymptotic result derived for the observer error dynamics, since the observer is based on the IE/HI-based gain and SEG $S(t)$; of course the discrete-time results would be equivalent and accurate for small sampling interval.

(b) Next, a nonlinear continuous-time system is considered,

$$\dot{X}_1(t) = 2.5\cos(t) - 0.68X_1(t) - X_2(t) - 0.0195X_2^3(t) \tag{12.233}$$

$$\dot{X}_2(t) = X_1(t) \tag{12.234}$$

The model error is determined by eliminating the terms from eqn. (12.233) in turn: (i) X_2^3, (ii) X_1, X_2, X_2^3, and utilizing the deficient models in the corresponding estimation algorithm; then,

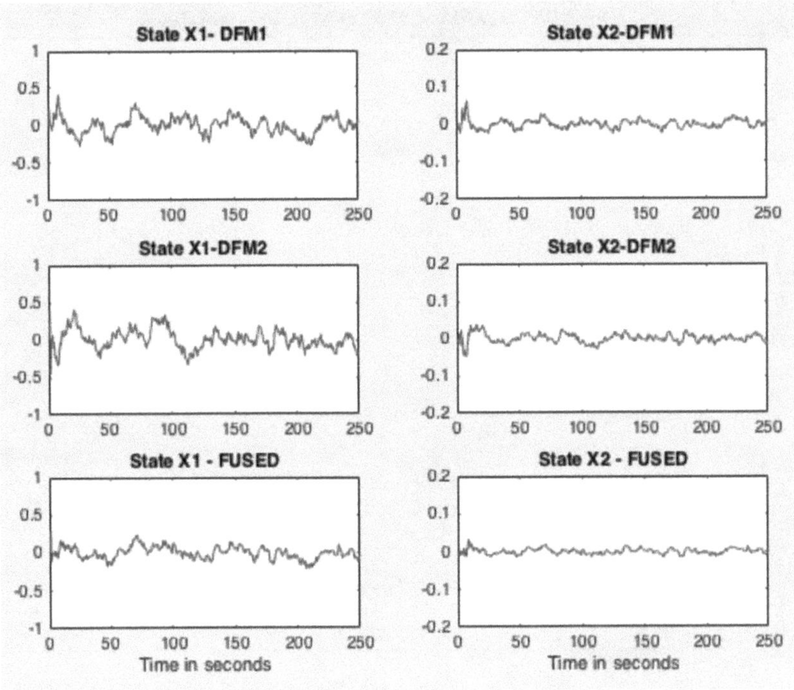

FIGURE 12.9 States $X1$ and $X2$ and their respective fused states – DTS.

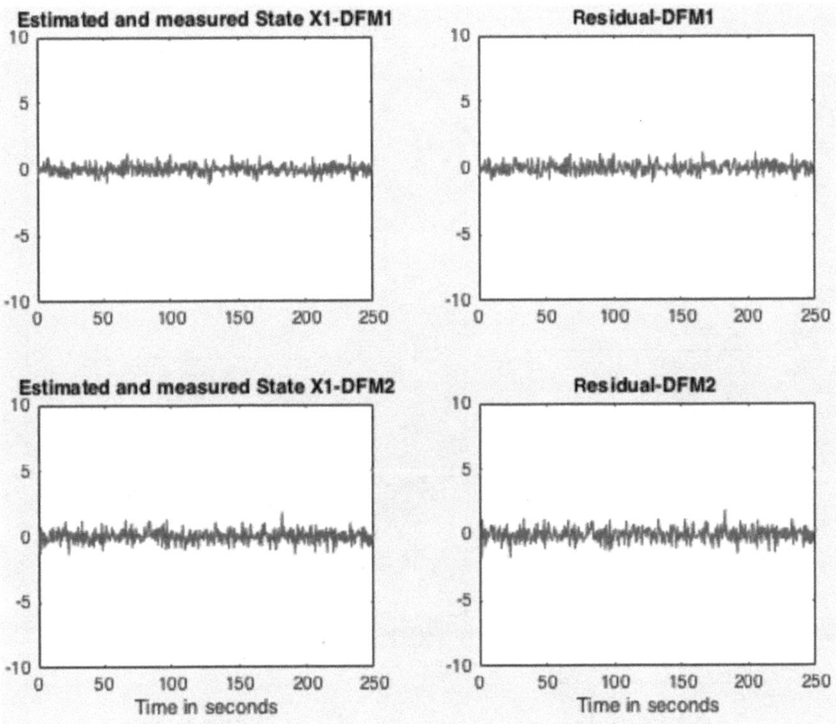

FIGURE 12.10 Estimated and measured states with residuals – DTS.

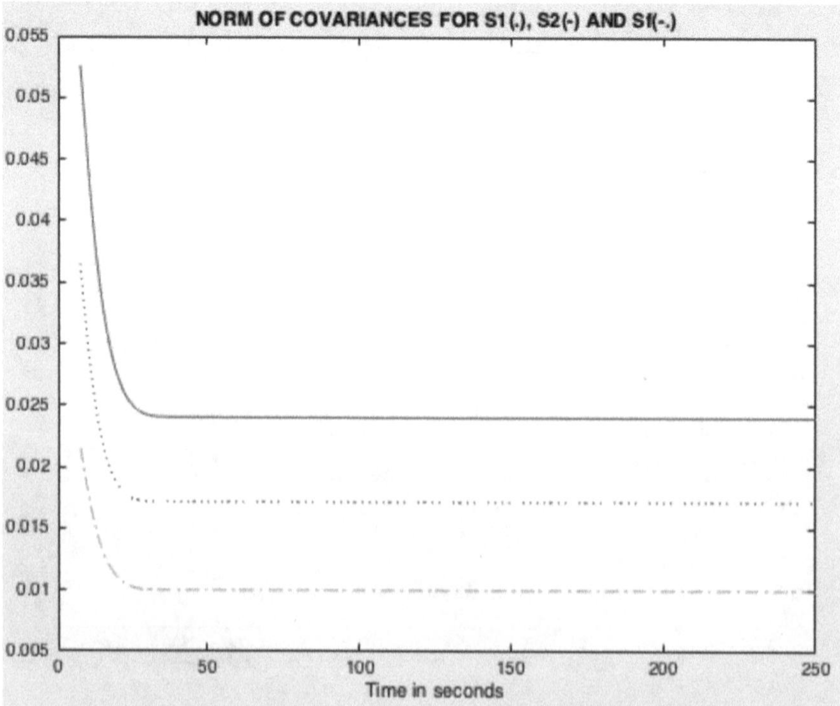

FIGURE 12.11 Norm of state error Gramians/IE method – DTS.

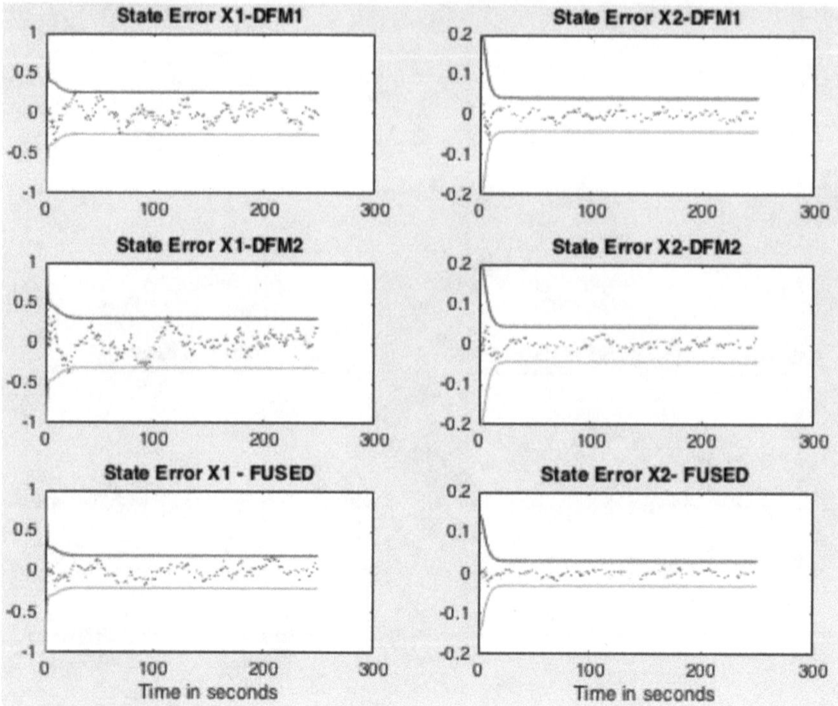

FIGURE 12.12 State errors with their bounds – DTS.

TABLE 12.4
Parameter Estimation of the Nonlinear DTS-IE/HI Method

Case	Parameter / True values	a_1x_1 0.8	$a_1x_1^2$ 0	$a_1x_1^3$ -0.05	a_1x_2 0.223	Terms removed -
DFM1	IE method	*(0.8)*	-1.03e-5	-0.0497	*(0.223)*	X_1^3, X_1^2
	IE+HI method $\gamma = 5$	*(0.8)*	-0.0000	-0.05	*(0.223)*	
DFM2	IE method	0.7961	-8.3e-6	-0.0498	*(0.223)*	X_1, X_1^3
	IE+HI method $\gamma = 7.5$	0.8	-0.00	-0.0499	*(0.223)*	
DFM3	IE method	0.8000	-3.07e-7	-0.05	0.2224	X_1, X_2, X_1^3
	IE+HI method $\gamma = 0.2$	0.8	-0.00	-0.05	0.223	

(.): true values used in the model; DFM: deficient model.

TABLE 12.5
Percentage Residual Fit Errors – DTS

Method	Model	Measurement
IE	DFM1	0.5391
	DFM2	0.8495
IE/HI	DFM1	0.4291
$\gamma = 0.5$	DFM2	0.7395

TABLE 12.6
Percentage State Errors – DTS

Method	Def Model	State X_1	State X_2
IE	DFM1	0.0257	0.9230
	DFM2	0.0271	1.0851
	Fusion	0.0180	0.6679
IE/HI $\gamma = 0.5$	DFM1	0.0234	0.8391
	DFM2	0.0247	0.9864
	Fusion	0.0164	0.6072

a regression model is fitted to the discrepancy $d(t) = a_1X_1(t) + a_2X_2(t) + a_3X_2^3(t)$ to estimate the parameters of the CTS. The values for R and Q for obtaining convergence are Q = diag. (0.001,30) and R = 18. The term γ is used that would generate a robust estimator in finding the states of the system and the model error.

Table 12.8 depicts the state estimates with the true value. Tables 12.9 to 12.11 show the PFEs, the state errors and HI norm for the continuous-time case, which are reasonably low.

Figures 12.13 to 12.16 depict performance graphs for the continuous-time case for the DF, from which it can be seen that the DF scheme that uses (at a time any) two DFMs, gives very satisfactory state-vector fusion results. The norm of the SEG ($S(t)$, studied in the context of DF using IE/HI estimators) is shown in typical graph in Figure 12.15, which again corroborates the asymptotic result derived for the observer error dynamics, since the observer is based on the IE/HI-based gain and SEG $S(t)$.

TABLE 12.7
H-Infinity Norm – DTS

Case	Method	Norm
DFM1 & DFM2	IE	0.0770
	IE+HI ($\gamma = 0.5$)	0.0765
DFM2 & DFM3	IE	0.0783
	IE+HI($\gamma = 7.5$)	0.0721
DFM3 & DFM1	IE	0.0743
	IE+HI($\gamma = 0.2$)	0.0698

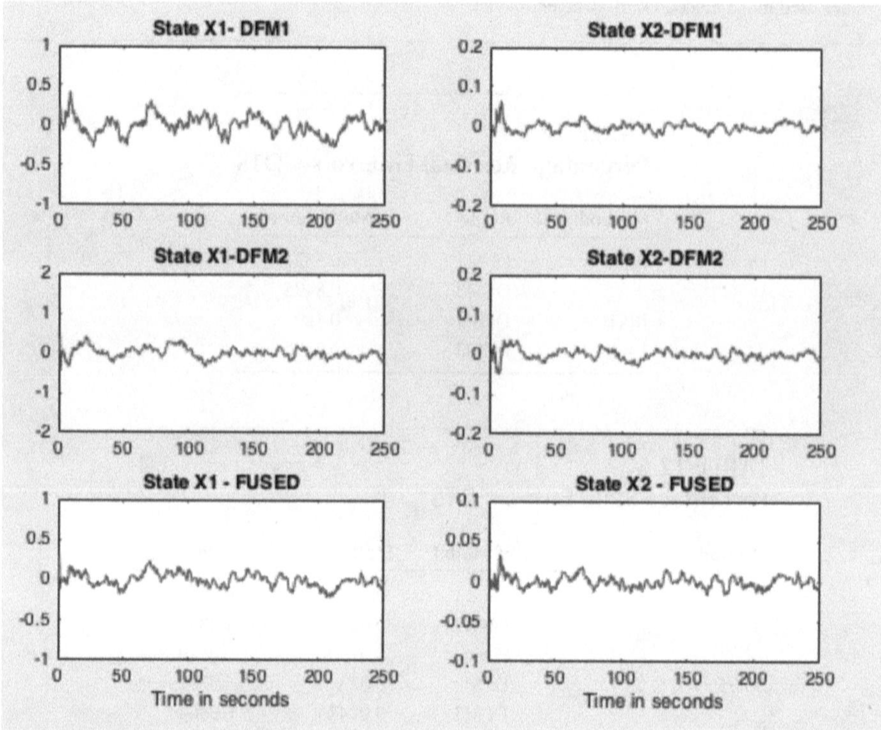

FIGURE 12.13 Estimated/fused states IE-HI method – CTS.

A DF fusion scheme that utilizes the IE and IE/HI-based ME estimators for two deficient models (DFMs at a time, in turn) has been presented in a SVF mode, as if these two information-results are coming from two independent (fictitious sensors) channels. The various metrics evaluated for continuous- and discrete-time systems using MATLAB-based simulations and the performance graphs validate the efficacy of the scheme.

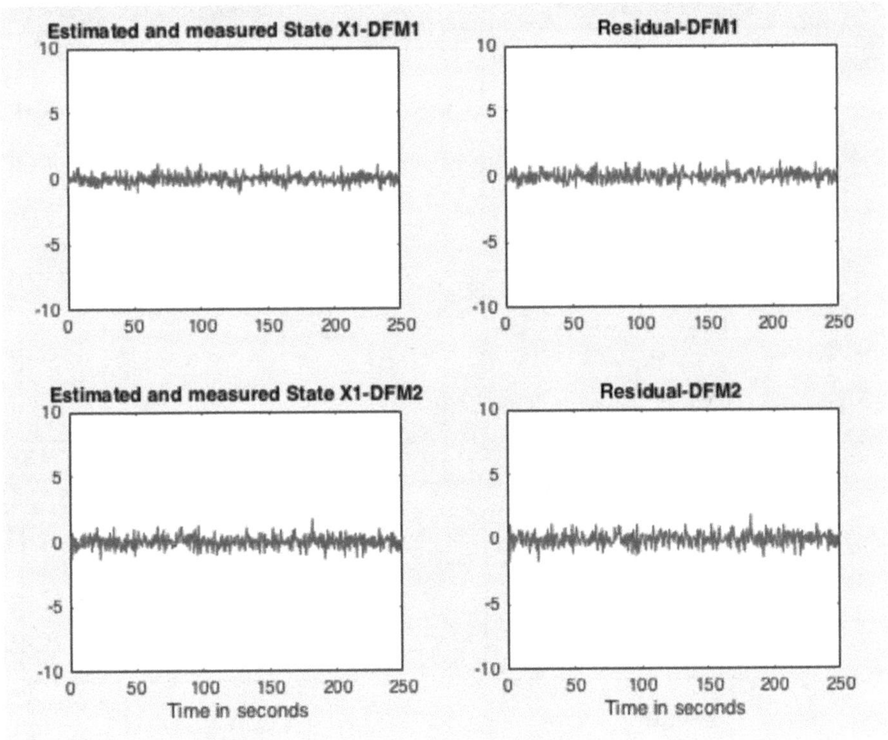

FIGURE 12.14 Estimated/measured states and residuals – CTS.

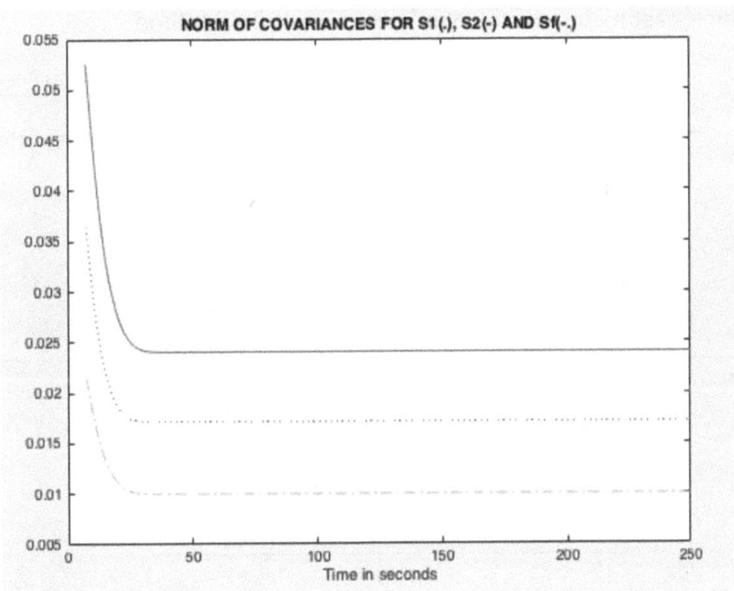

FIGURE 12.15 Norm of state error Gramians/IE method – CTS.

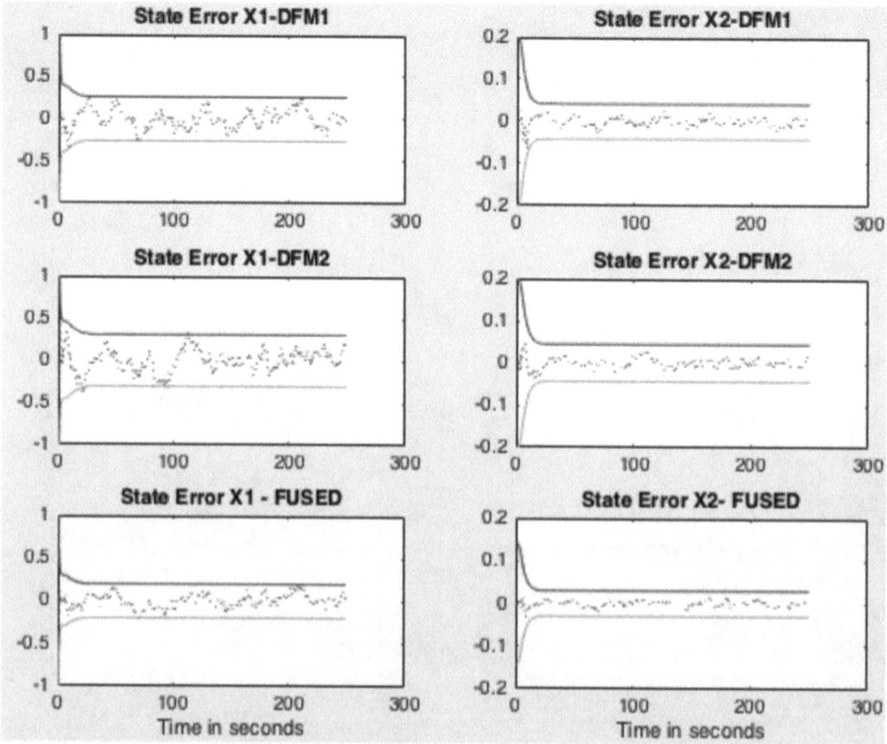

FIGURE 12.16 State errors with their bound – CTS.

TABLE 12.8
Parameter Estimation of the Nonlinear CTS-IE/HI Method

	Parameter	$a_1 x_1$	$a_1 x_1^2$	$a_1 x_2^3$	Terms Removed
Case	True Value	0.68	1	0.0195	-
DFM1	IE method	*(0.68)*	*(1)*	0.0187	X_2^3
	IE + HI method	*(0.68)*	*(1)*	0.0196	
	$\gamma = 22.902$				
DFM2	IE method	0.5576	0.9647	0.0198	X_1, X_2, X_2^3
	IE + HI method	0.68	1.08	0.0197	
	$\gamma = 0.3524$				

TABLE 12.9
Percentage Residual Fit Errors – CTS

Method	Model	Measurement
IE	DFM1	0.4290
	DFM2	0.7391
IE/HI	DFM1	0.3990
$\gamma = 0.3524$		
	DFM2	0.7091

TABLE 12.10
Percentage State Errors – CTS

Method	Model	State X_1		State X_2
IE	DFM1	0.0214	0.7692	
	DFM2		0.0226	0.9042
	Fusion		0.0150	0.5566
IE/HI	DFM1	0.0198	0.7100	
γ=0.3524				
	DFM2		0.0209	0.8346
	Fusion		0.0138	0.5137

TABLE 12.11
H-Infinity Norm – CTS

Case	Method	Norm
DFM1	IE	0.0870
&	IE+HI	0.0846
DFM2	γ=0.3524	

REFERENCES

1. Swetha, A., Robust Estimation of Model Error Using H-Infinity Framework Ph.D. Thesis, Electronics Engineering, Jain University, Bangalore, India, pp. 1–90, Dec. 2017.
2. Raol, J.R., Girija, G., and Twala, B. *Nonlinear Filtering: Concepts and Engineering Applications*, CRC Press, Taylor & Francis, Florida, May 2017.
3. Raol, J.R., *Modelling and Parameter Estimation of Dynamic Systems*, IEE/IET Control Series Books, IEE/IET, London, UK, Vol. 65, Aug. 2004.
4. Raff, T., and Allgower, F., An EKF-based observer for nonlinear time-delay systems. In Proc. of the American Control Conference, Minneapolis, MN, USA, 2006, Vol. 4, pp. 3130–3133, July 2006, http://bibbase.org/network/publication/raff-allgwer-anekfbasedobserver fornonlineartimedelay-systems-2006, accessed January 2016.
5. Reif, K., Sonnemann, F., and Unbehauen, R., An EKF-based nonlinear observer with a prescribed degree of stability, *Automatica*, Vol. 34, pp. 1119–1123, Sept. 1998, www.researchgate.net/publication/220157426, accessed January 2017.
6. Vershinin, Y.A., A data fusion algorithm for multi-sensor systems. In Proc. of the 5th International Conference on Information Fusion, Annapolis, MD, USA, 2002, pp. 341–345, July 2002.
7. Hassibi, B., Sayad, A. H., and Kailath, T., Linear estimation in Krein spaces-Part II: Applications. *IEEE Transactions on Automatic Control*, Vol. 41, No. 1, pp. 34–49, Jan. 1996.
8. Raol, J.R., *Data Fusion Mathematics – Theory and Practice*, CRC Press, Taylor & Francis, Florida, July 2015.

13 Hybrid Global H-Infinity Fusion Filter

13.1 INTRODUCTION

Sensor data fusion (SDF) is defined as the process or procedure of integrating information, in fact derived knowledge, from multiple sensors to produce the most comprehensive unified data result about an entity, activity or event [1]. One basic operation in sensor data fusion is signified by the numeric procedure involving estimation techniques and statistical methods, and one conventionally uses the Kalman/extended Kalman filter [2]; however, these are next addressed in the context of HI filter theory [3]; another alternative is Luenberger's observer [4]; these two latter estimation approaches are deterministic.

First, a global HI fusion filter for a discrete-time linear dynamic system is briefly stated. This filter is then converted into a continuous-time filter using the same procedure as is often used in the theory of the Kalman filter [5]. It turns out that the resulting global HI fusion filter is in some ways like the discrete global HI fusion filter; very interestingly this new continuous time global HI fusion filter turns out to be a hybrid-type filter: (i) the main part of the filter is in the continuous-time domain, that is for the local sensors; and (ii) the remaining part, mainly the global fusion (of the local filters) for m sensors is a combination of continuous- and discrete-time domains. This is a new feature arising out of the conversion process of the existing discrete-time global HI fusion filter into a new continuous-time HI filter.

Then, a new observer is presented that utilizes the gain and the covariance equations from this hybrid continuous-discrete time global HI fusion filter (HGFF). A suitable Lyapunov energy functional (LEF) is utilized to derive the conditions for the stability of the observer's error dynamics. The performance of the new hybrid global fusion filter is evaluated using implementations in MATLAB. The results establish the stability convergence of the new hybrid fusion filter, and that of the new observer. This (inherent composite) evaluation result is a novel contribution to the deterministic estimation theories.

13.2 GLOBAL H-INFINITY FUSION FILTER – DGHIFF IN DISCRETE TIME

We consider a linear discrete-time dynamic system:

$$y(k+1) = \phi y(k) + w(k) \tag{13.1}$$

$$z(k) = Hy(k) + v(k) \tag{13.2}$$

Variables $y(.)$, and $z(.)$ are the states and observations, respectively, and $w(.)$, and $v(.)$ are, respectively, the process noise and observation noise, which are assumed to be white Gaussian noise (WGN) sequences with zero mean; and Q, and R as their variances.

DOI: 10.1201/9781003620211-13

We make a direct recourse to the existing HI filtering theory, which does not require any assumptions on these noise processes, even if the actual data/observations contain stochastic noise [3]. The global HI fusion filter (DGHIFF in discrete time) is based on [6]. The local filters are given for each sensor (i = 1,2,...,m). In the HI theory, all the variables are considered as generalized 'stochastic' variables, and the associated 'covariance' like matrices are called Gramians, even though one tends to use the Kalman filter like terminologies (by force of habit). Thus, the noise processes are in fact the deterministic unknown disturbances.

13.2.1 At the Local Sensors

Part I – local filters: time propagations of the states and the state-error Gramian (SEG), $G(.)$ for the local sensors are given as follows:

$$\tilde{y}_i(k+1) = \phi\hat{y}_i(k) \tag{13.3}$$

$$\tilde{G}_i(k+1) = \phi\hat{G}_i(k)\phi^T + Q(k) \tag{13.4}$$

Part II – local filters: the measurement updates of the states and SEG for the local sensors are given as follows.

First, we have the information Gramian as $Y(.) = G^{-1}$, then the SEIG (state-error information Gramian) update is given as (here, in fact $\tilde{Y}_i(k+1) = \tilde{G}_i^{-1}(k+1)$, from eqn. (13.4)):

$$\hat{Y}_i\,(k+1) \Leftarrow \tilde{Y}_i\,(k+1) + [H_i^T\; L_i^T]\begin{bmatrix} I & 0 \\ 0 & -\gamma^2 I \end{bmatrix}^{-1}\begin{bmatrix} H_i \\ L_i \end{bmatrix} \tag{13.5}$$

The filter gains for the local sensors are given as (here use $\hat{G}_i\,(k+1) = \{\hat{Y}_i\,(k+1)\}^{-1}$, from eqn. (13.5)):

$$B_i = I + \frac{1}{\gamma^2}\hat{G}_i\,(k+1)L_i^T L_i\;\;;\quad K_i = B_i^{-1}\hat{G}_i(k+1)H_i^T \tag{13.6}$$

Then, the measurements/data update of the states for the local sensors is given as

$$\hat{y}_i(k+1) = \tilde{y}_i(k+1) + K_i(z_i(k+1) - H_i\tilde{y}_i(k+1)) \tag{13.7}$$

13.2.2 At the Fusion Centre

Part III – fusion filter: time propagations of the states and the state-error Gramians (SEG) for the fusion filter are given as follows.

The time propagation of the fused states and the SEG matrix is given as

$$\tilde{y}^f(k+1) = \phi\hat{y}^f(k) \tag{13.8}$$

$$\tilde{G}^f(k+1) = \phi\hat{G}^f(k)\phi + Q(k) \tag{13.9}$$

Part IV – fusion filter: the measurement update of the states and SEIG for the fusion filter.

The measurements updates of the SEIG is given as (here use $\tilde{Y}^f(k+1) = (\tilde{G}^f(k+1))^{-1}$, from eqn. (13.9)):

$$\hat{Y}^f(k+1) \Leftarrow \tilde{Y}^f(k+1) + \sum_{i=1}^{m} \{\hat{Y}_i(k+1) - \tilde{Y}_i(k+1)\} + \frac{m-1}{\gamma^2} L^T L \tag{13.10}$$

The required SEIGs, for the right hand side (RHS) of eqn. (13.10), in the curly brackets, are obtained from eqn. (13.5), and for two sensors – scenario $m = 2$.

The global fusion gain is given as (here use $\hat{G}^f(k+1) = (\hat{Y}^f(k+1))^{-1}$, from eqn. (13.10)):

$$B^f = I + \frac{1}{\gamma^2} \hat{G}^f(k+1) L^T L \tag{13.11}$$

Then finally global measurement update of the fused state is given, (here $\hat{Y}_i(k+1)$is taken from eqn. (13.5)) as

$$\begin{aligned} \hat{y}^f(k+1) &= [I - (B^f)^{-1} \hat{G}^f(k+1)(H^f)^T H^f]\tilde{y}^f(k+1) \\ &+ (B^f)^{-1} \hat{G}^f(k+1)[\sum_{i=1}^{m} \{\hat{Y}_i(k+1) B_i \hat{y}_i(k+1) \\ &- (\hat{Y}_i(k+1) B_i + H_i^T H)\breve{y}_i(k)\}] \end{aligned} \tag{13.12}$$

The state estimates needed on the RHS of eqn. (13.12) are taken from eqn. (13.8) for the fused state, and eqn. (13.7) for the individual states ($i = 1,2,\ldots,m$) for the first term in the square brackets.

In eqn. (13.12) the last term is given from eqn. (13.3) as

$$\breve{y}_i(k) = \phi \hat{y}_i(k) = \tilde{y}_i(k+1) \tag{13.13}$$

For this DGHIFF, the state-vector fusion formulae are not required, since the fusion is achieved using eqn. (13.10), and eqn. (13.12). The factor γ specifies the upper bound on the error 'variance'/ energy gain from the input energies due to the disturbances ($w(.)$, and $v(.)$); and the input error in the initial state condition ($x(0)$) to the output state (estimate) error energy, directly following the HI norm.

13.3 DERIVATION OF THE HYBRID GLOBAL HI FUSION FILTER – HGFF

Much of HI filter theory is in the domain of discrete-time systems. However, a continuous-time development offers a more favourable chance to derive and study easily the analytical results of these discrete-time filters; that is the continuous-time results can be easily taken as validation for the equivalent discrete time filters, if the discrete time step is very (very) small.

Next, the discrete time deterministic global HI fusion filter of Section 12.2 is converted to continuous-time filter.

A continuous-time linear dynamic system is

$$\dot{y}(t) = Ay(t) + gw(t) \tag{13.14}$$

$$z(t) = Hy(t) + v(t) \tag{13.15}$$

Although, in eqn. (12.14), and eqn. (12.15), the random variables $w(.)$, and $v(.)$ have the same symbols as in eqn. (12.1), and eqn. (12.2); in essence, for continuous-time systems, these noise processes have the spectral densities ($Q(t)$, and $R(t)$) as their strengths and for discrete-time systems, they are variance matrices.

Next, the main idea is to convert the discrete time GHFF (DGHIFF) to continuous time HGFF filter. To proceed with this, the following substitutions are used as a first order approximation [5]:

$$\phi(k) = I + A\Delta t \left(\text{transition matrix approximation}\right) \tag{13.16}$$

$$Q(k) = gQ(t)g^T \Delta t; \quad \text{and} \quad R(k) = \frac{R(t)}{\Delta t} \tag{13.17}$$

Then, eqn. (13.3), and eqn. (13.4) are used (from DGHIFF filter, Part I, local filters/Section 12.1) to derive a new and equivalent Part I for CGFF (continuous global fusion filter, and later on HGFF). These difference equations are used (after and using the substitutions from eqn. (13.16), and eqn. (13.17)), then the behaviour of these equations is studied as the discrete-time step $\Delta t \rightarrow 0$.

Local filters: the continuous-time equivalent equations from eqns. (13.3) and (13.4) are obtained very simply as

$$\begin{aligned} &\tilde{y}_i(k+1) = (I + A\Delta t)\hat{y}_i(k) \Rightarrow \\ &\frac{\tilde{y}_i(k+1) - \hat{y}_i(k)}{\Delta t} \Rightarrow \dot{\hat{y}}_i(t) = A\hat{y}_i(t) \end{aligned} \tag{13.18}$$

With initial condition as $y_i(0) = \hat{y}_{i0} = \hat{y}_i(0)$, for $t = 0$. In a similar way, by using eqn. (13.16), and eqn. (13.17) in eqn. (13.4), and neglecting a second-order term (of the time step; usual and normal approximation), we get after simplification:

$$\dot{\hat{G}}_i(t) = A\hat{G}_i(t) + \hat{G}_i(t)A^T + gQ(t)g^T \tag{13.19}$$

Next, eqn. (13.5) is handled (Part II):

$$\frac{\hat{Y}_i(k+1) - \tilde{Y}_i(k+1)}{\Delta t} = [H_i^T \; L_i^T]\begin{bmatrix} I\Delta t & 0 \\ 0 & -\gamma^2 I\Delta t \end{bmatrix}^{-1}\begin{bmatrix} H_i \\ L_i \end{bmatrix} \tag{13.20}$$

Then eqn. (13.20) is written in the continuous-time format, by taking limit on both the sides as time step $\rightarrow 0$, as

$$\dot{\hat{Y}}_i(t) = [H_i^T \; L_i^T]\begin{bmatrix} R & 0 \\ 0 & -\gamma^2 R \end{bmatrix}^{-1}\begin{bmatrix} H_i \\ L_i \end{bmatrix} \tag{13.21}$$

Here, we can use $R(t)$ from eqn. (13.17), (instead of literally putting the timestep = 0), otherwise the measurement update part of the local sensors filters will be ill-conditioned; this artefact is based on the prudent engineering judgement, rather than strictly considering the theoretical basis of the equation. Of course, if the proper theoretical basis is used then, the RHS of eqn. (13.21) would yield a very large number (inverse of nearly a null matrix) when the timestep tends to 0; and then one can proceed with the continuous-time part of eqn. (13.21), but with a large constant on its RHS. This will also reflect that the SEIG varies nearly at a constant rate, which practically may not be a useful approach/solution.

Next, the gain equations in eqn. (13.6) are handled (Part II) by inserting the first equation into the second of equation:

$$K_i = \left(I + \frac{1}{\gamma^2}\hat{G}_i(k+1)L_i^T L_i\right)^{-1}\hat{G}_i(k+1)H_i^T \tag{13.22}$$

$$\lim_{\Delta t \to 0}\left(\frac{1}{\Delta t}K_i(k+1)\right) = \lim_{\Delta t \to 0}\left\{\frac{1}{\Delta t}\left(I + \frac{1}{\gamma^2}\widehat{G}_i(k+1)L_i^T L_i\right)^{-1}\widehat{G}_i(k+1)H_i^T\right\} \tag{13.23}$$

$$\lim_{\Delta t \to 0}\left(\frac{1}{\Delta t}K_i(k+1)\right) = \lim_{\Delta t \to 0}\left\{\left(I\Delta t + \Delta t\frac{1}{\gamma^2}\widehat{G}_i(k+1)L_i^T L_i\right)^{-1}\widehat{G}_i(k+1)H_i^T\right\} \tag{13.24}$$

Evaluating the limits, we obtain the following gain equation in the continuous-time domain:

$$K_i(t) = (R_i(t))^{-1}\widehat{G}_i H_i^T; \quad R = I\Delta t \tag{13.25}$$

Next, local measurement update eqn. (13.7) is taken and by using the same procedure of dividing by the step size and taking the limit yields, with the initial conditions appropriately chosen:

$$\frac{\hat{y}_i(k+1) - \tilde{y}_i(k)}{\Delta t} = \dot{\hat{y}}_i(t) = R_i^{-1}\widehat{G}_i H_i^T(z_i(t) - H_i\tilde{y}(t)) \tag{13.26}$$

Fusion filters: now the discrete-time fusion filters are handled (Section 13.2, Part III). Eqns. (13.8) and (13.9) are similar to the ones in eqn. (13.3), and eqn. (13.4), and hence we directly obtain the continuous-time equivalent equation as follows:

$$\dot{\tilde{y}}^f(t) = A\hat{y}^f(k) \tag{13.27}$$

$$\dot{\tilde{G}}^f(t) = A\widehat{G}^f(t) + \widehat{G}^f(t)A^T + (gQg^T)^f \tag{13.28}$$

Now, the measurement update equations (Part IV) of the discrete-time filter (13.10)–(13.11) are converted to the continuous-time case. First eqn. (13.10) is handled and it can be easily expressed as the equivalent continuous-time equation as given below:

$$\frac{\hat{Y}^f(k+1) - \tilde{Y}^f(k+1)}{\Delta t} = \sum_{i=1}^{m}\left(\frac{\hat{Y}_i(k+1) - \tilde{Y}_i(k+1)}{\Delta t}\right) + \left(\frac{m-1}{\gamma^2\Delta t}L^T L\right) \tag{13.29}$$

Taking the usual limits, eqn. (13.29), is expressed in a compact form as the SEIG

$$\dot{\hat{Y}}^f(t) = \sum_{i=1}^{m}\dot{\hat{Y}}_i(t) + \left(\frac{m-1}{\gamma^2\Delta t}L^T L\right) \tag{13.30}$$

One can easily see that eqn. (13.30) would be ill-conditioned, if the step size is set to zero. Hence, practical judgment is to use a small number for the step size, then the last term can still be retained, but it could be a large number.

Now, eqn. (13.12) is handled in conjunction with eqn. (13.11) as follows:

$$\begin{aligned}\hat{y}^f(k+1) - \tilde{y}^f(k+1) &= -(B^f)^{-1}\hat{G}^f(k+1)(H^f)^T H^f\tilde{y}^f(k+1)\\ &+ (B^f)^{-1}\hat{G}^f(k+1)[\sum_{i=1}^{m}\{\hat{Y}_i(k+1)B_i\hat{y}_i(k+1)\\ &- (\hat{Y}_i(k+1)B_i + H_i^T H)\tilde{y}_i(k)\}]\end{aligned} \tag{13.31}$$

Since we need $(B^f)^{-1}$ in eqn. (13.31), we simplify it using the procedure adopted earlier, and get

$$(B^f) = \Delta t I = R^f \tag{13.32}$$

Utilizing eqn. (13.32), and eqn. (13.11) in eqn. (13.31) we obtain

$$\begin{aligned}\frac{\hat{y}^f(k+1)-\tilde{y}^f(k+1)}{\Delta t} &= \dot{\hat{y}}^f(t) = -(R^f)^{-1}\hat{G}^f(H^f)^T H^f \tilde{y}^f(t)\\ &+(R^f)^{-1}\hat{G}^f[\sum_{i=1}^{m}\{\hat{Y}_i(k+1)B_i(\hat{y}_i(k+1)\\ &-\tilde{y}_i(k+1))-H_i^T H \tilde{y}_i(k+1)\}]\end{aligned} \tag{13.33}$$

In eqn. (13.33), eqn. (13.3) is used for simplification, and some terms are combined. One can easily see that the step size can be effectively used only in the inverse of the gain terms and cannot be utilized in the last term in the square bracket, since it is a term appearing as a product. Hence, finally we have the following equation for the global data update of the states for the fusion:

$$\dot{\hat{y}}^f(t) = -F^f \tilde{y}^f(t) + (R^f)^{-1}\hat{G}^f D^f \tag{13.34}$$

In eqn. (13.34) we have the following equivalent expressions used:

$$\begin{aligned}F^f &= (R^f)^{-1}\hat{G}^f(H^f)^T H^f\\ D^f &= [\sum_{i=1}^{m}\{\hat{Y}_i(k+1)B_i(\hat{y}_i(k+1)-\tilde{y}_i(k+1))-H_i^T H_i \tilde{y}_i(k+1)\}]\end{aligned} \tag{13.35}$$

$$B_i = I + \frac{1}{\gamma^2}\hat{G}_i(k+1)L_i^T L_i \tag{13.36}$$

This completes the derivation of the continuous-time global HI-based fusion filter (CGFF) from the discrete-time global HI-based fusion filter (DGHIFF). One can observe from eqns. (13.34) to (13.36) that this new global HI-based fusion filter is as such a hybrid filter (HGFF) because eqn. (13.34) is hybrid, since it utilizes the continuous part, the first term of eqn. (13.34), whereas the expressions for D^f, and the gain B_i are the same as in the discrete-time global HI-based fusion filter as can be easily observed from eqn. (13.6), and eqn. (13.12). This is true since, the step size could not be carried into these expressions, and no limiting operation was feasible. Hence, the global HI fusion filter's measurement part is hybrid. Although, the other equations for the local filters, and time-propagation part of the global fusion filter are in fact in continuous-time domain (CGFF), since the global data update part is in continuous as well as discrete-time domains, the entire new filter can be termed as a hybrid global HI-based fusion filter (HGFF).

13.4 HYBRID GLOBAL HI-BASED FUSION FILTER: HGFF COLLECTED

In this section all the equations for the newly derived hybrid global HI-based fusion filter of Section 12.3 are put together for easy reference.

13.4.1 At the Local Sensors

Part I – local filters: time propagations of the states and the state-error Gramian (SEG) for the local sensors are given as follows (proper initial conditions should be used on the RHS):

$$\dot{\tilde{y}}_i(t) = A\hat{y}_i(t) \tag{13.37}$$

$$\dot{\hat{G}}_i(t) = A\hat{G}_i(t) + \hat{G}_i(t)A^T + gQg^T \tag{13.38}$$

Part II – local filters: the measurement updates of the states and SEG for the local sensors are given next. First, we have $\tilde{Y}_i(t) = (\tilde{G}_i(t))^{-1}$, this is used as the initial condition for the following equation of the SEIG update:

$$\dot{\hat{Y}}_i(t) = [H_i^T \; L_i^T]\begin{bmatrix} R_i & 0 \\ 0 & -\gamma^2 R_i \end{bmatrix}^{-1}\begin{bmatrix} H_i \\ L_i \end{bmatrix}; \quad R = I\Delta t \tag{13.39}$$

Then, we have to use $\hat{G}_i(t) = (\hat{Y}_i(t))^{-1}$ and $K_i(t) = R_i^{-1}\hat{G}_i H_i^T$ in the following state update:

$$\dot{\hat{y}}_i(t) = R_i^{-1}\hat{G}_i H_i^T(z_i(t) - H_i\tilde{y}_i(t)); \text{ (with the previous value from eqn. (13.37) as the initial condition} \tag{13.40}$$

13.4.2 At the Fusion Centre

Part III – fusion filter: time propagations of the states and the state-error Gramians (SEG) for the fusion filter are given next. The time propagation of the fused states and the SEG matrix is given as (with proper initial conditions used):

$$\dot{\hat{y}}^f(t) = A\hat{y}^f(t); \text{ to start with one can use initial condition as } \hat{y}^f(0) = 0.5\times(\hat{y}_1 + \hat{y}_2) \tag{13.41}$$

$$\dot{\hat{G}}^f(t) = A\hat{G}^f(t) + \hat{G}^f(t)A^T + (gQg^T)^f;$$

$$\hat{G}^f(0) = \frac{\hat{G}_1 \times \hat{G}_2}{(\hat{G}_1 + \hat{G}_2)} \Rightarrow \hat{G}_1 \times (\hat{G}_1 + \hat{G}_2)^{-1} \times \hat{G}_2 \tag{13.42}$$

Part IV – fusion filter: the measurement update of the states and SEIG for the fusion filter are given next. First, we have $\hat{Y}_i(t)$ from eqn. (39), use previous values from local sensors, and $m = 2$.

$$\dot{\hat{Y}}^f(t) = \sum_{i=1}^{m} \dot{\hat{Y}}_i(t) + \left(\frac{m-1}{\gamma^2\Delta t}L^T L\right); \quad \text{then } \hat{G}^f(t) = (\hat{Y}^f(t))^{-1}; \tag{13.43}$$

$$\hat{G}_i(t) = (\hat{Y}_i(t))^{-1}; \; B_i = I + \frac{1}{\gamma^2}\hat{G}_i(t)L_i^T L_i;$$

$$D^f = [\sum_{i=1}^{m} \{\hat{Y}_i(k+1)B_i(\hat{y}_i(k+1) - \tilde{y}_i(k+1)) - H_i^T H_i \tilde{y}_i(k+1)\}]$$

$$F^f = (R^f)^{-1}\hat{G}^f(H^f)^T H^f;$$

$$\dot{\hat{y}}^f(t) = -F^f\tilde{y}^f(t) + (R^f)^{-1}\hat{G}^f D^f \tag{13.44}$$

The state and the SEIG are taken from the local filters (Part II).

13.5 OBSERVER BASED ON THE GLOBAL HYBRID HI FUSION FILTER (HGFF)

Next, a new observer that uses the gain and Gramian formulae from the new hybrid fusion filter is proposed.

A nonlinear continuous-time system is represented, time being the specific symbolism that specifies the individual sensor/s or fusion (or even estimates) are omitted:

$$\dot{y} = f(y,t) \tag{13.45}$$

$$z = Hy \tag{13.46}$$

A nonlinear observer for the system of eqn. (13.45) is given for each individual sensor as (subscript '*i*' omitted):

$$\begin{aligned}\dot{\hat{y}}(t) &= f(\hat{y}(t),t) + L(t)(z(t) - \hat{z}(t)) \\ \hat{z}(t) &= H\hat{y}(t)\end{aligned} \tag{13.47}$$

In eqn. (13.47), *L*(*t*) is the observer gain vector/matrix with appropriate dimensions, and it is from the HGFF filter as

$$L(t) = R^{-1}\widehat{G}H^T \tag{13.48}$$

The matrix *G*(*t*) is obtained as the solution of the observer Riccati-type differential (RTD) equation, which is also directly used from the HGFF filter as

$$\dot{\widehat{G}}_i(t) = A\widehat{G}_i(t) + \widehat{G}_i(t)A^T + gQg^T \tag{13.49}$$

Interestingly for eqn. (13.49), the Jacobian is obtained as

$$A(t) = \frac{\partial f(.)}{\partial \hat{y}(t)} \tag{13.50}$$

By subtracting eqn. (13.47) from eqn. (13.45), the observer error dynamics are obtained as

$$\begin{aligned}\dot{e}(t) &= A(t)e(t) - L(t)H(y(t) - \hat{y}(t)) + \phi(.) \\ &= A(t)e(t) - L(t)He(t) + \phi(.)\end{aligned} \tag{13.51}$$

In eqn. (13.51), the nonlinear function is given as

$$\phi(.) = -A(t)e(t) + f(y,t) - f(\hat{y},t) \tag{13.52}$$

This new nonlinear function in eqn. (13.52) is intentionally devised to add *A*(*t*)*e*(*t*), in order to make error dynamics look like normal differential equations. The state errors are given as

$$e(t) = y(t) - \hat{y}(t) \tag{13.53}$$

13.5.1 Analytical Results

The local/asymptotic behaviour of the observer error dynamics of eqn. (13.49) is studied, for which certain conditions are now specified [7,8]:

1. The SEG which is a solution of the matrix RTD eqn. (13.49) needs to be bounded

$$p_l I \leq G(t) \leq p_u I \tag{13.54}$$

In eqn. (13.54), $p_l, p_u > 0$ are positive constants, because the SEG, G is theoretically, positive definite and symmetrical matrix; and these constants are the lower and upper bounds, respectively.

2. The nonlinearity, eqn. (13.52), in the error dynamics is required to be bounded

$$\|\phi(.)\| \leq \rho\|y(t) - \hat{y}(t)\|^2 \tag{13.55}$$

More conditions would follow as the development proceeds further. Now, in order to derive the conditions for the convergence of the observer, one needs to consider LEF:

$$V(t) = e^T(t)Y(t)e(t) \tag{13.56}$$

The LE functional (13.56) is positive definite because of the condition 1 that is using the inequality of eqn. (13.54), so, one obtains

$$\frac{1}{p_u}\|e(t)\|^2 \leq e^T(t)Y(t)e(t) \leq \frac{1}{p_l}\|e(t)\|^2 \tag{13.57}$$

The main condition for the stability/convergence of the observer is that the time derivative of the LEF, eqn. (13.56), should be negative definite with the constraints of: (a) error dynamics eqn. (13.51); (b) gain eqn. (13.48); and (c) the RTD of the SEG eqn .(13.49).

The time derivative of the LEF is obtained as

$$\dot{V}(t) = e^T(t)\dot{Y}(t)e(t) + e^T(t)Y(t)\dot{e}(t) + \dot{e}^T(t)Y(t)e(t) \tag{13.58}$$

Since, $\dot{Y} = -Y\dot{G}Y$, one can rewrite eqn. (13.58) as

$$\dot{V}(t) = -e^T(t)Y(t)\dot{G}(t)Y(t)e(t) + e^T(t)Y(t)\dot{e}(t) + \dot{e}^T(t)Y(t)e(t) \tag{13.59}$$

Substitute, RTD of the SEG eqn. (13.49) in eqn. (13.59) to obtain the following equation:

$$\begin{aligned}\dot{V}(t) = &-e^T(t)Y(t)[AG(t) + G(t)A^T + gQg^T]Y(t)e(t)\\ &+ e^T(t)Y(t)\dot{e}(t) + \dot{e}^T(t)Y(t)e(t)\end{aligned} \tag{13.60}$$

$$\begin{aligned}\dot{V}(t) = &-e^T(t)Y(t)AG(t)Y(t)e(t) - e^T(t)Y(t)G(t)A^TY(t)e(t)\\ &- e^T(t)Y(t)gQg^TY(t)e(t)\\ &+ e^T(t)Y(t)\dot{e}(t) + \dot{e}^T(t)Y(t)e(t)\end{aligned} \tag{13.61}$$

Simplifying eqn. (13.61), since $GY = YG = I$, one obtains

$$\begin{aligned}\dot{V}(t) = &-e^T(t)Y(t)Ae(t) - e^T(t)A^TY(t)e(t) - e^T(t)Y(t)gQg^TY(t)e(t)\\ &+ e^T(t)Y(t)\dot{e}(t) + \dot{e}^T(t)Y(t)e(t)\end{aligned} \tag{13.62}$$

Now, the expression for error dynamics eqn. (13.51) is substituted in eqn. (13.62) to obtain

$$\begin{aligned}\dot{V}(t) = &-e^T(t)Y(t)Ae(t) - e^T(t)A^TY(t)e(t) - e^T(t)Y(t)gQg^TY(t)e(t)\\ &+e^T(t)Y(t)\{A(t)e(t) - L(t)He(t) + \phi(.)\}\\ &+\{A(t)e(t) - L(t)He(t) + \phi(.)\}^T Y(t)e(t)\end{aligned} \tag{13.63}$$

Next, substitute the expression of the gain L, eqn. (13.48) in eqn. (13.63) to obtain the following expression:

$$\begin{aligned}\dot{V}(t) = &-e^T(t)Y(t)Ae(t) - e^T(t)A^TY(t)e(t) - e^T(t)Y(t)gQg^TY(t)e(t)\\ &+e^T(t)Y(t)\{A(t)e(t) - R^{-1}GH^T He(t) + \phi(.)\}\\ &+\{e^T(t)A(t) - e^T(t)H^T HGR^{-1} + \phi^T(.)\}Y(t)e(t)\end{aligned} \tag{13.64}$$

$$\begin{aligned}\dot{V}(t) = &-e^T(t)Y(t)Ae(t) - e^T(t)A^TY(t)e(t) - e^T(t)Y(t)gQg^TY(t)e(t)\\ &+e^T(t)Y(t)A(t)e(t) - e^T(t)Y(t)R^{-1}GH^T He(t) + e^T(t)Y(t)\phi(.)\\ &+e^T(t)A^T(t)Y(t)e(t) - e^T(t)H^T HGR^{-1}Y(t)e(t) + \phi^T(.)Y(t)e(t)\end{aligned} \tag{13.65}$$

Further simplifying we get

$$\begin{aligned}\dot{V}(t) = &-e^T(t)Y(t)gQg^TY(t)e(t) - e^T(t)Y(t)R^{-1}GH^T He(t) + e^T(t)Y(t)\phi(.)\\ &-e^T(t)H^T HGR^{-1}Y(t)e(t) + \phi^T(.)Y(t)e(t)\end{aligned} \tag{13.66}$$

$$\begin{aligned}\dot{V}(t) = &-e^T(t)Y(t)gQg^TY(t)e(t) - e^T(t)Y(t)R^{-1}GH^T He(t) - e^T(t)H^T HGR^{-1}Y(t)e(t)\\ &+\phi^T(.)Y(t)e(t) + e^T(t)Y(t)\phi(.)\end{aligned} \tag{13.67}$$

Now, the additional bounds need to be defined are

(i) It is assumed that $\|R^{-1}\| \le (1/r); \|H^T H\| \le h^2$, ($r$ and h are positive constants);

$$\|e(t)\|^2 \le \varepsilon^2 \tag{13.68}$$

(ii) $\varepsilon = \|e\| = \mathrm{k}$ (13.69)

After substituting these bounds and those from conditions 1 and 2 of eqns. (13.54), (13.55), one gets the following form of eqn. (13.67) in terms of the norms:

$$\dot{V}(t) = -\frac{q_l}{p_u^2}\|e(t)\|^2 + \frac{2\rho\varepsilon}{p_u}\|e(t)\|^2 - \frac{2p_l h^2}{rp_u}\|e(t)\|^2 \tag{13.70}$$

In eqn. (13.70), q_l is the smallest (positive) eigenvalue of the matrix gQg^T which itself is positive definite and symmetric. Combining certain terms in eqn. (13.70), one gets

$$\dot{V}(t) = -\frac{q_l}{p_u^2}\|e(t)\|^2 - \{\frac{2p_l h^2}{rp_u} - \frac{2\rho\varepsilon}{p_u}\}\|e(t)\|^2 \tag{13.71}$$

Since, the first term is a positive number with a –ve sign, one can simplify eqn. (13.71) to get

$$\dot{V}(t) \le -\{\frac{p_l h^2}{r} - \rho\varepsilon\}\|e(t)\|^2 \tag{13.72}$$

From eqn. (13.72), the final condition for the derivative of LE functional is obtained as

$$\frac{h^2 p_l}{r\rho} > k \tag{13.73}$$

Because all the constants and bounds (as defined earlier) in eqn. (13.73) are positive, it is seen that the time derivative of the LEF is negative definite as in eqn. (13.72); and hence, the error dynamics, eqn. (13.51) of the observer, with the gain and state-error Gramian utilized from the continuous-time GHI fusion filter, are asymptotically stable; and simultaneously this result also establishes that the newly derived global HI fusion filter would be asymptotically stable (HGFF). This is a novel interpretation, since it gives a procedure for establishing the convergence result of any linear/linearized estimator/filter by way of establishing the convergence of the associated observer, the structure of which directly utilizes the gain and state error (information) Gramian expressions from the newly derived filter. In this way numerical validation of the filter (structured observer) would establish the convergence of the associated observer (original filter).

Similarly, one can propose a new nonlinear observer, which would utilize the gains and state-error Gramians for the fusion filter given in Section 4.2 (Part III and IV).

13.6 PERFORMANCE EVALUATION OF THE NEW HYBRID FUSION FILTER (HGFF)

The newly derived hybrid algorithm is implemented in MATLAB and validated using simulated data for a linear dynamic system:

$$\dot{x}(t) = Ax(t) + Bu + w \tag{13.74}$$

$$A = \begin{bmatrix} 0.06 & -2.0 \\ 0.8 & -0.8 \end{bmatrix}; \; B = \begin{bmatrix} -0.6 \\ 1.5 \end{bmatrix} \tag{13.75}$$

The measurement model is given as

$$z(t) = Hx(t) + v \tag{13.76}$$

The simulated data for 5 sec are generated using the models, eqn. (13.74), and eqn. (13.75), with a sampling interval of 0.01 sec. Appropriate additive process noise in states and measurement noise are used (SNR = variance of signal/variance of noise = 20). Control input signal $u(.)$ is a doublet that excites the modes of the system eqn. (13.74). The % fit errors (metrics) of the measurement residuals are computed as PFE = 100×cov(of measurements)/cov(of true or actual meas.). If covariance measure is found to be ill-conditioned, then 'norm' can be used instead. Also, the H-infinity norm is evaluated as

$$\mathrm{HI}(\text{norm}) = \frac{\sum_{k=0}^{N} (\hat{y}(k) - y(k))^T (\hat{y}(k) - y(k))}{(\hat{y}_0 - y_0)^T P_0 (\hat{y}_0 - y_0) + \sum_{k=0}^{N} w^T(k)w(k) + \sum_{k=0}^{N} v^T(k)v(k)} \tag{13.77}$$

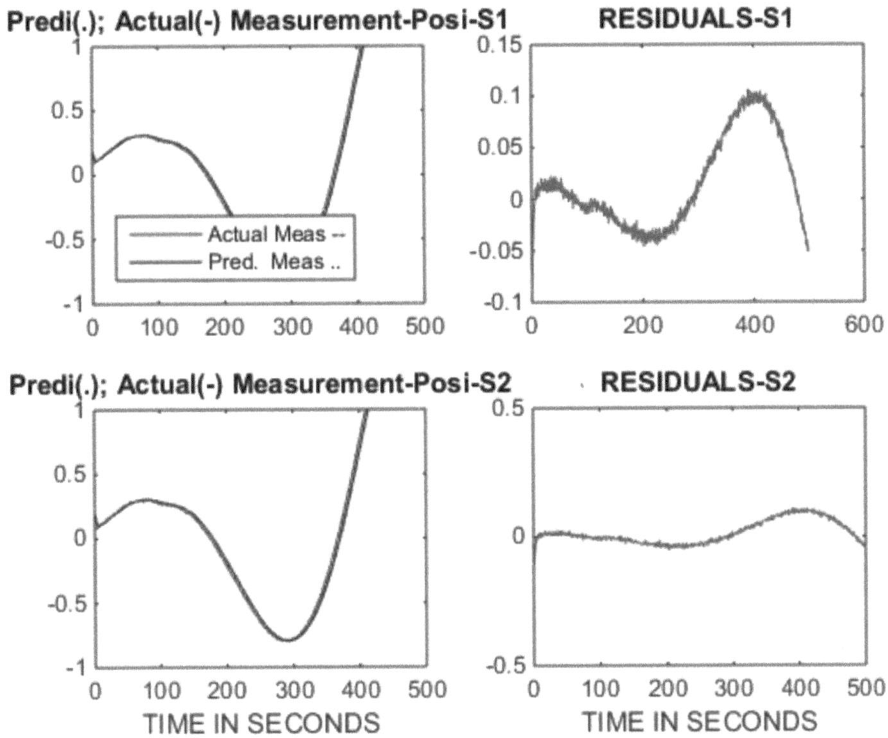

FIGURE 13.1 Time history match of the measurements and residuals (X-axis no. of samples).

The dynamic equations for simulation as well as filtering algorithm are solved by RK4 (or Euler) method of integration. The RTD eqn. (13.38) is solved by using the transition matrix method [5]. The performance metrics are HI norm = 0.262, PFE1 = 0.2629, PFE2 = 0.2773. Figure 13.1 shows the performance which shows very satisfactory trajectory matching; Figure 13.2 depicts the time history of the norm of $P(t)$, the state-error Gramian, solution of the RTD equation, which corroborates the asymptotic convergence of the observer as well as the inherent continuous time H-infinity filtering algorithm. Further extensions of the presented results to the applications in the domain of multisensory data fusion for more complex dynamics can be easily pursued [1,9,10].

A continuous-time *a posteriori* H-infinity filter (CGFF) has been derived from an existing discrete-time *a posteriori* H-infinity filter (DGHIFF). It turns out that the resultant HI filter is a hybrid one (HGFF) that has the features of the new continuous-time filter as well as the existing discrete-time filter. Then, a nonlinear observer is presented, which uses the gain and the state-error Gramian from the new filter. The conditions of the asymptotic stability result of the observer's error dynamics are derived using the Lyapunov energy functional. The performance of the HGFF algorithm is evaluated using numerical simulations carried out in MATLAB. These results establish the asymptotic stability/convergence of the observer and the new filter in one go. This type of result is novel in the literature on the HI theory of filters and the deterministic observers. Such studies impart confidence in utilization and applications of the HGFF algorithms and observers in communication systems, wireless sensor networks, mechanical/aerospace engineering (aircraft trajectory and parameter estimation and target tracking), and robotics.

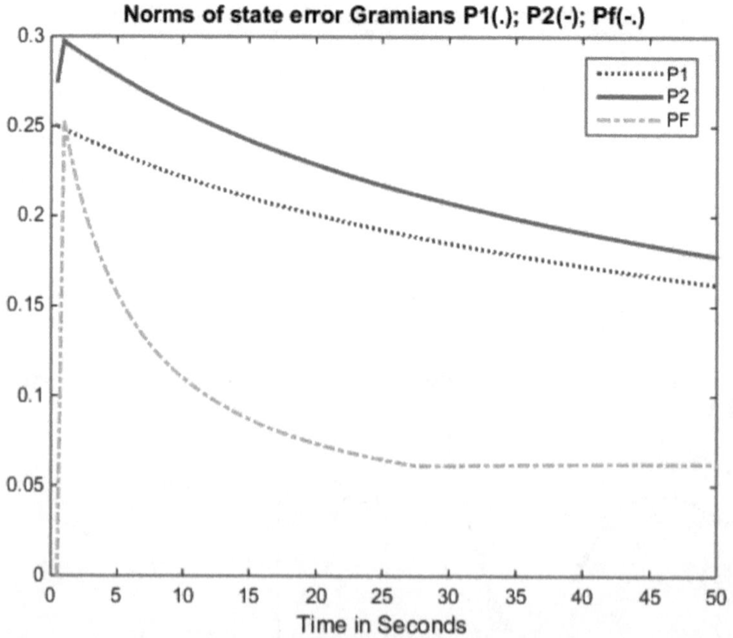

FIGURE 13.2 Norm of the state-error Gramians showing the convergence of the HGFF (*X*-axis no. of samples).

REFERENCES

1. Raol, J.R., *Data Fusion Mathematics – Theory and Practice*, CRC Press, Taylor & Francis, Florida, July 2015.
2. Raol, J.R., Girija, G., and Twala, B., *Nonlinear Filtering: Concepts and Engineering Applications*, CRC Press, Taylor & Francis, Florida, May 2017.
3. Hassibi, B., Sayad, A.H., and Kailath, T., Linear estimation in Krein spaces – Part II: Applications, *IEEE Transactions on Automatic Control*, Vol. 41, No. 1, pp. 34–49, Jan. 1996.
4. Kailath, T., *Linear Systems*, Prentice-Hall, New York, 1980.
5. Gelb, A. (Ed.), *Applied Optimal Estimation*, The M.I.T. Press, USA, 1974.
6. Park, B.S., and Oh, J.H., Gain fusion algorithm for decentralized parallel Kalman filters. *IEE Proceedings – Control Theory and Applications*, Vol. 147, No. 1, pp. 97–103, 2000.
7. Raff, T., and Allgower, F., An EKF-based observer for nonlinear time-delay systems. In *Proc. of the American Control Conference, Minneapolis, MN, USA*, 2006, Vol. 4, pp. 3130–3133, July 2006, http://bibbase.org/network/publication/raff-allgwer-anekfbasedobserverfornonlineartimedelay-systems-2006, accessed January 2016.
8. Reif, K., Sonnemann, F., and Unbehauen, R., An EKF-based nonlinear observer with a prescribed degree of stability, *Automatica*, Vol. 34, pp. 1119–1123, Sept. 1998, www.researchgate.net/publication/220157426, accessed January 2017.
9. Vershinin, Y.A., A data fusion algorithm for multi-sensor systems. In Proc. of the 5th International Conference on Information Fusion, Annapolis, MD, USA, pp. 341–345, 2002, July 2002.
10. Raol, J.R., Girija, G., and Singh, J., *Modelling and Parameter Estimation of Dynamic Systems.* IET/IEE Control Series Books, Vol. 65, IET/IEE Society, London, UK, 2004.

14 Neural Networks for Parameter Estimation with Lyapunov Stability

14.1 INTRODUCTION

The process of formulating a mathematical model that effectively describes and represents the characteristics of a dynamic system by using measurements from the system is known as parameter estimation [1,2]. Estimation of parameters and system identification are very important for the design and development of aircraft control systems [3]. The practice of determining the optimal values of coefficients and their partial derivatives of an airplane is critical as these coefficients help in-flight testing-based data analysis and evaluation of the overall aircraft performance [2]. System performance under all regimes may be tested by examining an airplane prototype and also by performing wind-tunnel tests [4–6]. Improved sensors, faster computational power, improvements in filtering and control theory have further contributed to successful flight vehicle system identification [7,8]. The Kalman filter is the optimal algorithm for a linear state-space model with Gaussian noise assumption [2]. However, aircraft parameter estimation is a nonlinear filtering challenge due to the fact that the aerodynamic models are nonlinear and the measurements from the sensors are noisy and biased.

The extended Kalman filter (EKF) is a commonly used method for dealing with nonlinear filtering and is frequently used to estimate aircraft aerodynamic parameters [9–11]. The key strategy used to develop EKF was to linearize the nonlinear process or measurement model around the current mean value. However, this might lead to linearization errors. Unlike EKF, the unscented Kalman filter (UKF) put forward an intuitive method by which the Gaussian density, that represents the prior and posterior pdfs, is represented using a set of deterministic points and corresponding weights [12]. Several other algorithms for parameter estimation, such as iterated EKF and adaptive UKF, have also been developed [11,13].

The application of artificial neural networks (ANNs) in parameter estimation has been receiving a great deal of attention in recent times [14–16]. Neural networks have emerged as powerful tools in the domain of parameter estimation, enabling the extraction of intricate patterns and relationships from complex datasets. However, their usage for aircraft parameter estimation is relatively less. ANNs constitute neurons and connections between nodes/neurons where each connection has its own weight/coefficient. Appropriate choice of ANN and proper training of weights can help in achieving good parameter estimation. The training of conventional ANN is mainly based on the theory of optimization. The ability of ANN to estimate any given function to a desired level of accuracy by selecting the right number of hidden layers and neurons makes them a viable tool for parameter estimation [17–20]. Feedforward neural networks (FFNNs) and recurrent neural networks (RNNs) are two forms of neural networks that can be used for the estimation of parameters.

DOI: 10.1201/9781003620211-14

FFNN is distinguished by its unidirectional signal flow [21]. FFNN estimates parameters by identifying the optimal weights that provide the best fit for a particular training data set [22,23]. FFNN uses backpropagation, which is a technique that is based on the classical gradient algorithm that is used to minimize the mean-squared error between the estimated output and the actual output for a given input to the network. RNN is a class of neural networks in which the output from the previous step is fed as input to the current step. RNNs have the ability to learn features and long-term dependencies from data that evolve with time as well as ordered sequences. The RNNs consist of a stack of nonlinear units connected by at least one directed cycle [24]. RNN techniques have been used for solving various types of state and parameter estimation problems [21,25–29]. FFNN and RNN have been used in various control and identification problems [30–33].

The requirement of stability analysis of the parameter estimation algorithms is very important. In the case of aircraft parameter estimation, stability analysis is critical, especially if the determined aircraft mathematical models are to be used for fault identification and other safety-critical applications. There are various methods by which stability analysis can be performed [34]. The Lyapunov method is an established technique for studying the stability conditions of nonlinear systems [35].

In this chapter, we delve into two distinct types of FFNNs – the conventional backpropagation method and a more advanced approach incorporating the recursive least-squares algorithm (RLS). Additionally, our exploration extends to two variants of RNNs – the weight-bias method and the gradient-based method. Our primary focus is on the algorithmic convergence of these architectures, assessed through the lens of the normalized Lyapunov energy functional (LEF). The chapter meticulously examines the convergence behaviour of the discussed FFNNs and RNNs, emphasizing the utilization of the normalized LEF as a guiding metric. The LEF provides a normalized measure of the convergence process, offering insights into the stability and efficiency of the training algorithms.

14.2 FEEDFORWARD NEURAL NETWORKS

FFNN is formed by neurons that are arranged into an input layer, one or more hidden layers and an output layer. The number of input and output variables in the given estimation problem determines the number of neurons in the input and output layers. There are various algorithms proposed to decide on the number of neurons in the hidden layer [36]. FFNN is normally used along with back propagation (BPN) to train neural networks. The basic goal of a FFNN-BPN is to learn and map the input–output relationship from the presented data for training. The forward step involves the computation of input weights, whereas the backward step deals with updating weights and computing errors. First, the training algorithms for the FFNNs are presented for the sake of completion, then the analytical results for deriving the conditions for the stability using the Lyapunov energy functional are derived.

14.2.1 FFNN-BPN (Back-Propagation) Training Algorithm

The state-space representation of a dynamical system is as given in eqn. (14.1).

$$\dot{x} = Ax + Bu \tag{14.1}$$

The main objective of parameter estimation is to find the numerical values of the elements of matrices A and B provided $\dot{x}, x$ and u are known.

$v_0 = [x, u]$ are the known inputs.

The estimation of parameters procedure using the FFNN-BPN can be divided into two steps. The first step is to train the network and hence obtain the weights of the network. The second step is to perform the delta method on the responses from the trained network to estimate the parameters.

Training of the network is achieved by giving the measured data to the network, comparing it with the predicted response and backpropagating the error of the output layer to estimate the weights.

The series of steps involved in training the network is as follows:

I. Initialize the weights randomly.
$W_1 = [w_{11}, w_{12}, \ldots, w_{1n}]$, and $W_2 = [w_{21}, w_{22}, \ldots, w_{2n}]$ are the guesstimates of the initial weights. These weights are later computed based on the recursive weight update rule as given in eqn. (14.2) [1]

$$W(j+1) = W(j) + \mu e v^T + \Omega\big(W(j) - W(j-1)\big) \tag{14.2}$$

where μ is the learning rate, e is the error computed in each step and Ω is the momentum constant which is used to expedite the convergence of the algorithm by smoothing the weight changes.

II. Calculate the intermediate vectors using available data.
y_1 is an intermediate vector between the input and hidden layer, which is a combination of the weights and the known inputs as expressed in eqn. (14.3). Here, W_1 is the weight matrix between the input layer and hidden layer and W_{1b} is the bias weight vector.

$$y_1 = W_1 v_0 + W_{1b} \tag{14.3}$$

The input to the hidden layer v_1 is obtained by applying the activation function on the intermediate value y_1 as given in eqn. (14.4).

$$v_1 = f(y_1) \tag{14.4}$$

The function $f(y_1)$ is a nonlinear sigmoid activation function as given in eqn. (14.5).

$$f(y_1) = \frac{1 - e^{-\lambda y_1}}{1 + e^{-\lambda y_1}} \tag{14.5}$$

λ represents the sigmoid slope parameter.

Intermediate vector y_2 between the hidden and output layers, and the output layer vector v_2 is computed as given in eqn. (14.6) and eqn. (14.7), respectively.

$$y_2 = W_2 v_1 + W_{2b} \tag{14.6}$$

$$v_2 = f(y_2) \tag{14.7}$$

III. Evaluate the error based on data from steps I and II.
The quadratic cost function is as given in eqn. (14.8)

$$E = \frac{1}{2} e e^T \tag{14.8}$$

where $e = \dot{x} - v_2$

The rate of change in the values of parameters with respect to time represented by $\frac{dW}{dt}$ is in the negative direction of the gradient of the cost function with respect to the parameter, i.e. $\frac{\partial E(W)}{\partial W}$. This is mathematically expressed as in eqn. (14.9).

$$\frac{dW}{dt} = -\mu(t)\frac{\partial E(W)}{\partial W} \tag{14.9}$$

where μ is the learning rate.

$$\frac{\partial E}{\partial W_2} = -v_2' e v_1^T \tag{14.10}$$

Eqn. (14.10) is obtained from eqns. (14.6), (14.7) and (14.8).

v_2' is the derivative of the activation function and is computed as given in eqn. (14.11)

$$v_2' = \frac{2\lambda e^{-\lambda y_2}}{\left(1+e^{-\lambda y_2}\right)^2} \tag{14.11}$$

The error in output layer is denoted by e_{2b} and is as given in eqn. (14.12).

$$e_{2b} = v_2' e \tag{14.12}$$

IV. Update the weights.
From eqn. (14.2), the weight update for the output layer is as given in eqn. (14.13).

$$W_2(j+1) = W_2(j) + \mu e_{2b} v_1^T + \Omega\left(W_2(j) - W_2(j-1)\right) \tag{14.13}$$

Eqn. (14.14) shows the back-propagated error of the hidden layer and the weight update for W_1.

$$e_{1b} = v_1' W_2^T e_{2b} \tag{14.14}$$

where v_1' is the derivative of the activation function with respect to y_1.

Similarly, the weight update rule for W_1 is as given in eqn. (14.15).

$$W_1(j+1) = W_1(j) + \mu e_{1b} v_0^T + \Omega\left(W_1(j) - W_1(j-1)\right) \tag{14.15}$$

V. The data is then presented again and the updated weights from the previous iteration is used. This process continues until convergence is reached.

Once the network is trained, the predicted responses from the network are perturbed one by one in order to estimate each parameter. This is called the Delta method.

14.2.2 FFNN-RLS (Recursive Least-Squares) Training Algorithm

Feedforward neural networks with backpropagation using recursive least squares (FFNN-RLS) uses an enhanced form of the backpropagation technique. The conventional approach (Section 14.2.1) minimizes the mean-squared error with respect to the weights. Here, the algorithm minimizes the mean-squared error between the expected output and observed output with respect to the summation outputs [1,37]; and Kalman filter gain and covariance are computed in each layer to update the weights.

The first two steps, I and II, in the algorithms are the same as FFNN-BPN. Next, the Kalman filter equations are computed as follows.

The updates for filter gain K_j and covariance matrix P_j (j for each layer) are calculated as in eqns. (14.16) and (14.17).

$$K_j = P_j v_{j-1}\left(f_j + v_{j-1} P_j v_{j-1}\right)^{-1} \tag{14.16}$$

$$P_j = \frac{P_j - K_j v_{j-1} P_j}{f_j} \tag{14.17}$$

Here f_j represents the forgetting factor of the jth layer.

The modified output error e_{2b} and the back-propagation inner-layer error e_{1b} is calculated in a similar way as FFNN-BPN eqns. (14.12) and (14.14).

The weight update for the output layer is as given in eqn. (14.18).

$$W_2(j+1) = W_2(j) + (d - y_2)K_2^T \tag{14.18}$$

where d is the summation of output and is calculated by using the inverse function as shown in eqn. (14.19).

$$d = \frac{1}{\lambda} ln \frac{1+\dot{x}}{1-\dot{x}} \tag{14.19}$$

λ is the sigmoid slope parameter and $\dot{x}$ is known as given in eqn. (14.1).

The weights of the hidden layer are updated as given in eqn. (14.20).

$$W_1(j+1) = W_1(j) + \mu e_{1b} K_1^T \tag{14.20}$$

The updated weights are used and the algorithm is repeated till the desired convergence is achieved.

As explained in Section 14.2.1, the delta method is used to obtain the parameters.

The major difference between the two algorithms is in the method of back propagation of errors. As FFNN-RLS uses Kalman gain and covariance to update the weights, it is expected to achieve better network parameters in lesser number of iterations as compared to FFNN-BPN.

14.3 LYAPUNOV STABILITY ANALYSIS

Lyapunov stability provides a strong assurance on the convergence of the system for any state. It is well known that a system is Lyapunov stable if and only if there exists a LEF that is strictly positive definite and the derivative of the LEF is strictly negative definite. In order to establish the conditions for the convergence of the algorithms, the Lyapunov–Krasovskii technique is used.

The error state is defined as given in eqn. (14.21).

$$e(\kappa+1) = Ge(\kappa) \tag{14.21}$$

A feasible normalized LEF is described as in eqn. (14.22).

$$V(e(\kappa)) = e^T(\kappa) P^{-1}(\kappa) e(\kappa) \tag{14.22}$$

Eqn. (14.22) is a positive definite matrix, and the derivative of the Lyapunov energy function is obtained from eqns. (14.23)–(14.25).

$$\Delta\{V(e(\kappa))\} = V(e(\kappa+1)) - V(e(\kappa)) \tag{14.23}$$

$$\begin{aligned}\Delta\{V(e(\kappa))\} &= V(e(\kappa+1)) - V(e(\kappa)) \\ &= e^T(\kappa+1)Ye(\kappa+1) - e^T(\kappa)Ye(\kappa) \\ &= e^T(\kappa)G^TYGe(\kappa) - e^T(\kappa)Ye(\kappa) \\ &= e^T(\kappa)[G^TYG - Y]e(\kappa)\end{aligned} \tag{14.24}$$

$$\Delta\{V(e(\kappa))\} = -e^T(\kappa)[-(G^TYG - Y)]e(\kappa) \tag{14.25}$$

The derivative term in eqn. (14.25) should be negative definite for convergence. Therefore, the condition for convergence of error dynamics is stated as in eqn. (14.26).

$$-(G^TYG - Y) = Y - G^TYG$$

$$|Y - G^TYG| > 0$$

$$Y - G^TYG > 0 \Rightarrow G^2 < 1 \tag{14.26}$$

14.3.1 For FFNN-BPN

The network error is represented as $e(\kappa)$, i.e. the difference between the desired output and the actual network output is as shown in eqn. (14.27).

$$e(\kappa+1) = d(\kappa+1) - y(\kappa+1) \tag{14.27}$$

In eqn. (14.27), $d(\kappa+1)$ is the desired output and $y(\kappa+1)$ is the output of the neural network at $(\kappa+1)^{th}$ time instant as shown in eqns. (14.28) and (14.29), respectively.

$$d(\kappa+1) = F_\kappa v_{0,\kappa+1} \tag{14.28}$$

Here v_0 is the input to the input layer of the neural network and F_κ is the transition matrix.

$$y(\kappa+1) = f(y_{2,\kappa+1}) \tag{14.29}$$

where $y_{2,\kappa+1}$ is the intermediate signal between the hidden layer and the output layer and is expressed as shown in eqn. (14.30). $f(\cdot)$ is the nonlinear sigmoidal activation function as mentioned in eqn. (14.5).

$$y_{2,\kappa+1} = \sum w_{2,\kappa+1} v_{1,\kappa+1} \tag{14.30}$$

In eqn. (14.30), $v_1 = f(y_1)$, where v_1 is the input to the hidden layer. y_1 is a vector of intermediate values and is as given in eqn. (14.31).

$$y_{1\kappa} = \sum w_{1,\kappa+1} v_{0,\kappa+1} \tag{14.31}$$

The network error is as given in eqn. (14.32) and is obtained by substituting eqns. (14.28) and (14.29) in eqn. (14.27).

$$e_{\kappa+1} = F_\kappa v_{0,\kappa+1} - f\left(y_{2,\kappa+1}\right) \tag{14.32}$$

$e_{\kappa+1}$ is further elaborated in eqn. (14.33) by substituting eqns. (14.30) and (14.31) in eqn. (14.32)

$$e_{\kappa+1} = F_\kappa v_{0,\kappa+1} - f\left(\sum w_{2,\kappa+1} \times f\left(\sum w_{1,\kappa+1} v_{0,\kappa+1}\right)\right) \tag{14.33}$$

$w_{2,\kappa+1}$ can be rewritten in terms of $w_{2,\kappa}$ as in eqn. (14.13), and eqn. (14.34) is obtained.

$$w_{2,\kappa+1} = w_{2,\kappa} + \mu e_{2,\kappa} v_{1,\kappa}^T + \Omega\left(\Delta w_{2,\kappa}\right) \tag{14.34}$$

where output layer error $e_{2b,\kappa}$ can be derived from eqn. (14.12), as in eqn. (14.35).

$$e_{2b,\kappa} = f'\left(y_{2,\kappa}\right)\left(d_\kappa - y_\kappa\right) = f'\left(y_{2,\kappa}\right) e_\kappa \tag{14.35}$$

Similarly, $w_{1,\kappa+1}$ and $e_{1,\kappa}$ are obtained as in eqns. (14.36) and (14.37) by substituting eqns. (14.14) and (14.15).

$$w_{1,\kappa+1} = w_{1,\kappa} + \mu e_{1,\kappa} v_{0,\kappa}^T + \Omega\left(\Delta w_{1,\kappa}\right) \tag{14.36}$$

$$e_{1,\kappa} = f'\left(y_{1,\kappa}\right) w_{2,\kappa+1}^T e_{2,\kappa} = f'\left(y_{1,\kappa}\right)\left[w_{2,\kappa} + \mu e_{2,\kappa} v_{1,\kappa}^T + \Omega\left(\Delta w_{2,\kappa}\right)\right]^T f'\left(y_{2,\kappa}\right) e_\kappa \tag{14.37}$$

Therefore, eqn. (14.33) can be rewritten as in eqn. (5.33).

$$\begin{aligned} \left\|e(\kappa+1)\right\| = \left\|F_\kappa v_{0,\kappa+1}\right\| - \Big\| f\Big(w_{2,\kappa} + \mu f'\left(y_{2,\kappa}\right) e_\kappa v_{1,\kappa}^T + \Omega\left(\Delta w_{2,\kappa}\right) \times f\Big\{w_{1,\kappa} + \mu f'\left(y_{1,\kappa}\right) \\ \left[w_{2,\kappa} + \mu f'\left(y_{2,\kappa}\right) e_\kappa v_{1,\kappa}^T + \Omega\left(\Delta w_{2,\kappa}\right)\right] f'\left(y_{2,\kappa}\right) e_\kappa v_{0,\kappa}^T + \Omega\left(\Delta w_{1,\kappa}\right)\Big\} v_{0,\kappa+1}\Big)\Big\| \end{aligned} \tag{14.38}$$

To simplify the error dynamics, the following bounds on the norms of the vectors are required as given in eqn. (14.39).

$$\left\|F v_0\right\| = \rho \left\|e\right\|^2$$

$$\left\|e_\kappa\right\| \le \varepsilon;\ \left\|f'(y)\right\| \le h;\ \left\|w\right\| \le \omega;\ \left\|v\right\| \le \upsilon \tag{14.39}$$

Substituting, $1+\Omega = k$ and solving eqn. (14.38), eqn. (14.40) is obtained as

$$\left\| e(\kappa+1) \right\| = \rho\varepsilon^2 - f\left(\omega k + \mu h \upsilon \varepsilon \times f\left(\omega \upsilon k + \mu h^2 \upsilon^2 \varepsilon\left[\omega k + \mu h \upsilon \varepsilon\right]\right)\right) \tag{14.40}$$

ε can be factored out from eqn. (14.40) to get eqn. (14.41).

$$\left\| e(\kappa+1) \right\| = \varepsilon\left(\rho\varepsilon - \frac{1}{\varepsilon} f\left(\omega k + \mu h \upsilon \varepsilon \times f\left(\omega \upsilon k + \mu h^2 \upsilon^2 \varepsilon\left[\omega k + \mu h \upsilon \varepsilon\right]\right)\right)\right) \tag{14.41}$$

Eqn. (14.42) is obtained by comparing eqns. (14.21) and (14.41).

$$G^2 \equiv \left(\rho\varepsilon - \frac{1}{\varepsilon} f\left(\omega k + \mu h \upsilon \varepsilon \times f\left(\omega \upsilon k + \mu h^2 \upsilon^2 \varepsilon\left[\omega k + \mu h \upsilon \varepsilon\right]\right)\right)\right)^2 < 1 \tag{14.42}$$

Now applying the condition of eqn. (14.26) in eqn. (14.42) and taking into consideration the positive square root of 1, eqns. (14.43) and (14.44) are obtained.

$$\rho\varepsilon - \frac{1}{\varepsilon} f\left(\omega k + \mu h \upsilon \varepsilon \times f\left(\omega \upsilon k + \mu h^2 \upsilon^2 \varepsilon\left[\omega k + \mu h \upsilon \varepsilon\right]\right)\right) < \mathrm{sqrt}(1) = \pm 1 \tag{14.43}$$

$$\rho\varepsilon < 1 + \frac{1}{\varepsilon} f\left(\omega k + \mu h \upsilon \varepsilon \times f\left(\omega \upsilon k + \mu h^2 \upsilon^2 \varepsilon\left[\omega k + \mu h \upsilon \varepsilon\right]\right)\right) \tag{14.44}$$

By defining

$$f_{\max} = \max\left(\omega k + \mu h \upsilon \varepsilon \times f\left(\omega \upsilon k + \mu h^2 \upsilon^2 \varepsilon\left[\omega k + \mu h \upsilon \varepsilon\right]\right)\right) \tag{14.45}$$

$$\rho\varepsilon < 1 + \frac{f_{\max}}{\varepsilon} \tag{14.46}$$

All the coefficients appearing in eqn. (14.46) are precisely defined as these are the constants with bounds with positive values. If the condition is satisfied, the time derivative of Lyapunov energy function, as in eqn. (5.18), will be negative definite, and FFNN-BPN will converge.

14.3.2 For FFNN-RLS

The normalized LEF is defined as in eqn. (14.22) and the output error state as in eqn. (14.21).

The steady state solution is defined as $P(\infty)$ and this is expected to be in the range as in eqn. (14.47).

$$p_l I \le P \le p_u I \tag{14.47}$$

Here, $p_u, p_l > 0$ are positive numbers, $P(\circ)$ is a positive definite matrix by definition and the bounds are specified by the constants.

The bounds in eqn. (14.48) are defined to simplify the derivative.

$$\left\| G^T G \right\| \le g^2; \left\| e_\kappa \right\|^2 \le \varepsilon^2; \text{ and } f_1 \text{ is the forgetting factor} \tag{14.48}$$

Eqn. (14.49) defines the derivative of Lyapunov energy function

$$\Delta\{V(e(\kappa))\} = V(e(\kappa+1)) - V(e(\kappa))$$
$$= e^T(\kappa+1)P^{-1}(\kappa+1)e(\kappa+1) - e^T(\kappa)P^{-1}(\kappa)e(\kappa) \quad (14.49)$$

Eqns. (14.50) and (14.51) gives the covariance and Kalman gain for the $\kappa+1^{th}$ time instant.

$$P(\kappa+1) = \frac{P(\kappa) - K(\kappa+1)v_1(\kappa)P(\kappa)}{f_1} \quad (14.50)$$

$$K(\kappa+1) = P(\kappa)v_1(\kappa)\left(f_1 + v_1^T(\kappa)P(\kappa)v_1(\kappa)\right)^{-1} \quad (14.51)$$

v_1 is the input to the hidden layer which is obtained after transforming y through a nonlinear activation function which is the summation of the product of weights and input vector as given in eqn. (14.52).

$$v_1 = f(y_\kappa) = \frac{1 - e^{-\lambda y_k}}{1 + e^{-\lambda y_k}} \quad (14.52)$$

λ is the sigmoid slope parameter of the hidden layer.

This can be rewritten as $v_1 = f\left(\sum_{j=1}^{n} w_{1,j}v_0\right) = f(y_k)$ where n is the number of hidden nodes; $w_{1,j}$ are the weights between input and hidden layer; and v_0 is the input vector to the neural network.

Eqn. (14.53) is obtained by substituting eqns. (14.21), (14.50), (14.51) into eqn. (14.49).

$$\Delta\{V(e(\kappa))\} = e_k^T G^T \left(\frac{P_k - \left[P_k f(y_\kappa)\left(f_1 + f(y_\kappa)^T P_k f(y_\kappa)\right)^{-1}\right] f(y_\kappa) P_\kappa}{f_1} \right)^{-1}$$
$$Ge_\kappa - e_\kappa^T P_\kappa^{-1} e_\kappa \quad (14.53)$$

Eqn. (14.54) is derived by substituting the bounds defined in eqns. (14.47), (14.48) and $\|f(y_k)\| = y_m$ into eqn. (14.53).

$$\Delta\{V(e(\kappa))\} = \varepsilon^2 g^2 \left(\frac{p_l - \dfrac{p_l^2 y_m^2}{f_1 + y_m^2 p_l}}{f_1} \right)^{-1} - \frac{\varepsilon^2}{p_l}$$
$$= \varepsilon^2 \left[g^2 \frac{f_1 + y_m^2 p_l}{p_l} - \frac{1}{p_l} \right] \quad (14.54)$$

The condition for stability is $\nabla\{V(e(\kappa))\}< 0$. Eqns. (14.55) and (14.56) are obtained by applying the condition for stability to eqn. (14.54).

$$\left[g^2\frac{f_1+y_m^2 p_l}{p_l}-\frac{1}{p_l}\right]<0 \tag{14.55}$$

$$g^2\left(f_1+y_m^2 p_l\right)<1 \tag{14.56}$$

All the coefficients appearing in eqn. (14.56) are defined as these are the constants with bounds of positive values. If the conditions of eqn. (14.56) are satisfied, the time derivative of Lyapunov energy function will be negative definite, and FFNN-RLS will converge.

14.4 AIRCRAFT PARAMETER ESTIMATION WITH FFNNS

The following postulated aircraft model is used to depict aircraft longitudinal motion. [1,2]:

$$\dot{u}=\frac{\overline{q}S}{m}C_X-qw-g\sin\theta$$

$$\dot{w}=\frac{\overline{q}S}{mV}C_Z+qu+g\cos\theta \tag{14.57}$$

$$\dot{q}=\frac{\overline{q}S\overline{c}}{I_{yy}}C_m$$

$$\dot{\theta}=q$$

In eqn. (14.58), C_x, C_z and C_m are the nondimensional aerodynamic coefficients expressed as

$$C_Z=C_{Z_0}+C_{Z_\alpha}\alpha++C_{Z_q}\frac{q\overline{c}}{2V}+C_{Z_{\delta_e}}\delta_e$$

$$C_X=C_{X_0}+C_{X_\alpha}\alpha+C_{X_{\alpha^2}}\alpha^2$$

$$C_m=C_{m_0}+C_{m_\alpha}\alpha++C_{m_{\alpha^2}}\alpha^2+C_{m_q}\frac{q\overline{c}}{2V}+C_{m_{\delta_e}}\delta_e \tag{14.58}$$

Here, the state variables are the velocity components in the longitudinal and vertical axis represented by u and w respectively, pitch angle θ, and pitch rate $_q$. The various parameters used for aircraft data simulation are listed in Table 14.1. The control input to aircraft dynamics is elevator deflection and is denoted by δ_e. β is the vector of unknown parameters that is to be estimated and it comprises of the aerodynamic derivatives:

$$\beta=\left[C_{x0},C_{x\alpha},C_{x\alpha^2},C_{z0},C_{z\alpha},C_{z\delta e},C_{m0},C_{m\alpha},C_{m\alpha^2},C_{mq},C_{m\delta e}\right] \tag{14.59}$$

MATLAB implementation is performed to obtain the short period data of a light transport aeroplane and nondimensional longitudinal characteristics are computed to assess the efficacy of

TABLE 14.1
Aircraft Parameters used for Simulation

Parameter	Values Used
Mass m	2280 kg
Wing chord $\overline{c}$	1.5875 m
Wing span S	23.23 m^2
Dynamic Pressure $\overline{q} = \frac{1}{2}\rho V^2$	6.1042 Pa
Initial velocity in x axis $u0$	36 m/s
Initial velocity in z axis $w0$	7 m/s
True air speed $V = \sqrt{u^2 + w^2}$	36.6742 m/s

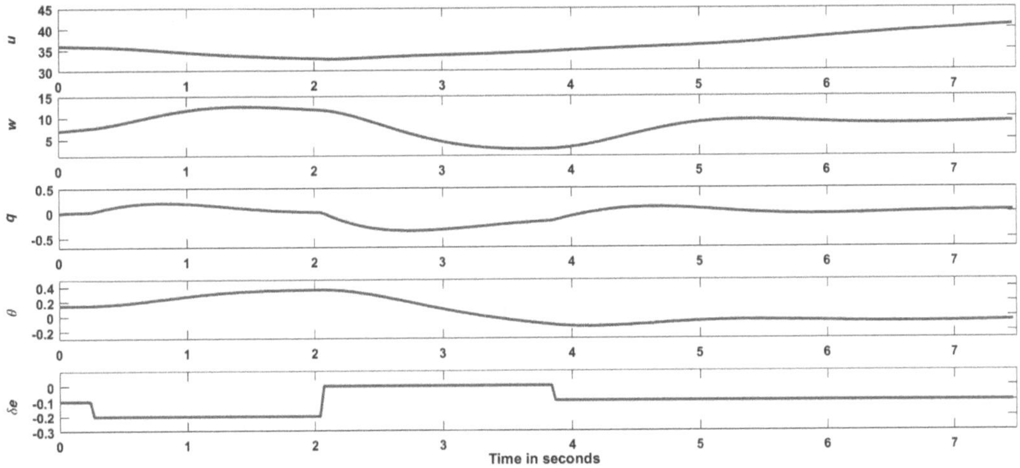

FIGURE 14.1 Time series plot of inputs (Ex1).

FFNN-BPN and FFNN-RLS. Eqns. (14.57) and (14.58) are utilized to estimate the 4-degrees of freedom (4-DOF) longitudinal axis model. In order to validate the algorithms for aircraft parameter estimation, two datasets/examples were simulated.

The two examples are denoted as Ex1 and Ex2. The training dataset for Ex1 is generated for a duration of nearly 7 sec with a sampling interval of 0.03 sec by giving a doublet input to the elevator control surface. The second dataset (for Ex2) is also generated for a duration of approximately 7 sec with a sampling interval of 0.03 sec by giving a 3211 input to the aircraft control surface. The initial values for the states $\left[u, w, q, \theta\right]$ are chosen as $\left[36, 7, 0, 0.15\right]$ for both examples [1]. The time history of input states and control input for Ex1 and Ex2 are shown in Figure 14.1 and Figure 14.3, respectively. Aircraft parameters used to generate both the data sets are given in Table 14.1.

A fourth-order Runge-Kutta integration method is used to obtain the longitudinal flight variables u, w, q and θ from eqn. (14.57). The performance metric used is parameter estimation error norm (PEEN). PEEN is defined as

$$\text{PEEN} = \frac{\text{norm}\left(\beta_{\text{true}} - \beta_{\text{est}}\right)}{\text{norm}\left(\beta_{\text{true}}\right)} \times 100 \tag{14.60}$$

TABLE 14.2
Estimated Aerodynamic Parameters

Parameter	True Values	FFNN-BPN Ex1	FFNN-RLS Ex1	FFNN-BPN Ex2	FFNN-RLS Ex2
C_{X0}	-0.0540	-0.0576	-0.0562	-0.05242	-0.0548
$C_{X\pm}$	0.2330	0.3079	0.2548	0.2221	0.2476
$C_{X\pm^2}$	3.6089	3.5278	3.5533	3.7262	3.5178
C_{Z0}	-0.1200	-0.1226	-0.1197	-0.1291	-0.1205
$C_{Z\pm}$	-5.6800	-5.6799	-5.6683	-5.6767	-5.6783
$C_{Z'e}$	-0.4070	-0.4231	-0.3918	-0.3236	-0.4177
C_{m0}	0.0550	0.0483	0.0481	0.05153	0.05601
$C_{m\pm}$	-0.7290	-0.6835	-0.6819	-0.6786	-0.7455
$C_{m\pm^2}$	-1.7150	-1.8463	-1.8183	-1.9538	-1.5419
C_{mq}	-16.300	-16.5556	-16.0477	-16.4078	-16.2233
$C_{m'e}$	-1.9400	-1.9663	-1.9374	-1.9376	-1.9387
PEEN(%)		**1.8935**	**1.718**	**1.8378**	**1.2801**

where β_{true} is the vector of true values of the aircraft parameters and β_{est} is the vector of estimated values of the aircraft parameters. PEEN is calculated for each method. Table 14.2 presents the estimated parameter values for Ex1 and Ex2. The comparison of the computational times and the PEENs is depicted in Table 14.4.

In order to train the network, simulations are carried out with specifications as in Table 14.1 to generate time histories for the variables$\alpha, \alpha^2, q, \delta_e$ and coefficients C_x, C_m and C_z using the

Eqns. (14.57) and (14.58). The data$\alpha, \alpha^2, q, \delta_e$ is given as the input to the network and C_x, C_m and C_z is presented as the output. The algorithms are used to train the network. FFNN-BPN and FFNN-RLS are used to obtain the updated weights, which can be used to estimate the parameters. Once the network is trained, the delta method is used to estimate the aerodynamic parameters such as $C_{x\alpha}, C_{x\alpha^2}, C_{z\alpha}, C_{z\delta_e}, C_{m\alpha}, C_{m\alpha^2}, C_{m\delta_e}, C_{mq}$. In the delta method, the input variables are perturbed to cause a change in the output variable. For example, the value of q is changed to $q+\Delta q$ for all the data points and a corresponding change in C_m^+ is obtained. Similarly, q is changed to $q-\Delta q$ for all the data points and corresponding change C_m^- is obtained. The derivative C_{mq} will be $C_{mq} = \frac{C_m^+ - C_m^-}{2\Delta q}$. The remaining three parameters C_{x0}, C_{m0} and C_{z0} can be calculated by substituting the parameters estimated using the delta method and the available coefficients, i.e. C_x, C_m and C_z into eqn. (14.58). The estimated parameter values are compared with the true values that are given in the second column of Table 14.2. It is observed that the estimated values are quite close to the true values even when there is noise of SNR = 10 in the data. The PEEN related to Ex1 is 1.89% for FFNN-BPN and 1.71% for FFNN-RLS, which is reasonably small. Similarly, simulations were performed for Ex2 and the PEEN was observed to be 1.97% and 1.35% for the FFNN-BPN and FFNN-RLS algorithms, respectively.

The tuning parameters that have been used for training the network are listed in Table 14.3.

The time histories of the estimated parameters in Figure 14.2 (for doublet control input) and Figure 14.4 (for 3211 control input) also show that the estimated values are significantly close to the true values and the estimates using FFNN-RLS have approximately 10% better PEEN than FFNN-BPN.

TABLE 14.3
Tuning Parameters for FFNN-BPN/FFNN-RLS

Tuning Parameters	FFNN-BPN	FFNN-RLS
No. of hidden layers	1	1
No. of nodes in hidden layer	6	6
Sigmoid slope parameter λ_1	0.8	0.8
Sigmoid slope parameter λ_2	0.75	0.75
Learning parameter	0.2	0.2
Momentum parameter	0.4	-
No. of training iterations	1000	400

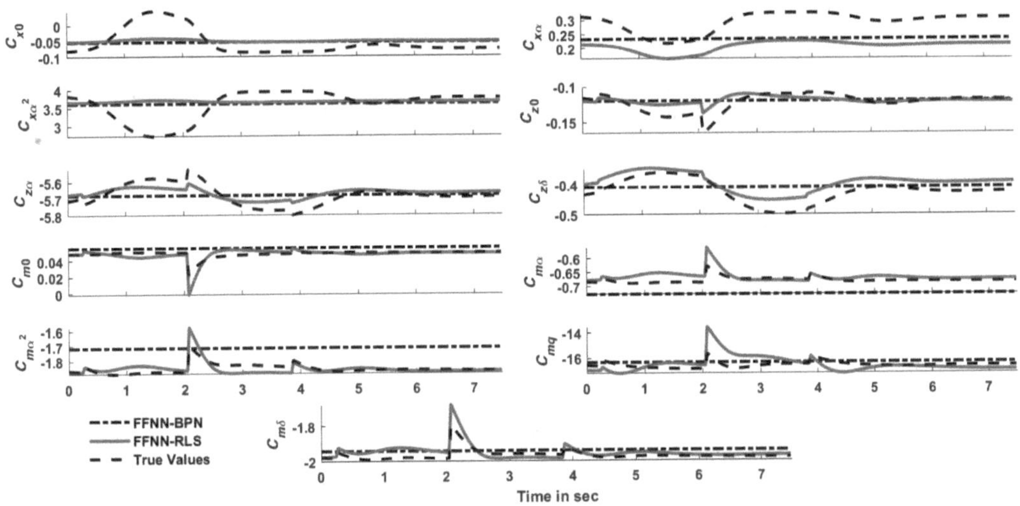

FIGURE 14.2 Time history match of parameters estimated using FFNN-BPN/FFNN-RLS (Ex1).

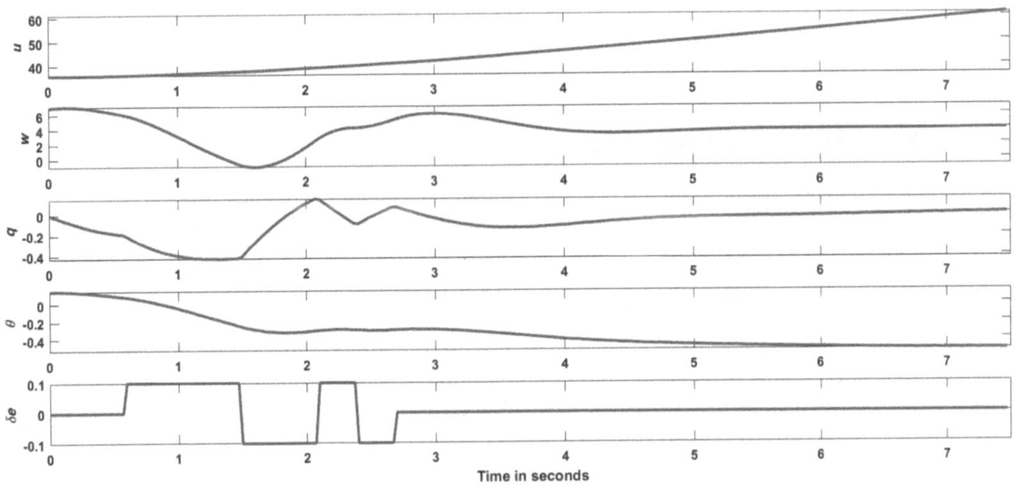

FIGURE 14.3 Time series plot of inputs (Ex2).

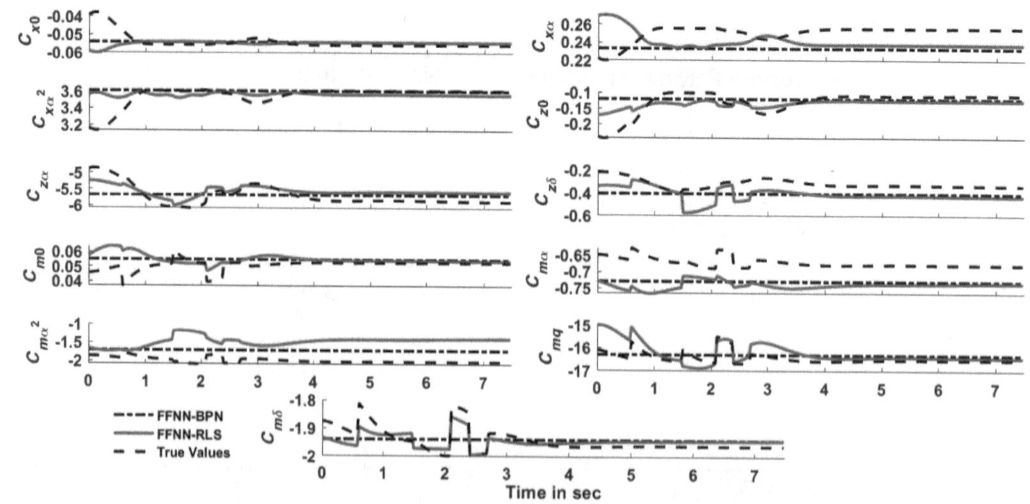

FIGURE 14.4 Time history match of parameters estimated using FFNN-BPN/FFNN-RLS for 3211 input.

TABLE 14.4
Comparison of Execution Times and PEENs

	PEEN in %		Execution Time in Seconds	
	Ex1	Ex2	Ex1	Ex2
FFNN-BPN	1.894	1.8378	3.912173	4.364223
FFNN-RLS	1.718	1.2801	1.129653	1.15643

The comparative CPU time requirements for the two methods are shown in Table 14.4. Computation of the Kalman gain is needed in FFNN-RLS. However, this method, which uses an improved form of back propagation using the Kalman gain (and covariance) requires a lesser number of training iterations to converge. Hence, the overall computational time is seen to be about 60% lesser in FFNN-RLS as compared to FFNN-BPN.

14.5 RECURRENT NEURAL NETWORKS

RNN extracts the nonlinear dynamics of the system by feeding current inputs as well as previous network outputs back into the input layer. These cyclic, time-delayed connections, which can be used as memory for previous events, makes RNN an enhancement over the regular neural networks. RNNs can be programmed to perform the estimation of parameters in dynamic systems [30]. Two variants of RNNs namely the weight-bias method and the gradient-based method are presented. The stability conditions of both the algorithms are derived using the LEF. The performance of proposed RNN-based algorithms is evaluated and compared with true values using MATLAB implementations with realistic simulated data for aircraft parameter estimation for two different control input conditions.

14.5.1 RNN-Weight-Bias Method

Consider the state-space representation of a dynamical system as given in eqn. (14.61).

$$\dot{x} = Ax + Bu \tag{14.61}$$

The aim of parameter estimation is to find the numerical values of the elements of matrices A and B provided $\dot{x}$, x and u are known.

$v_0 = [\dot{x}, x, u]$ are the known quantities.

$\beta = [A, B]$ represent the parameters to be estimated.

The RNN-weight-bias method (RNN-WB) for parameter estimation is based on the Hopfield neural network (HNN).

The dynamic representation of a Hopfield neural network is given as in eqn. (14.62).

$$\dot{x}_i(t) = -x_i(t)R^{-1} + \sum_{j=1}^{n} w_{ij}\beta_j(t) + b_i \tag{14.62}$$

where $\beta_j = f(x_j)$ is the output state of the network and it is the parameters to be estimated in eqn. (14.61). Here, the nonlinear activation function $f(.)$ has been applied to the states x_j of the network. w_{ij} is the neuron weights, b is the bias input to the neurons, R is the neuron impedance and n is the total number of parameters to be estimated.

To neglect the internal state dynamics, R is assumed to be very high and hence eqn. (14.62) is rewritten as in eqn. (14.63).

$$\dot{x}_i(t) = \sum_{j=1}^{n} w_{ij}\beta_j(t) + b_i \tag{14.63}$$

The cost function is as given in eqn. (14.64).

$$E(\beta) = \frac{1}{2}\sum_{k=1}^{N} e^T(k)e(k) \tag{14.64}$$

where $e(k)$ is the error, i.e. $e(\kappa) = \dot{x} - Ax - Bu$

From optimization theory,

$$\frac{d\beta}{dt} = -\frac{\partial E(\beta)}{\partial \beta} = -\frac{1}{2}\frac{\partial \sum_{k=1}^{N} e^T(\kappa)e(\kappa)}{\partial \beta} \tag{14.65}$$

The energy function of Hopfield network is expressed as in eqn. (14.66) **Invalid source specified.**.

$$E = -\frac{1}{2}\sum_i\sum_j W_{ij}\beta_i\beta_j - \sum_i b_i\beta_i \tag{14.66}$$

Eqn. (14.67) is obtained by differentiating eqn. (14.66).

$$\frac{\partial E}{\partial \beta_i} = -\sum_{j=1}^{n} w_{ij}\beta_j - b_i \tag{14.67}$$

Eqns. (14.63) and (14.67) are compared to obtain eqn. (14.68).

$$\frac{\partial E}{\partial \beta_i} = -\dot{x}_i \tag{14.68}$$

The weight and bias matrices are determined through eqn. (14.65) and comparing the same with eqn. (14.67).

$$\text{Consider } \beta_i = f(x_i) \tag{14.69}$$

$$\dot{x}_i = (f^{-1})' \dot{\beta}_i \tag{14.70}$$

Eqn. (14.71) is obtained by combining eqns. (14.68) and (14.70)

$$(f^{-1})' \dot{\beta}_i = -\frac{\partial E}{\partial \beta_i} \tag{14.71}$$

Further solving eqn. (14.71), eqn. (14.72) is obtained.

$$\dot{\beta}_i = -\frac{1}{(f^{-1})' \beta_i} \frac{\partial E}{\partial \beta_i} \tag{14.72}$$

Substituting eqn. (14.67) in eqn. (14.72) gives eqn. (14.73).

$$\dot{\beta}_i = \frac{1}{(f^{-1})' \beta_i} \sum_{j=1}^{n} w_{ij}\beta_j - b_i \tag{14.73}$$

Here f is the nonlinear sigmoid nonlinearity $f(x_i) = \rho\left(\frac{1-e^{-\lambda x_i}}{1+e^{-\lambda x_i}}\right)$

Eqn. (14.74) is obtained by taking inverse and differentiating with respect to the non-linearity function.

$$(f^{-1})' \beta_i = \frac{2\rho}{\lambda(\rho^2 - \beta_i^2)} \tag{14.74}$$

Combining eqns. (14.73) and (14.74), eqn. (14.75) is derived as

$$\dot{\beta}_i = \frac{\lambda(\rho^2 - \beta_i^2)}{2\rho} \sum_{j=1}^{n} w_{ij}\beta_j - b_i \tag{14.75}$$

The initial values of β_i are chosen randomly. Then by integrating eqn. (14.75), the solution to the parameter estimation problem is obtained. The final estimate of the parameters is obtained by performing iterations of eqn. (14.75) and updating the value of parameter in each iteration until convergence is obtained.

14.5.2 RNN-Gradient-Based Method

In the RNN-gradient (RNN-G) method presented in this chapter, the nonlinearity directly operates on the equation error. The nonlinear function does not enter into the neuron dynamic equation. However, it does affect the error by way of quantizing them and thereby reducing the effect of measurement outliers **Invalid source specified.**.

The cost function is the same as in the RNN-WB method as given in eqn (14.64). The gradient of the cost function with respect to the parameters $\frac{\partial E}{\partial \beta}$ is calculated as given in eqn. (14.76).

$$\frac{\partial E}{\partial \beta} = f\left(e(k)\right)\frac{\partial f\left(e(k)\right)}{\partial \beta} \tag{14.76}$$

As mentioned above, in this algorithm nonlinearity is applied on the error and hence $f\left(e(k)\right)$is as in eqn. (14.77).

$$f\left(e(k)\right) = \rho\left(\frac{1-e^{-\lambda e(k)}}{1+e^{-\lambda e(k)}}\right) \tag{14.77}$$

Since $e(\kappa) = \dot{x} - Ax - Bu$ as mentioned in eqn. (14.64), the derivative of $e(\kappa)$is expressed as a combination of weights w, inputs v and bias b as in eqn. (14.78).

$$\frac{\partial e(\kappa)}{\partial \beta} = \sum_{j=1}^{n} w_{ij} v_j + b_i = Wv + b \tag{14.78}$$

Eqn. (14.79) is obtained by differentiating eqn. (14.77) and using eqn. (14.78).

$$\frac{\partial f\left(e(k)\right)}{\partial \beta} = \frac{2\lambda\rho e^{-\lambda e_i}}{\left(1+e^{-\lambda e_i}\right)^2}\left(Wv+b\right) \tag{14.79}$$

Parameters are updated such that it changes with respect to a proportion of the gradient as shown in eqn. (14.80).

$$\beta_{\kappa+1} = \beta_{\kappa} - \mu\frac{\partial E}{\partial \beta} \tag{14.80}$$

Here, μ is the step size.

The initial values of β are chosen randomly. The error is calculated based on the data that is available and the parameters. The nonlinearity is applied on the error. The gradient is calculated, and parameters are updated. Iterations are performed with the updated parameters until convergence.

14.6 LYAPUNOV STABILITY ANALYSIS

In the study of dynamical systems, derivation of stability conditions is one of the most crucial issues. The stability characteristics of RNN directly dictate its engineering applications. Lyapunov approach is employed to derive the conditions for convergence of the algorithms.

The error state is described as given in eqn. (14.81).

$$e(\kappa+1) = Ge(\kappa) \tag{14.81}$$

A feasible normalized LEF is defined as in eqn. (14.82).

$$V(e(\kappa)) = \frac{1}{2}e^T(\kappa)e(\kappa) \tag{14.82}$$

Eqn. (14.82) is a positive definite matrix, and the expression in eqn. (14.83) is an equivalent or analogous derivative of LEF.

$$\Delta\nabla\{V(e(k))\} = V(e(k+1)) - V(e(k)) = \frac{1}{2}e^T(k+1)e(k+1) - \frac{1}{2}e^T(k)e(k)$$

$$= \frac{1}{2}e^T(k)G^T Ge(k) - \frac{1}{2}e^T(k)e(k)$$

$$= \frac{1}{2}e^T(k)[G^T G - 1]e(k)$$

$$= -\frac{1}{2}e^T(k)[-(G^T G - 1)]e(k) \tag{14.83}$$

For convergence of error dynamics, the condition is as in eqn. (14.84)

$$-(G^T G - 1) = 1 - G^T G$$

$$|1 - G^T G| > 0$$

$$\|1 - G^T G\| > 0 \Rightarrow \|G\|^2 < 1 \tag{14.84}$$

14.6.1 For RNN-Weight-Bias Method

Let $e(\kappa)$ represent the network error in an RNN-WB network.

Eqn. (14.85) is obtained by differentiating eqn. (14.82).

$$\frac{\partial V(e(\kappa))}{\partial e(\kappa)} = e(\kappa) \tag{14.85}$$

The Taylor series expansion of $e(k+1)$ and approximation is as in eqn. (14.86) and (14.87).

$$e(\kappa+1) = e(\kappa) + \frac{\partial e(\kappa)}{\partial \beta}\Delta\beta + \ldots \approx e(\kappa) + \frac{\partial e(\kappa)}{\partial \beta}\Delta\beta$$

$$e(\kappa+1) = e(\kappa) + \frac{\partial e(\kappa)}{\partial \beta}\Delta\beta \tag{14.86}$$

$$= e(\kappa) + \frac{\partial e(\kappa)}{\partial E}\frac{\partial E}{\partial \beta}\Delta\beta \tag{14.87}$$

Since the cost function in RNN-WB and LEF are the same, substituting eqn. (14.68) into eqn. (14.87), eqn. (14.88) is obtained.

$$e(\kappa+1) = e(\kappa) - \frac{\dot{x}(\kappa)\Delta\beta}{e(\kappa)} \tag{14.88}$$

Eqn. (14.63) is written as in eqn. (14.89).

$$\dot{x}(t) = W\{f(x(t))\} + b = W\beta + b \tag{14.89}$$

Substituting eqns. (14.75) and (14.89) into eqn. (14.88)

$$e(\kappa+1) = e(\kappa) - \frac{1}{e(\kappa)}(W\beta+b)\left(\frac{\lambda(\rho^2-\beta_i^2)}{2\rho}(W\beta+b)\right) \tag{14.90}$$

Eqn. (14.91) is obtained by simplifying eqn. (14.90).

$$e(\kappa+1) = e(\kappa) - \frac{\lambda(\rho^2-\beta^2)}{2\rho e(\kappa)}(W\beta+b)^2 \tag{14.91}$$

The bounds on the norms of the vectors in eqn. (14.92) are needed to simplify the error dynamics.

$$\|e_k\| \le \varepsilon; \quad \|\beta\| \le l; \quad \|W\| \le w; \quad \|b\| \le \delta \tag{14.92}$$

Eqn. (14.93) is obtained by substituting eqn. (14.81) and the bounds in eqn. (14.92) into eqn. (14.91) and simplifying

$$G = 1 - \frac{\lambda(\rho^2-l^2)}{2\rho\varepsilon^2}(wl+\delta)^2 \tag{14.93}$$

Squaring eqn. (14.93) and applying the condition in eqn. (14.84), eqn. (14.94) is obtained.

$$\|G\|^2 \equiv \left\{1 - \frac{\lambda(\rho^2-l^2)}{2\rho\varepsilon^2}(wl+\delta)^2\right\}^2 < 1 \tag{14.94}$$

Considering the positive square root of 1, eqns. (14.95) and (14.96) are obtained.

$$1 - \frac{\lambda(\rho^2-l^2)}{2\rho\varepsilon^2}(wl+\delta)^2 < 1 \tag{14.95}$$

$$\frac{\lambda(\rho^2-l^2)}{2\rho\varepsilon^2}(wl+\delta)^2 > 0 \tag{14.96}$$

As all the terms in eqn. (14.96) are positive, the final condition for the term to be greater than 0 is that $\rho^2 > l^2$, implying that the magnitude of the activation function should be greater than the magnitude of the parameters, and this is intuitively appealing as if the nonlinear activation function is acting as a band pass filter for the parameters, and at the same time it acts as a saturation/cut-off for the large variations of the parameters as well as noise.

14.6.2 For RNN-Gradient-Based method

The normalized LEF is defined as in eqn. (14.97).

$$V(e(k)) = \frac{1}{2} e^T(k) e(k) \tag{14.97}$$

where $e(k)$ represents the network error in RNN-G network. Since the LEF and the cost function are the same, eqn. (14.80) can be rewritten as in eqns. (14.98) and (14.99).

$$\beta_{\kappa+1} = \beta_\kappa - \mu \frac{\partial V(e(\kappa))}{\partial \beta} \tag{14.98}$$

$$\Delta\beta = -\mu \frac{\partial V}{\partial \beta} \tag{14.99}$$

Eqn. (14.100) is obtained by substituting eqn. (14.76) in eqn. (14.99).

$$\Delta\beta = -\mu f(e(\kappa)) \frac{\partial f(e(\kappa))}{\partial \beta} \tag{14.100}$$

Substituting eqn. (14.79) and $f(e(k)) = \gamma = \rho\left(\frac{1-e^{-\lambda e_i}}{1+e^{-\lambda e_i}}\right)$ in eqn. (14.100), eqn. (14.101) is obtained.

$$\Delta\beta = -\frac{2\mu\gamma\lambda\rho e^{-\lambda e(\kappa)}}{\left(1+e^{-\lambda e(\kappa)}\right)^2}(Wv+b) \tag{14.101}$$

The Taylor series expansion of $e(k+1)$ and approximation used for the derivation is as in eqn. (14.102)

$$e(k+1) = e(k) + \frac{\partial e(k)}{\partial \beta}\Delta\beta + \ldots \approx e(k) + \frac{\partial e(k)}{\partial \beta}\Delta\beta$$

$$e(k+1) = e(k) + \frac{\partial e(k)}{\partial \beta}\Delta\beta \tag{14.102}$$

Eqn. (14.103) is obtained by substituting eqns. (14.78) and (14.101) into eqn. (14.102).

$$e(k+1) = e(k) - \frac{2\mu\gamma\lambda\rho e^{-\lambda e(k)}}{\left(1+e^{-\lambda e(k)}\right)^2}(Wv+b)^2 \tag{14.103}$$

Also, the bounds expressed in eqn. (14.104) are needed to simplify the derivative.

$$\|e(k)\| \le \varepsilon,\ \|W\| \le w,\ \|\gamma\| \le \rho, \left\| \frac{e^{-\lambda e_i}}{\left(1+e^{-\lambda e_i}\right)^2} \right\| \le \frac{1}{4},\ \|b\| \le \delta \tag{14.104}$$

Applying the bounds and simplifying eqn. (14.103), we get eqn. (14.105)

$$G\varepsilon = \varepsilon - \frac{\mu\lambda\rho^2}{2}(wv+\delta)^2 \tag{14.105}$$

Applying the condition of eqn. (14.84) and considering the positive square root of 1, eqn. (14.106) is given as

$$\|G\|^2 \equiv \left\{1 - \frac{\mu\lambda\rho^2}{2\varepsilon}(wv+\delta)^2\right\}^2 < 1 \tag{14.106}$$

$$\left\{1 - \frac{\mu\lambda\rho^2}{2\varepsilon}(wv+\delta)^2\right\} < sqrt(1) = \pm 11.1.1$$

$$1 - \frac{\mu\lambda\rho^2}{2\varepsilon}(wv+\delta)^2 < 11.1.1$$

Thus, the convergence condition for the error dynamics is derived as in eqn. (14.107).

$$\frac{\mu\lambda\rho^2}{2\varepsilon}(wv+\delta)^2 > 0 \tag{14.107}$$

The bracketed term is always positive, and the stability of the RNN-gradient-based method for aircraft parameter estimation is assured.

All the coefficients appearing in eqn. (14.107) are well defined as these are positive bound constants. The condition derived in eqn. (14.107) means that if this is satisfied then, the time derivative, analogously the differential of LEF in eqn. (14.83) is negative definite and RNN converges. This is a novel way of establishing the asymptotic stability of the RNN-G algorithm as, if these conditions are satisfied, then the algorithms using the RNN converges.

14.7 AIRCRAFT PARAMETER ESTIMATION WITH RNNS

In this section, the performance of RNN-WB and RNN-G are tested and validated for aerodynamic parameter estimation with two different control inputs. The aerodynamic parameter estimation problem mentioned in Section 14.4 is considered for evaluating the efficacy of the algorithms. MATLAB implementation is performed to obtain short period data of an LTA and nondimensional longitudinal characteristics are estimated using RNN-WB and RNN-G. Two examples based on the choice of the control input, namely doublet control input and 3211 control input are considered. The evaluation metric considered for analysing the aerodynamic parameter estimation efficiency is PEEN and computation time. The 4-degrees of freedom (4-DOF) longitudinal body axis model is used for estimation. The initial states for $[u, w, q, \theta]$ are chosen as [36, 7, 0, 0.15]. Random Gaussian noise with a standard deviation 0.001 (as the process noise) is added to the states u, w, q and θ. A fourth-order Runge-Kutta (RK) integration is used to obtain the longitudinal flight variables

u, w, q and θ from eqn. (14.57). The tuning parameters used for training the networks are listed in Table 14.5. Initial values of parameters are chosen randomly. The same initial parameters are used for both the algorithms. The network is trained using time histories of $\alpha, \alpha^2, q, \delta$ to obtain the coefficients C_x, C_m and C_z using eqns. (14.57) and (14.58).

The steps taken to train the network and estimate the parameters are explained below.

In the RNN-WB method, parameters are nonlinear functions of state and the following steps are followed.

(i) The cost function and then the derivative of cost function with respect to each aerodynamic parameter is computed using eqns. (14.64) and (14.65).
(ii) This derivative is then compared with eqn. (14.68) to obtain the weights and biases.
(iii) The parameter values to be estimated are chosen randomly.
(iv) Iterations are performed by substituting the computed weights/biases in eqn. (14.75) and integrating to estimate these parameters.

In the case of the RNN-G method, the following steps are followed.

(i) The initial values of parameters β are chosen randomly.
(ii) The error is calculated based on the data that is available and the parameters.

TABLE 14.5
Tuning Parameters

Tuning Parameters	RNN-WB	RNN-Gradient
Sigmoid slope parameter λ	0.5	0.5
Learning parameter μ	-	0.01
Sigmoid magnitude parameter ρ	40	1
No. of training iterations	15000	6000

TABLE 14.6
Estimated Parameters for RNN-WB/RNN-Gradient

Parameter	True Values	RNN-WB	RNN-G
C_{X0}*	-0.0540	-0.054	-0.054
$C_X\alpha$	0.2330	0.2328	0.2325
$C_X\alpha^2$	3.6089	3.609	3.6098
C_{Z0}	-0.1200	-0.1200	-0.1200
$C_Z\alpha$	-5.6800	-5.6797	-5.6799
$C_Z\delta$	-0.4070	-0.4068	-0.4069
C_{m0}	0.0550	0.0621	0.0601
$C_m\alpha$	-0.7290	-0.792	-0.7706
$C_m\alpha^2$	-1.7150	-1.571	-1.6166
C_{mq}	-16.300	-16.122	-16.171
$C_m\delta$	-1.9400	-1.9318	-1.932
PEEN(%)		**1.331**	**0.937**

(*) The aerodynamic derivatives are expressed in the form of the text for convenience.

(iii) The nonlinearity is applied on the error as in eqn. (14.80).
(iv) The gradient is calculated, and parameter is updated.
(v) Iterations are performed with the updated parameters until convergence.

In this scenario, doublet control input is provided to the elevator control surface and data is generated using eqns. (14.57)–(14.59), for a span of approximately 7 sec with a sampling interval of 0.03 sec. The time series plot of variation of input states when this control is applied is shown in Figure 14.1. The aerodynamic parameters are estimated using RNN-WB and RNN-G algorithms.

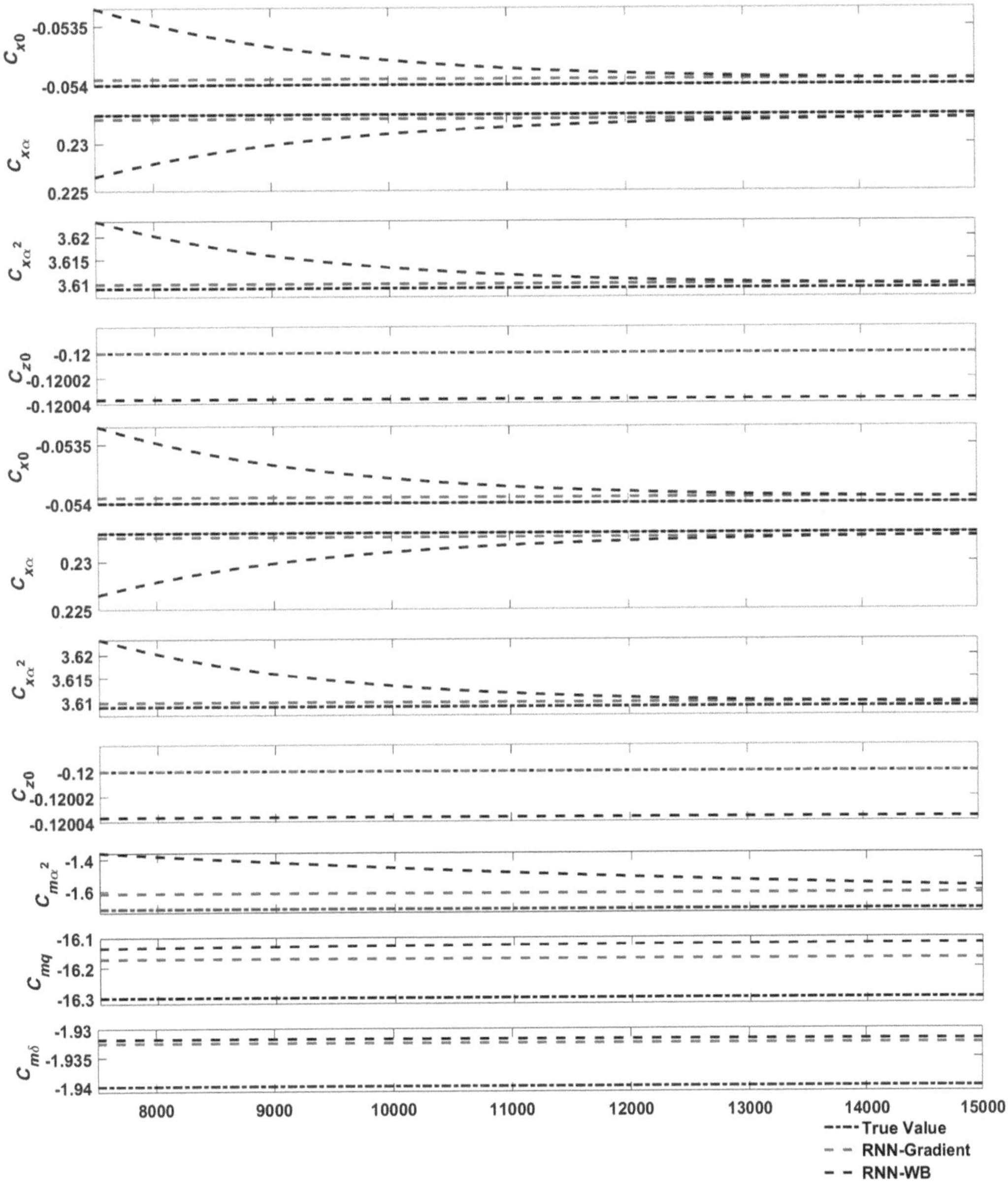

FIGURE 14.5 Estimated parameters from 7500 to 15,000 iterations for doublet control input.

Initial values for the parameters are chosen randomly with the same values for both the algorithms. The estimated aerodynamic parameter values from the final iteration along with the true values are presented in Table 14.6. It is observed that PEEN for RNN-WB and RNN-G is 1.331% and 0.937%. From the PEEN values at every iteration, it is also observed that RNN-G converged at approximately 6000 iterations while RNN-WB converged at 15,000 iterations. However, for the sake of comparing the algorithms graphically, RNN-G was continued for 15,000 iterations. It was also noted that towards convergence the estimates from both the algorithms cannot be distinguished from one another when all the iterations were plotted as the initial variations are very high due to randomly selected values for the parameters to be estimated. Hence, the estimation plot is presented in Figure 14.5 for 7500 to 15,000 iterations. Figure 14.5 also shows that RNN-G converges faster than RNN-WB.

Next, 3211 control input is provided to the elevator control surface and data is generated using eqns. (14.57)–(14.59), for a span of approximately 7 sec with a sampling interval of 0.03 sec. The time series plot of variation of input states when this control is applied is shown in Figure 14.3. The aerodynamic parameters are estimated using RNN-WB and RNN-G algorithms. The initial values for the parameters are chosen randomly and the same values are presented to both the algorithms. The estimated aerodynamic parameter values from the final iteration along with the true values are presented in Table 14.7.

It can be observed that PEEN for RNN-WB and RNN-G is 1.5152% and 1.0938%. The RNN-G converged at roughly 6000 iterations, but RNN-WB required over 15,000 iterations. However, for the sake of comparing the algorithms graphically RNN-G is continued for 15,000 iterations. The plots are presented in Figure 14.6 for 7500 to 15,000 iterations.

The comparative CPU time requirements and PEEN are shown in Table 14.8.

It is observed that the performance of RNN-G is slightly better than RNN-WB. Further, in order to verify the stability analysis conditions of the algorithms, the aircraft parameter estimation problem with doublet control input is simulated and time-steps from 200 to 250 are considered to obtain the bounds. The bounds of the various terms obtained from the simulations and the conditions of stability are listed in Table 14.9. It is observed that the stability conditions are satisfied for all the algorithms.

TABLE 14.7
Estimated Parameters for RNN-WB/RNN-Gradient for 3-2-1-1 Control

Parameter	True Values	RNN-WB	RNN-G
C_{X0}	-0.0540	-0.0539	-0.05402
$C_x\alpha$	0.2330	0.2326	0.2336
$C_x\alpha^2$	3.6089	3.6097	3.6054
C_{Z0}	-0.1200	-0.1201	-0.1200
$C_z\alpha$	-5.6800	-5.6797	1.1. -5.6799
$C_z\delta$	-0.4070	-0.4068	-0.4069
C_{m0}	0.0550	0.0639	1.2. 0.0557
$C_m\alpha$	-0.7290	-0.8102	-0.7443
$C_m\alpha^2$	-1.7150	-1.5314	-1.6603
C_{mq}	-16.300	-16.1193	-16.4866
$C_m\delta$	-1.9400	-1.9315	-1.9462
PEEN(%)		**1.516**	**1.093**

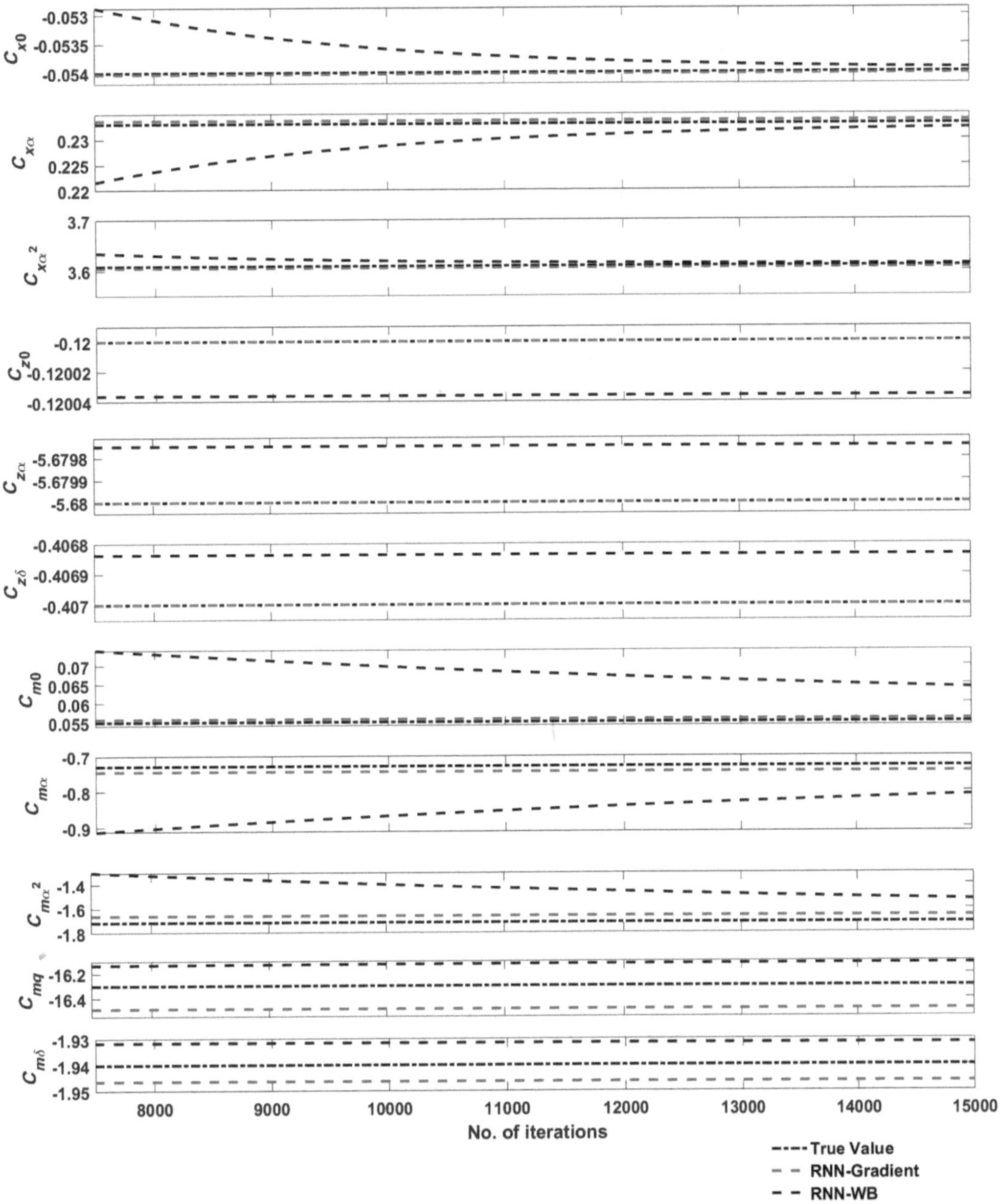

FIGURE 14.6 Estimated parameters from 7500 to 15,000 iterations for 3211 control input.

14.8 CONCLUSION

Two variations each of feedforward neural network and recurrent neural network namely FFNN-BPN, FFNN-RLS, RNN-WB and RNN-G are discussed in this chapter and the performance of these algorithms is evaluated for aircraft parameter estimation with two different elevator control inputs. The major difference between the two FFNN is in the method by which the errors are propagated back to update the network weights. The simulation results show that the estimation accuracy of FFNN-RLS is better than FFNN-BPN by approximately 10%. Although the number of steps in

TABLE 14.8
Comparison of PEENs and Computation for Doublet Control

	PEEN in %	Comp. Time (seconds)
RNN-WB	1.331	2.788
RNN-Gradient	0.937	2.125

TABLE 14.9
Stability Conditions for Doublet Control

Algorithms	Formula	Bounds obtained by simulation	Substitution
RNN-WB	$\frac{\lambda(\rho^2 - l^2)}{2\rho\varepsilon^2}(wl+\delta)^2 > 0$	$\lambda = 0.5$ $A = 20$ $l = 18$ $\mu = 0.4$ $w = 1$ $\delta = 1$	**2143 > 0**
RNN-G	$\frac{\mu\lambda\rho^2}{2\varepsilon}(wk+\delta)^2 > 0$	$\mu = 0.01$ $\lambda = 0.5$ $A = 1$ $\mu = 0.4$ $k = 17$ $w = 1$ $\delta = 1$	**2.025 > 0**

the RLS algorithm is more than the BPN algorithm due to the computation of the Kalman gain and covariance, iterations required for convergence has been observed to be lesser in FFNN-RLS making the algorithm faster on the whole. RNN-WB is based on Hopfield neural network and relies on pre-computation of weight and bias information. Here the nonlinearity is applied to the states of the network. On the other hand, RNN-G is based on computation of direct gradient and nonlinearity is applied on the error. The asymptotic stability conditions for all the algorithms are derived using the Lyapunov approach via proposition of a normalized LEF. The simulation results show that the estimation accuracy obtained by both the neural-network-based algorithms are encouraging, and this paves the way to further consider these techniques for parameter estimation of other aircrafts, and even for rotorcraft system identification.

EXERCISES

Exercise 14.1

Compare and contrast the back-propagation method with the recursive least-squares algorithm (RLS) in the context of parameter estimation.

Exercise 14.2

Describe the steps involved in the convergence analysis of feedforward neural networks (FFNNs) using the normalized Lyapunov energy functional (LEF).

Exercise 14.3

How can knowledge of convergence behaviour impact the practical implementation of neural network models in complex systems?

Exercise 14.4

Explain the RNN-weight-bias method (RNN-WB) for parameter estimation in dynamic systems. Outline the key steps involved in this method.

Exercise 14.5

Explore the applicability of the convergence analysis using Lyapunov stability in domains beyond parameter estimation, such as classification or regression problems.

REFERENCES

1. Raol, J.R., Girija, G., and Singh, J., *Modelling and Parameter Estimation of Dynamic Systems*. The IEE/IET Control Series, 2004, 65.
2. Sara, M.G. Approximations in Nonlinear Filtering: Some Novel Approaches. Ph.D. Thesis, Electrical and Electronics Engineering Sciences, Visvesvaraya Technological University (VTU), Jnana Sangama, VTU Main Road, Machhe, Belagavi, Karnataka, 2023.
3. Iliff, K.W., Parameter estimation for flight vehicles, AIAA Journal of Guidance and Control, Vol. 12, No. 5, pp. 609–622, 1989.
4. Milenković-Babić, M., Samardžić, M., Antonić, V., Marjanović, M., and Stefanović-Gobeljić, V., Longitudinal stability characteristics of the LASTA airplane, *Aircraft Engineering and Aerospace Technology*, Vol. 89, No. 6, pp. 911–919, 2017.
5. Wolowicz, H., and Yancey, R.B., Longitudinal aerodynamic characteristics of light, twin-engine, propeller-driven airplanes, NASA Technical Note, 1972.
6. Nicolosi, F., Corcione, S., and Della Vecchia, P., Commuter aircraft aerodynamic characteristics through wind tunnel tests, *Aircraft Engineering and Aerospace Technology: An International Journal*, Vol. 88, No. 4, pp. 523–534, 2016.
7. Chauhan, K., and Singh, S., Review of aerodynamic parameter estimation techniques. In *International Conference on Infocom Technologies and Unmanned Systems (Trends and Future Directions) (ICTUS)*, pp. 864–869, 2017.
8. Chen, R.T.N., Eulrich, B.J., and Lebacqz, J.V., *Development of Advanced Techniques for the Identification of V/STOL Aircraft Stability and Control Parameters*, Cornell Aeronautical Lab., Inc., USA, Report No. BM-2820-F-1, 1971.
9. Jategaonkar, R.V., and Plaetschke, E., Algorithms for aircraft parameter estimation accounting for process and measurement noise, Journal of Aircraft, Vol. 26, No. 4, pp. 360–372, 1989.
10. Garcfa-Velo, J., and Walker, B.K., Aerodynamic parameter estimation for high performance aircraft using extended Kalman filtering, *Journal of Guidance, Control, and Dynamics*, Vol. 20, No. 6, pp. 1257–1259, 1997.
11. Chowdhary, G., and Jategaonkar, R., Aerodynamic parameter estimation from flight data applying extended and unscented Kalman filter, *Aerospace Science and Technology*, Vol. 14, No. 2, pp. 106–117, 2010.
12. Julier, S.J., and Uhlmann, J.K., Unscented filtering and nonlinear estimation, *Proceedings of the IEEE*, Vol. 92, No. 3, pp. 401–422, 2004.
13. Chaabane, M., Baklouti, I., Mansouri, M., Jaoua, N., Nounou, H., Nounou, M., and Destain, M.F., Nonlinear state and parameter estimation using iterated sigma point kalman filter: Comparative studies, *Nonlinear Systems – Design, Analysis, Estimation and Control*, IntechOpen, 2016.
14. Gutierrez-Villalobos, J.M., Rodríguez-Reséndiz, J., Rivas-Araiza, E.A., and Mucino, V.H., A review of parameter estimators and controllers for induction motors based on artificial neural networks, *Neurocomputing*, Vol. 118, pp. 87–100, 2013.

15. Sadeghi-Goughari, M., Mojra, A., and Sadeghi, S., Parameter estimation of brain tumors using intraoperative thermal imaging based on artificial tactile sensing in conjunction with artificial neural network, *Journal of Physics D: Applied Physics*, Vol. 49, No. 7, 075404, 2016.
16. Jamili, E., and Dua, V., Parameter estimation of partial differential equations using artificial neural network, *Computers & Chemical Engineering*, Vol. 147, 107221, 2021.
17. Agrawal, S., Gobiha, D., and Sinha, N., Nonlinear parameter estimation of airship using modular neural network, *The Aeronautical Journal*, Vol. 124, No. 1273, pp. 409–428, 2020. doi:10.1017/aer.2019.125
18. Gabrielli, L., Tomassetti, S., Squartini, S., and Zinato, C., Introducing deep machine learning for parameter estimation in physical modelling. In *Proceedings of the 20th International Conference on Digital Audio Effects (DAFx-17)*, 2017.
19. Sinha, M., Kalra, P.K., and Kumar, K., Parameter estimation using compensatory neural networks, *Sadhana*, Vol. 25, pp. 193–203, 2000.
20. Yang, Y., and Shi, Y., State and parameter estimation algorithm for state space model based on linear neural network and Kalman filter. In *IEEE International Conference on Mechatronics and Automation (ICMA)*, pp. 2314–2318, 2019.
21. Raol, J.R., and Jategaonkar, R.V., Aircraft parameter estimation using recurrent neural networks – a critical appraisal. In *AIAA Atmospheric Flight Mechanics Conference*, 1995, Chapter No AIAA 95-3504.
22. Alessandri, A., Cirimele, G., Cuneo, M., Pagnan, S., and Sanguineti, M., EKF learning for feed-forward neural networks. In European Control Conference (ECC), pp. 1990–1995, 2003.
23. Narendra, K.S., and Parthasarathy, K., Identification and control of dynamical systems using neural networks, *IEEE Transactions on Neural Networks*, 1990, Vol. 1, No. 1, pp. 4–27.
24. Salehinejad, H., Sankar, S., Barfett, J., Colak, E., Valaee, S., Recent advances in recurrent neural networks, ArXiv preprint arXiv:1801.01078, 2017.
25. Astha S., Dynamics estimation using recurrent neural network, Arxiv, vol. Preprint Arxiv:1809.06148, 2018.
26. Harika, P., Sonam, S.K., et al., State estimation and tracking using recurrent neural networks, *International Journal of Engineering Research and Technology*, Vol. 6, No. 5, pp. 545–549, 2017.
27. Behera, S.K., & Rana, D., System identification using recurrent neural network, *International Journal of Advanced Research in Electrical, Electronics, Instrumentation Engineering*, Vol. 3, No. 3, 2014.
28. Mohajerin, N., *Identification and Predictive Control Using Recurrent Neural Networks*, Orebro University, 2012.
29. Kaur, J., Mahajan, H.K., Singh, S., Banerjee, S., Recurrent neural network for aircraft parameter estimation. In *IEEE 6th International Conference on Computing, Communication and Automation (ICCCA)*, Arad, Romania, pp. 604–608, 2021.
30. Raol, J.R., and Madhuranath, H., Neural network architectures for parameter estimation of dynamical systems, *IEE Proceedings – Control Theory and Applications*, Vol. 143, No. 4, pp. 387–394, 1996.
31. Raol, J.R., Parameter estimation of state space models by recurrent neural networks, *IEE Proceedings – Control Theory and Applications*, Vol. 142, No. 2, pp. 114–118, 1995. doi:10.1049/ip-cta:19951733
32. Ogunmolu, O., Gu, X., Jiang, S., and Gans, N.R., Nonlinear systems identification using deep dynamic neural networks, ArXiv, 2016.
33. Rajagopal, R.B., Harika, P., BhanuPrasad, C., and Sonam, S.K., State estimation and tracking using recurrent neural networks, *International Journal of Engineering Research and Technology*, Vol. 6, No. 5, pp. 545–549, 2017.
34. Rhudy, M.B., *Sensitivity and Stability Analysis of Nonlinear Kalman Filters with Application to Aircraft Attitude Estimation*, West Virginia, West Virginia University, 2013.
35. Khalil, H.K., Lyapunov stability, Control Systems, Robotics and Automation, Vol. 12, p. 115, 2009.
36. Sheela, K.G., and Subramaniam, N.D., Review on methods to fix number of hidden neurons in neural networks, *Mathematical Problems in Engineering*, Vol. 1, Article ID 425740, 2013.
37. Scalero, R.S., and Tepedelenlioglu, N., A fast new algorithm for training feedforward neural networks, *IEEE Transactions on Signal Processing*, Vol. 40, No. 1, pp. 202–210, 1992.

15 Interactive Multiple Modelling for Target Tracking with New Algorithms

15.1 INTRODUCTION

Target tracking is a pivotal aspect of various technological domains, including surveillance, robotics, and autonomous systems [1]. The ability to accurately monitor and predict the movements of dynamic entities is crucial for applications ranging from military operations to traffic management and beyond [2]. Early tracking systems relied on simple algorithms and sensors, often struggling to cope with dynamic and unpredictable scenarios [3]. The advent of radar and sensor technologies prompted the development of more sophisticated tracking approaches. Various estimation techniques can be applied to target tracking applications. Many target tracking algorithms are proposed based on so many researches, such as the alpha-beta filter, Kalman filter, extended Kalman filter, cubature Kalman filter and so on [3–5]. In conventional target tracking, the most common method is the standard Kalman filter algorithm [1,6]. When the target state and observation equations are linear equations and the observation noise is Gaussian white noise, the Kalman filter algorithm is the optimal algorithm in the sense of minimum variance [7]. As technology continued to advance, the limitations of these traditional approaches became apparent in scenarios involving maneuvering targets. In maneuvering target tracking, where the target may change its motion dynamically, the motion model changes and hence Kalman filtering accuracy will be reduced or even diverged due to the model mismatch and noise characteristics that cannot be known exactly. In this case, the detection and tracking of motion model changes are crucial to get accurate estimation accuracy [1]. Recognizing the limitations of conventional tracking methodologies, researchers have turned to more adaptive and sophisticated techniques. One such innovative approach that has gained considerable attention is interactive multiple modelling (IMM) [8]. The motivation behind the adoption of IMM lies in its potential to overcome the shortcomings of single-model tracking approaches. IMM, as a multi-model framework, introduces the concept of using multiple dynamic models, each representing a different potential motion pattern of the target. The adaptability of IMM is a key motivator for its widespread adoption. The ability to seamlessly switch between models based on observed data allows the tracking system to dynamically respond to changes in the target's behaviour. This adaptability is particularly relevant in scenarios where targets may undergo rapid maneuvers or exhibit complex motion patterns. This chapter provides an in-depth exploration of IMM, aiming to elucidate its fundamental principles, showcase its application in tracking maneuvering targets, and rigorously evaluate its performance using advanced filtering techniques.

DOI: 10.1201/9781003620211-15

15.2 PROCESS OF TARGET TRACKING

In the pursuit of accurate target tracking, especially in dynamic environments with maneuvering targets, employing multiple dynamic models has become imperative. Section 15.2.1 delves into the intricacies of incorporating multiple models within the IMM framework, emphasizing their significance in handling the challenges posed by maneuvering targets.

The process of target tracking within the IMM framework is a sophisticated and dynamic system designed to predict and monitor the trajectories of moving objects in diverse scenarios. At the core of this process is the integration of various dynamic models, each tailored to represent specific aspects of a target's potential motion patterns. These models collectively contribute to a comprehensive suite that allows the tracking system to adapt to a broad spectrum of dynamic behaviours exhibited by targets.

The selection and design of these models are pivotal elements of the target tracking process. Different models may include representations of linear motion, constant velocity, or even more complex maneuvers. Each model is strategically chosen to address specific characteristics of the tracked targets and the environmental conditions in which they operate. For instance, a model capturing linear motion might be effective for tracking objects with consistent speed, while a model accommodating abrupt changes may handle dynamic scenarios more adeptly.

The process of target tracking within the IMM framework is a nuanced and adaptive system. By strategically integrating different dynamic models, the tracking system gains the flexibility to handle diverse motion patterns exhibited by targets. The selection, design and adaptability of these models are pivotal elements that collectively empower the system to provide accurate and reliable tracking in real-world scenarios. The utilization of various models reflects the system's versatility, enabling it to tackle the complexities posed by different targets and environmental conditions.

15.2.1 Multiple Models for Maneuvering Targets

One notable aspect of the target tracking process is the adaptability facilitated by model switching. The tracking system dynamically transitions between these multiple models based on observed data, ensuring responsiveness to changes in the target's behaviour. The criteria for model switching involve evaluating the likelihood of each model given the observed data. This adaptability ensures that the tracking system can make real-time adjustments to evolving dynamics, maintaining accurate and robust tracking performance.

The choice of different models in the target tracking process depends on the nature of the tracked objects and the complexity of their motion patterns. Models may include but are not limited to constant velocity models, constant acceleration models, coordinated turn or more intricate representations accommodating sudden changes in direction or speed. Each model contributes to the system's ability to handle a diverse array of tracking scenarios, offering versatility in adapting to various target behaviours.

15.2.2 Process of Interaction

The process of interaction within the IMM framework is an important element that sets it apart from traditional tracking methodologies. This dynamic and adaptive system leverages the interaction between multiple models to seamlessly switch and adapt to changing target behaviours, ensuring robust and accurate tracking in complex scenarios. The IMM estimator is a sub-optimal hybrid filter that is a cost-effective hybrid state estimation scheme. The main feature of this algorithm is its ability to estimate the state of a dynamic system with several behaviour modes, which can switch from one to another. As mentioned earlier, noise can occur either in the form of measurement noise or process noise. If there are sufficient models in the IMM model domain then the algorithm will give more credence to a correct model, so that the process noise will get reduced.

In multiple model estimation, it is assumed that the possible system behaviour patterns or structures, called system modes, can be represented by a set of models; a bank of filters runs in parallel at every time, each based on a particular model, to obtain the model-conditional estimates. The overall state estimate is a certain combination of these model-conditional estimates. The example implemented in this chapter considers an IMM filter with two models, they are the followinfg:

1. Constant velocity (CV) model
2. Coordinated turn (CT) model

In the case of maneuvering targets, the standard Kalman filter cannot provide good estimates. It is necessary for the tracking filter to adapt itself to the new model structure when tracking a maneuvering target. In order to yield satisfactory performance with maneuvering targets, the tracking algorithm needs to be capable of identifying the beginning and the end of maneuvers. The IMM estimator achieves this by updating the mode probabilities based on the received data.

Using the IMM approach, at time k the state estimate is computed under each possible current model using r filters, with each filter using a different combination of the previous model-conditioned estimates called the mixed initial condition [9].

One cycle of the algorithm consists of the following:

Step 1: Calculation of the mixing probabilities: the mixing probabilities, which can be written as

$$\mu_{i|j}(k-1|k-1) = \frac{1}{c_j} p_{ij}\mu_i(k-1) \quad i,j = 1,\ldots,r \tag{15.1}$$

where the normalizing constants are

$$\bar{c}_j = \sum_{i=1}^{r} p_{ij}\mu_i(k-1) \tag{15.2}$$

p_{ij} is the mode transition probability and μ_i is the mode probability.

Step 2: Mixing: starting with $\hat{x}^i(k-1|k-1)$ we compute the mixed initial condition for the filter matched to $M_j(k)$

$$\hat{x}^{0j}(k-1|k-1) = \sum_{i=1}^{r} \hat{x}^j(k-1|k-1)\mu_{i|j}(k-1|k-1) \quad j = 1,\ldots.,r \tag{15.3}$$

The covariance corresponding to the above is

$$P^{0j}(k-1|k-1) =$$

$$\sum_{i=1}^{r} \mu_{i|j}(k-1|k-1) \times \left\{ \begin{array}{c} P^i(k-1|k-1) + \\ \left[\hat{x}^i(k-1|k-1) - \hat{x}^{0j}(k-1|k-1)\right] \\ \left[\hat{x}^i(k-1|k-1) - \hat{x}^{0j}(k-1|k-1)'\right] \end{array} \right\} \quad j = 1,\ldots.,r \tag{15.4}$$

Step 3: Mode-matched filtering.

The estimate eqn. (15.3) and covariance eqn. (15.4) are used as an input to the filter matched to $M_j(k)$ which uses $z(k)$ to yield $\hat{x}^j(k \mid k)$ and $\hat{P}^j(k \mid k)$. The likelihood functions corresponding to r filters are computed using the mixed initial conditions and the associated covariance as

$$\Lambda_j(k) = N\left[x(k); \hat{z}^j\left[k|k-1; \hat{x}^{0j}(k-1|k-1)\right], S^j\left[k; P^{0j}(k-1|k-1)\right]\right] \quad j = 1,\ldots,r \tag{15.5}$$

Step 4: Mode probability update: this is done as follows:

$$\mu_j(k) = \frac{1}{c}\Lambda_j(k)\bar{c}_j \quad j = 1,..,r \tag{15.6}$$

where

$$c = \sum_{i=1}^{r}\Lambda_j(k)\bar{c}_j \tag{15.7}$$

is the normalization constant for (21).

Step 5: Estimate and covariance combination: combination of the model-conditioned estimates and covariance is done according to the mixture equations

$$\hat{x}(k|k) = \sum_{i=1}^{r}\hat{x}^j(k|k)\mu_j(k) \tag{15.8}$$

15.3 PERFORMANCE EVALUATION

A radar target tracking problem is considered where the goal is to precisely monitor the trajectory of an aircraft that manoeuvres at constant velocity and then takes a coordinated turn. This motion is subject to a small amount of noise modelled according to the relation in eqn. (15.9).

$$X_{k+1} = F_k(X_k) + \eta_k \tag{15.9}$$

where X_k is the state vector at time instant k and η_k is the process noise.

$$X = \left[x v_x y v_y\right]^T \tag{15.10}$$

In eqn. (15.10), (x, y) are the position coordinates, v_x, v_y are the velocities of the x and y components, respectively.

The state transition matrix of CV and CT models are as given in eqns. (15.11) and (15.12).

$$\begin{bmatrix} 1 & T & 0 & 0 \\ 0 & 1 & 0 & 0 \\ 0 & 0 & 1 & T \\ 0 & 0 & 0 & 1 \end{bmatrix} \tag{15.11}$$

$$
(14) \qquad \begin{bmatrix} 1 & \dfrac{\sin\omega_k T}{\omega_k} & 0 & \dfrac{-(1-\cos\omega_k T)}{\omega_k} \\ 0 & \cos\omega_k T & 0 & -\sin\omega_k T \\ 0 & \dfrac{(1-\cos\omega_k T)}{\omega_k} & 1 & \dfrac{\sin\omega_k T}{\omega_k} \\ 0 & \sin\omega_k T & 0 & \cos\omega_k T \end{bmatrix} \tag{15.12}
$$

ω_k is the turn rate at the kth instant. It is obtained from the rate of change of heading. $\omega_k = \dfrac{h_k - h_{k-1}}{T}$ where h_k is the heading at the kth instant and is calculated as $h_k = \tan^{-1}\dfrac{\dot{y}_k}{\dot{x}_k}$.

The covariance of the process noise is considered as in eqn. (15.13).

$$
Q = \begin{bmatrix} \hat{Q}_{2\times2} & 0_{2\times2} \\ 0_{2\times2} & \hat{Q}_{2\times2} \end{bmatrix} q_0 \tag{15.13}
$$

where $\hat{Q} = \begin{bmatrix} \dfrac{T^4}{4} & \dfrac{T^3}{2} \\ \dfrac{T^3}{2} & T^2 \end{bmatrix}$

Here σ_1 is the standard deviation of velocities in all the three dimensions and σ_2 is the standard deviation of the turn rate.

The measurement model is described as in eqn. (15.14). Here μ_k denotes the measurement noise.

$$
\chi_k = \gamma(X_k) + \mu_k \tag{15.14}
$$

The measurements χ_k received at time instant k are assumed to be range and azimuth from a radar located at the origin. Hence the observation matrix γ is defined as given in eqn. (15.15).

$$
\gamma = \begin{bmatrix} \sqrt{x_k^2 + y_k^2} \\ \tan^{-1}\left(\dfrac{y_k}{x_k}\right) \end{bmatrix} \tag{15.15}
$$

The covariance of the measurement noise is represented by R as in eqn. (15.16).

$$
R = \text{diag}\begin{bmatrix} \sigma_1^2 & \sigma_2^2 \end{bmatrix} \tag{15.16}
$$

where σ_1 and σ_2 are standard deviations of x and y, respectively In Figure 15.1, target trajectory is plotted.

Consider the scenario in which the target initially moves at nearly a constant velocity for 50 sec. Following this, the target executes a coordinated turn with turn angle 6 degrees/second for 50 sec. Again the target moves with constant velocity for 25 sec and then takes a turn with angle 10 degrees per second for the next 25 sec as shown in Figure 15.1. The target has initial state $[37000100\,35000100]^T$ and continues for 100 sec with constant velocity.

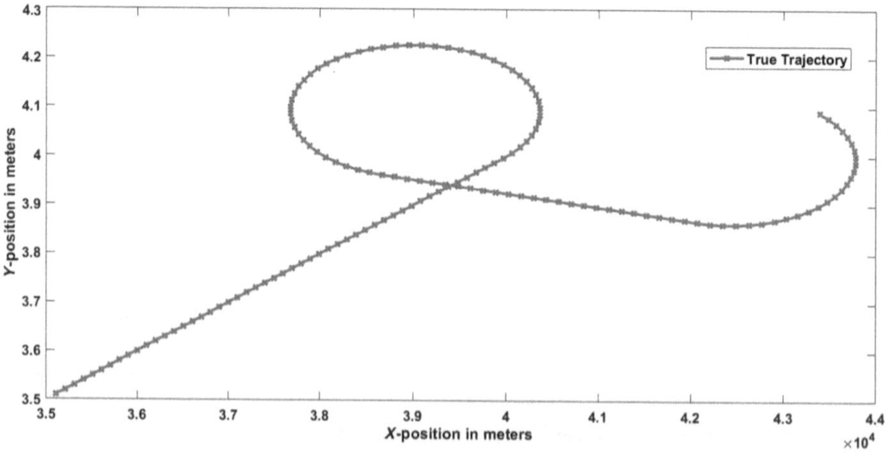

FIGURE 15.1 True trajectory of the target.

The measurement and process noise covariance is defined with parameters $\sigma_1 = \sigma_2 = 10\,\text{m/s}$. The sampling time is taken as $T = 2$ sec and measurements are taken for a period of 200 sec. The performance metric considered for this example is the root mean square error (RMSE). The standard deviation of the estimation error is provided by RMSE of states. Since, lower RMSE directly translates to higher accuracy, this metric is frequently used to assess the filtering accuracy of algorithms. Eqn. (15.17) calculates the RMSE at each time step over M Monte Carlo runs.

$$\text{RMSE}_k = \sqrt{\frac{1}{M}\sum_{j=1}^{M}\left(\text{X}_{j,k} - \hat{X}_{j,k}\right)^2} \tag{15.17}$$

Here $\text{X}_{j,k}$ and $\hat{\text{X}}_{j,k}$ is the true and estimated values respectively for the kth time step and M^{th} Monte Carlo run.

15.3.1 With GSKF

The IMM with Gaussian sum Kalman filter is a technique used for state estimation in dynamic systems characterized by multiple operating modes. This approach integrates two key components: the IMM algorithm and the Gaussian sum Kalman filter (GSKF). The IMM algorithm manages multiple models, each representing a distinct mode of system behaviour, and switches between them based on a probability distribution that evolves over time. On the other hand, the GSKF extends the traditional Kalman filter to handle non-Gaussian uncertainties by representing the state estimate as a sum of Gaussian components. In the IMM with GSKF framework, each mode of system behaviour is associated with a specific Kalman filter, possibly a GSKF, which estimates the state based on measurements and predictions. The IMM algorithm probabilistically selects the appropriate filter based on the current mode probability, and the state estimates from all filters are combined using a weighted sum to obtain the final estimate. This integration allows the approach to effectively handle complex systems with dynamic behaviours.

In this chapter, the IMM algorithm has been effectively implemented using both Kalman filter (KF) and GSKF methodologies, with a focus on radar tracking applications. Figure 15.2 shows the true trajectory of the target and the estimated trajectory using IMM-KF and IMM-GSKF.

Comparative analysis reveals that the IMM-GSKF framework outperforms the IMM-KF counterpart in terms of overall accuracy. While IMM-KF demonstrates superior tracking RMSE under normal operating conditions, it exhibits significant degradation in tracking accuracy when the target's

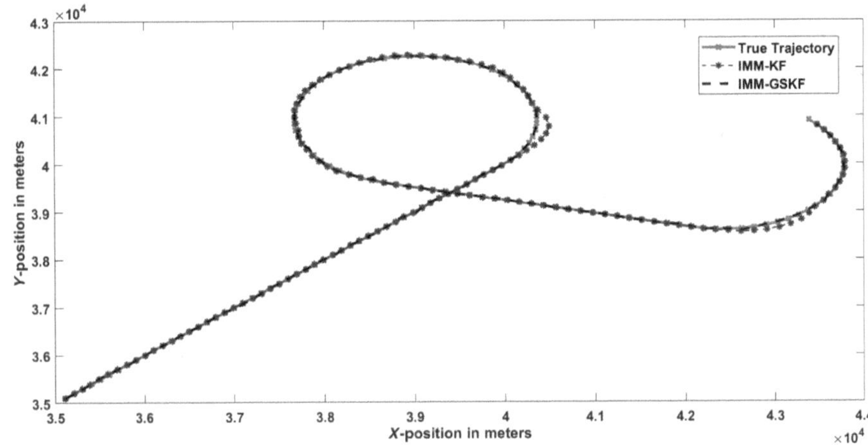

FIGURE 15.2 Comparison of the estimated trajectory by IMM-KF, IMM-GSKF and the true trajectory.

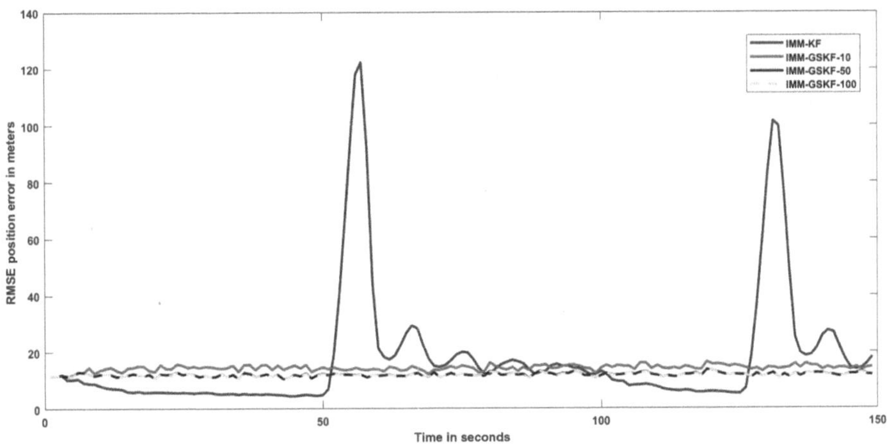

FIGURE 15.3 RMSE for position comparing IMM-KF and IMM-GSKF.

mode changes. This contrast is visually depicted in Figure 15.3, where IMM-GSKF maintains consistent RMSE values throughout the tracking process.

The number of components in GSF can be defined depending on the application. However, as this number increases, the computational complexity increases. The performance is analyzed for 10, 50 and 100 Gaussian components. The results indicate that employing 50 Gaussian components yields improved performance accuracy compared to utilizing only 10 components. This observation shows the delicate balance between computational efficiency and tracking precision in Gaussian sum Kalman filter-based systems, thereby highlighting the significance of appropriately tuning the number of components based on the application's demands.

15.3.2 With GSIF

The information filter (IF) is a variant of the Kalman filter that operates on the information form of the state estimate and covariance matrix, rather than the state itself [10,11]. It transforms the state and covariance matrices into their corresponding information matrices, allowing for more computationally efficient updates, particularly in high-dimensional state spaces. By working directly with the information matrices, the IF eliminates the need for matrix inversions, which can be computationally expensive, making it advantageous in scenarios with large state dimensions. The Gaussian sum information filter (GSIF) extends the information filter to handle multi-modal estimation problems by representing the state estimate as a mixture of Gaussian distributions. The interacting multiple model information filter (IMM-IF) combines the benefits of the IMM algorithm with the efficiency of the information filter for state estimation in dynamic systems. IMM-IF utilizes a multiple model approach to account for mode transitions in the system, allowing for better adaptation to changing dynamics. By integrating multiple models and their associated information filters, IMM-IF can effectively handle uncertainties and variations in system behaviour, leading to improved tracking accuracy.

The IMM-IF and IMM-GSIF algorithms have been deployed and assessed within the context of the radar tracking example. Figure 15.4 illustrates both the actual trajectory of the target and the estimated trajectory obtained through the implementations of IMM-IF and IMM-GSIF algorithms.

IMM-GSIF demonstrates superior accuracy compared to IMM-IF, primarily attributed to GSIF's utilization of a greater number of Gaussian components for state estimation. The findings depicted in Figure 15.5 highlight the superior accuracy achieved by the IMM-GSIF compared to the

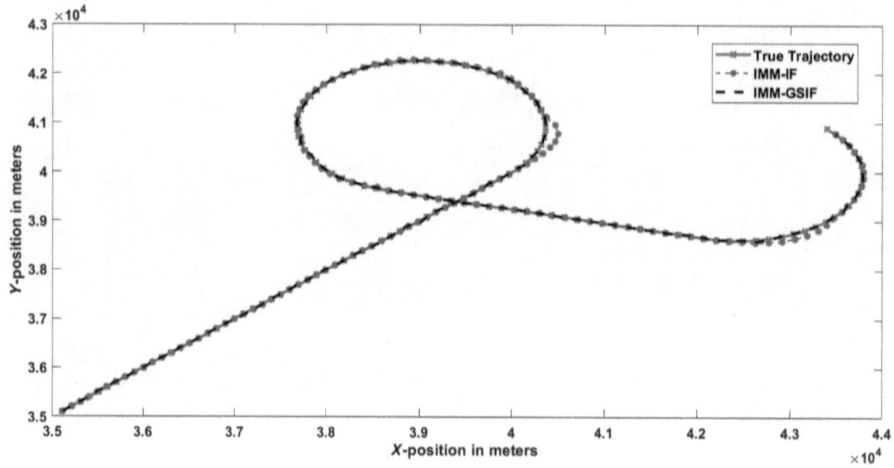

FIGURE 15.4 Comparison of the estimated trajectory by IMM-IF, IMM-GSIF and the true trajectory.

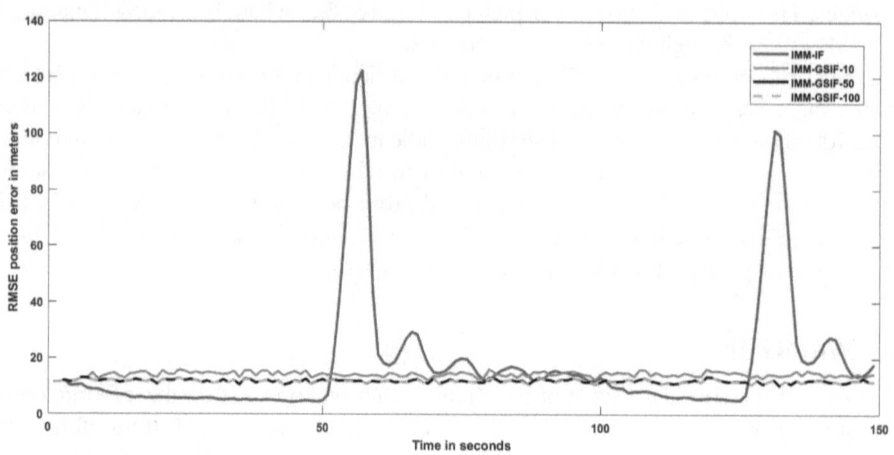

FIGURE 15.5 RMSE for position comparing IMM-IF and IMM-GSIF.

IMM-IF. Similar to previous implementations, experiments were conducted by varying the number of Gaussian components. The findings reveal consistent accuracy levels for scenarios employing 50 and 100 components, while a reduction in accuracy is observed with only 10 components employed in the estimation process.

15.4 DISCUSSION OF THE RESULTS

The comparison between the IMM-KF, IMM-GSKF, IMM-IF and IMM-GSIF sheds light on the efficacy of different tracking methodologies in radar target tracking scenarios. The results indicate that IMM-GSKF outperforms IMM-KF, primarily due to GSKF's ability to utilize a higher number of Gaussian components for estimation. This flexibility allows GSKF to better capture the complexity of target motion, resulting in improved accuracy in tracking trajectories. However, it is important to note that as the number of Gaussian components increases, so does the computational complexity of the algorithm. This trade-off between accuracy and computational efficiency needs to be carefully considered when designing tracking systems for real-world applications.

Further analysis of the IMM-GSIF algorithm highlights its superiority over IMM-IF. The reason being the same as mentioned above. However, increasing the number of components also leads to an escalation in computational complexity. The performance evaluation involving varying the number of components provides valuable insights into the impact of this parameter on tracking accuracy. Notably, the decrease in performance accuracy observed with 50 components compared to 10 Gaussian components suggests the presence of an optimal balance between accuracy and computational efficiency.

Table 15.1 shows the execution time for the different filters. The computational time required by both methods is nearly identical. The IF demonstrates superior computational efficiency particularly when dealing with large state matrices. However, in our specific scenario, where the size of the matrix is not substantial, this computational advantage is not observed.

Overall, these results demonstrate the importance of choosing the appropriate tracking algorithm and parameter settings based on the specific requirements of the application. While algorithms like GSKF and GSIF especially when combined with IMM offer superior accuracy, their increased computational demands necessitate careful consideration in resource-constrained environments. Additionally, the performance evaluation framework, utilizing metrics like RMSE, provides a quantitative assessment of tracking accuracy, facilitating informed decision-making in algorithm selection and parameter tuning.

TABLE 15.1
Comparison of Execution Times and PEENs

	Number of Components	Execution Time in Seconds
IMM-KF	-	0.070221
IMM-GSKF	10	0.246026
	50	0.419961
	100	0.804523
IMM-IF	-	0.069982
IMM-GSIF	10	0.246003
	50	0.419831
	100	0.799862

EXERCISES

EXERCISE 15.1

Explain the rationale behind employing multiple dynamic models in target tracking, particularly in scenarios involving maneuvering targets.

EXERCISE 15.2

Describe the process of model switching in the interactive multiple modelling (IMM) framework and its significance in adapting to changing target behaviours.

EXERCISE 15.3

Discuss the steps involved in the IMM estimator algorithm for tracking maneuvering targets, emphasizing the role of mode probability update and covariance combination.

EXERCISE 15.4

What are the key components of the radar target tracking problem considered in the chapter, and how are the state transition matrix and measurement model defined?

EXERCISE 15.5

Compare the performance of the Gaussian sum extended Kalman filter (GSKF) with other tracking algorithms discussed in the chapter, such as IMM KF and IMM-GSF, highlighting the impact of the number of Gaussian components on estimation accuracy.

REFERENCES

1. Bar-Shalom, Y., Li, X.R., & Kirubarajan, T., *Estimation with Applications to Tracking and Navigation: Theory Algorithms and Software*, John Wiley & Sons, New York, 2001.
2. Blackman, S., and Popoli, R., *Design and Analysis of Modern Tracking Systems*, Artech House, Boston, 1999.
3. Julier, S.J., and Uhlmann, J.K., Unscented filtering and nonlinear estimation, *Proceedings of the IEEE*, Vol. 92, No. 3, pp. 401–422, 2004.
4. Merwe, R.V.D., and Wan, E.A., Sigma-point Kalman filters for probabilistic inference in dynamic state-space models, *Proceedings of the Royal Society of London. Series A: Mathematical, Physical and Engineering Sciences*, Vol. 4, No. 2074, pp. 867–891, 2001.
5. Gustafsson, F., Gunnarsson, F., Bergman, N., Forssell, U., Jansson, J., Karlsson, R., and Holst, H., Particle filters for positioning, navigation, and tracking, *IEEE Transactions on Signal Processing*, Vol. 50, No. 2, pp. 425–437, 2002.
6. Vo, B.N., and Ma, W.K., The Gaussian mixture probability hypothesis density filter, *IEEE Transactions on Signal Processing*, Vol. 54, No. 11, pp. 4091–4104, 2006.
7. Grewal, M.S., and Andrews, A.P., *Kalman Filtering: Theory and Practice with MATLAB*, John Wiley & Sons, Hoboken, 2014.
8. Blom, H.A.P., and Bar-Shalom, Y., The interacting multiple model algorithm for systems with Markovian switching coefficients, *IEEE Transactions on Automatic Control*, Vol. 33, No. 8, pp. 780–783, 1988.
9. Kirubarajan, T., and Bar-Shalom, Y., Kalman filter versus IMM estimator: When do we need the latter?, *IEEE Transactions on Aerospace and Electronic Systems*, Vol. 39, No. 4, pp. 1452–1457, 2003.

10. Raol, J.R., Twala, B., and Gopalratnam, G., *Nonlinear Filtering: Concepts and Engineering Applications*, CRC Press, Taylor & Francis, Florida, 2017.
11. Sara, M.G., Approximations in Nonlinear Filtering: Some Novel Approaches. Ph.D. Thesis, Electrical and Electronics Engineering Sciences, Visvesvaraya Technological University (VTU), Jnana Sangama, VTU Main Road, Machhe, Belagavi, Karnataka, 2023.

16 Machine Learning for Estimation

16.1 INTRODUCTION

Operational data reveals that machine-learning algorithms are effective in capturing the dynamic behaviour of complex, nonlinear chemical processes. However, in chemical facilities, complete state data may not always be available. We propose two machine-learning-based approaches for state estimation. The first method integrates recurrent neural networks (RNNs) into the extended Luenberger observer architecture to develop data-driven state estimators. The second method employs a hybrid model that combines feed-forward neural networks (FFNNs) with first-principles models to capture process dynamics within the state estimator. Subsequently, a closed-loop system is stabilized at steady state using an output feedback model predictive controller, leveraging the state estimates obtained from the machine-learning-based estimators. The effectiveness of these proposed machine-learning-based state estimation and control techniques is demonstrated through a practical example involving a chemical process.

Estimation using machine learning (ML) involves utilizing ML methodologies to forecast or approximate unknown factors or variables using existing data. Estimation challenges emerge across diverse domains such as statistics, engineering, finance and healthcare. ML algorithms are adept at constructing predictive frameworks that discern patterns within data, facilitating precise estimations or forecasts. Here is a broad outline of the application of machine learning in estimation [1–4]:

- Problem formulation:
  Define the estimation problem and identify the target variable or quantity to be estimated. Determine the available data sources and features that may influence the estimation.
- Data collection and pre-processing:
  Collect pertinent data from a range of sources, such as sensors, databases, or historical archives. Prior to analysis, pre-process the data by addressing missing values, managing outliers, and standardizing or scaling features.
- Feature engineering:
  Choose and craft features that are both informative and pertinent to the estimation objective. This process might entail transforming raw data, generating novel features, or cherry-picking feature subsets based on domain expertise or feature selection methodologies.
- Model selection:
  Select the suitable machine-learning (ML) model or algorithm for the estimation undertaking. Frequently employed models for estimation encompass linear regression, logistic regression, decision trees, random forests, support vector machines (SVMs), and neural networks.

DOI: 10.1201/9781003620211-16

- Training the model:
  Segment the data set into training and validation sets. Proceed to train the selected ML model on the training data utilizing optimization algorithms such as gradient descent or stochastic gradient descent. Fine-tune hyper parameters to enhance model efficacy using methodologies like cross-validation or grid search.
- Model evaluation:
  Assess the performance of the trained model on the validation set by employing suitable evaluation metrics tailored to the task at hand. For regression tasks, common metrics include mean squared error (MSE), mean absolute error (MAE), or R-squared. For classification tasks, pertinent metrics encompass accuracy, precision, recall, or F1-score.
- Model deployment:
  Once satisfied with the model's performance, deploy it to make predictions or estimates on new, unseen data. Integrate the model into the production environment or application to automate the estimation process.
- Monitoring and maintenance:
  Continually oversee the deployed model's performance and adjust it as necessary to accommodate evolving data distributions or demands. Periodically retrain the model with fresh data to enhance accuracy and ensure its ongoing relevance over time.

  Machine learning for estimation provides robust tools for generating precise predictions and estimates across diverse applications. By leveraging data-driven approaches, ML techniques can help automate decision-making processes, optimize resource allocation, and improve overall efficiency and effectiveness in various domains.

16.2 EXTENDED LUENBERGER OBSERVER

The extended Luenberger observer (also known as the state observer or full-order observer) is a control system technique used for estimating the unmeasured states of a dynamic system based on available input and output measurements. It is an extension of the classical Luenberger observer and is commonly used in control engineering for state estimation, feedback control and fault diagnosis. Here is an overview of the extended Luenberger observer (ELO):

(a) Applications
  - State estimation in control systems where not all states are directly measurable.
  - Fault detection and diagnosis by comparing estimated states with measured outputs.
  - Observers are used in various industries, including aerospace, automotive, robotics and process control.

(b) Advantages:
  - Provides estimates of unmeasured states for feedback control purposes.
  - Can be implemented in real-time using computational hardware.
  - Robust to noise and disturbances in the system.

(c) Limitations:
  - Sensitivity to model uncertainties and parameter variations.
  - Requires knowledge of system dynamics and accurate model information.
  - Performance may degrade if the observer gain matrix is not properly tuned.

The ELO plays a crucial role in control systems engineering by estimating unmeasured states and improving system performance through feedback control. Its effective design and implementation necessitate thorough consideration of system dynamics, tuning of observer gain, and analysis of stability to guarantee dependable operation in real-world scenarios.

16.2.1 Notations

The magnitude of a vector is denoted by | · | in Euclidean space. The standard Lie derivative is expressed as $L_f h(x) = (\partial h(x)/\partial x) f(x)$. In set notation, the symbol \ denotes set subsection, where $A \backslash B = \{x \in Rn | x \in A, x \notin B\}$. A function $f(\cdot)$ is classified as C^1 if it exhibits continuous differentiability.

(d) Class of systems:

Let us examine the following collection of continuous-time nonlinear systems presented in state-space representation:

$$x = F(x,u) := \dot{f}(x) + g(x)u \tag{16.1}$$

$$y = h(x) \tag{16.2}$$

In this format, the state vector is represented as $x = [x1, \ldots, xn]^T \in R^n$, the output vector is denoted as $y = [y1, \ldots, yq]^T \in R^q$, and the input vector is $u = [u_1, \ldots, u_m]^T \in R^m$. The function $F(x,u)$ is nonlinear concerning both x and u. The constraints on control inputs are defined by $u \in U := \{u_{imin} \leq u_i \leq u_{imax}\}$. Furthermore, the functions $f(\cdot)$, $g(\cdot)$ and $h(\cdot)$ correspond to matrices of dimensions $n \times 1$, $n \times m$, and $q \times 1$, respectively.

The ELO has been developed for nonlinear systems as a natural extension of the Luenberger observer, which operates based on a linearized depiction of the system. The key aim of this state observer is to provide estimations of the unmeasured internal states within the system. This objective is achieved by leveraging the measured states derived from the system, along with the applied inputs. The formulation of the ELO for the nonlinear system outlined in eqn. (16.3) is presented as follows:

$$\hat{x} = F(\hat{x}, u) + K(y - h(\hat{x})) \tag{16.3}$$

In this context, $\hat{x}$ signifies the estimated state vector, and the observer gain is denoted by K. The observer gain is associated with the desired attributes of the state estimator, which will be comprehensively analyzed subsequently. Eqn. (16.2) reveals that the initial component represents the process model, whereas the concluding term, $K(y - h(\hat{x}))$, is identified as the output prediction error, serving as a corrective element.

The aim of the ELO is to minimize the estimation error (i.e., $e = x - \hat{x}$), with the dynamics of the error governed by the following equation:

$$\dot{e} = F(\hat{x} + e, u) - F(\hat{x}, u) - K(h(\hat{x} + e) - h(\hat{x})) \tag{16.4}$$

As illustrated in eqn. (16.3), the current task is to determine the conditions where e can converge to zero. Therefore, configuring K to achieve this goal becomes crucial. Simplifying eqn. (16.3) to the following equation by linearizing the process model at a particular fixed point facilitates the design of K.

$$\dot{e} = (A - KL)e \tag{16.5}$$

Here $A = \partial \left[\dfrac{F(x,u)}{\partial x} \right]_{x=\hat{x}}$ and $L = \left[\dfrac{\partial h(x,u)}{\partial x} \right]_{x=\hat{x}}$

The linearized terms of the nonlinear system are evaluated at a specific fixed point, typically the operational steady state. Ultimately, the selection of K is conducted to ensure that the eigenvalues of the matrix A-KL possess strictly negative real components.

Stabilization via control Lyapunov function relies on the presence of an observer and a feedback control law $u = \varphi(\hat{x}) \in U$ employing estimated states $\hat{x}$ from the observer to formulate an output feedback controller. This controller is capable of exponentially stabilizing the equilibrium point of the nonlinear system outlined in eqn. (16.1). This assumption of stabilizability suggests the existence of a C^1 control Lyapunov function $V(x)$ such that the following inequalities are satisfied for all x, $\hat{x}$ in an open neighbourhood D surrounding the equilibrium point.

$$c_1 |x|^2 \leq V(x) \leq c_2 |x|^2 \tag{16.6}$$

$$\frac{\partial V(x)}{\partial x} F(x, \varphi(\hat{x})) \leq -c_3 |x|^2 \tag{16.7}$$

$$|\partial V(x) / \partial x| \leq c_4 |x| \tag{16.8}$$

where c_1, c_2, c_3 and c_4 denote positive constants, and $F(x, u)$ represents the nonlinear system described in eqn. (16.1).

A prospective controller for $\varphi(\hat{x})$ is proposed utilizing the universal Sontag control law. Defining the closed-loop stability region Ω_{ρ^1} as a level set of the Lyapunov function within the domain D, where the time derivative of V becomes negative under the controller $\varphi(\hat{x}) \in U$. Specifically, Ω_{ρ^1} is characterized as the set of x values in D for which $\Omega_{\rho^1} : \{x \in D | V(x) \leq \rho^1\} > 0$, where $\rho^1 > 0$. Furthermore, considering the Lipschitz property of $F(x,u)$ and the boundedness of u, positive constants M, Lx, and Lx' exist, satisfying the following inequalities for all $x, x' \in D$, and $u \in U$.

$$|F(x,u)| \leq M \tag{16.9}$$

$$|F(x,u) - F(x',u)| \leq L_x |x - x'| \tag{16.10}$$

$$\left| \frac{\partial V(x)}{\partial(x)} F(x,u) - \frac{\partial V(x')}{\partial(x)} F(x',u) \right| \leq L'_x |x - x'| \tag{16.11}$$

For the existence of an output feedback controller meeting eqns. (16.6)–(16.8), it is essential that the observer states remain bounded within Ω_{ρ^1} and that the estimation error, $e = x - \hat{x}$, converges to zero within a finite time frame. In this investigation, the extended Luenberger observer (ELO) outlined in eqn. (16.2) and the Sontag control law are utilized as the observer and state feedback controller, respectively.

16.3 RECURRENT NEURAL NETWORK (RNN)-BASED STATE ESTIMATOR

The extended Luenberger observer specified ineqn. (16.3) necessitates a recurrent neural network (RNN) to serve as the process model. When a first-principles model is unavailable, the nonlinear system described in eqn. (16.1) is approximated using the subsequent RNN model, which is constructed based on operational data from the process:

$$\dot{\bar{x}} = F_{mn}(\bar{x}, u) := A\bar{x} + \Theta^T y \tag{16.12}$$

where $u = [u_1, \ldots u_m]$ is the altered input vector and $\bar{x} = [\bar{x}_1, \ldots \bar{x}_n]$ is the RNN state vector. $y = [y_1, \ldots y_n, y_{n+1}, \ldots y_{m+n}] = [\sigma(\bar{x}_1), \ldots \sigma(\bar{x}_n), u_1, \ldots u_m] \in R^{n+m}$. A vector of both $\bar{x}$ and u, is

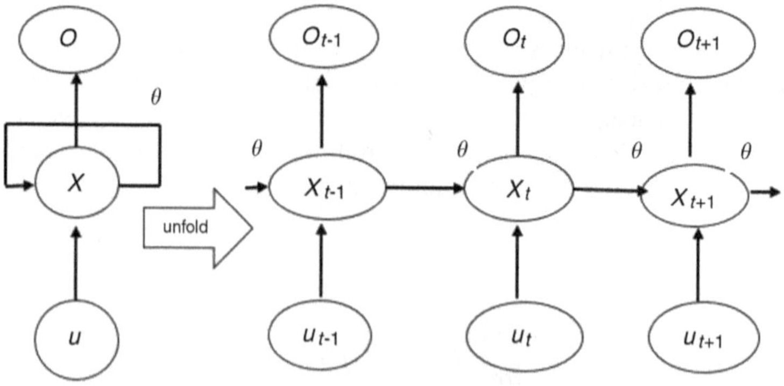

FIGURE 16.1 Structure of a recurrent neural network.

represented by R^{n+m}, where $\sigma(\cdot)$ represents the nonlinear activation function. $\Theta = \left[\theta_1, \ldots \theta_n\right] \epsilon R^{(m+n)xn}$ and A is a diagonal coefficient matrix. An optimised matrix containing the weights of a neural network is $\Theta = \left[\theta_1, \ldots \theta_n\right] \epsilon R^{(m+n)xn}$.

Figure 16.1 displays the structures of folded and unfolded RNNs. Weights of the neural network that will be optimized during training are contained in the coefficient matrices $A = \mathrm{diag}\left[-a_1, \ldots -a_n\right], a_i > 0,\ i = 1, \ldots n$ and Θ. When a neural network is placed into practice, its weights are adjusted to minimize the discrepancy between the training data set's actual and expected outputs, where $u = \left[u_1, \ldots u_m\right]$ is the modified state vector and $\bar{x} = \left[\bar{x}_1, \ldots \bar{x}_n\right]$ is the RNN state vector.

Once the architecture of the RNN, including the number of layers, neurons, and other hyper parameters, has been determined, the RNN undergoes training using standard learning procedures. Specifically, open-loop simulations of eqn. (16.1) are conducted under various initial conditions and control actions for a finite duration to generate training, validation, and testing datasets. Employing the explicit Euler method with a sufficiently small integration time step (hc), the continuous-time system of eqn. (16.1) is numerically integrated. Subsequently, the control actions (u) are applied in a sample-and-hold manner, defined as $u(t) = u\left(t_k, t_{k+1}\right)$ where $t_{k+1} := t_k + \Delta$ and Δ represents the sampling period. In this study, the RNN model is constructed using data from all integration time steps, allowing for the capture of the process's dynamic behaviour, as RNN models can effectively utilize time-series data.

For the RNN model to faithfully capture the dynamics of the process within the operating range under scrutiny, it is imperative that, throughout the training phase, it achieves a sufficiently low modelling error. This is expressed as $|v| = \left|F(x,u) - F_{\mathrm{RNN}}(\bar{x},u)\right| \leq \gamma|x| \leq v_m$. It is important to emphasize that the modelling error refers to the disparity in $\dot{x}$ between the RNN model's forecast and the $\dot{x}$ derived from the first-principles model under identical inputs and states. While the modelling error, 'v' varies when considering different inputs and states, it is not constant. However, since the operation is confined to the stability region Ω_p (characterized by bounded states and inputs), and due to the recurrent neural network model's training using data sets generated within Ω_p to achieve exceptionally high accuracy, the training process is effective.

16.3.1 Remarks

(i) The RNN models are used to introduce historical data into the network at hand. The suggested methodology can be expanded to encompass additional machine-learning techniques, like long short-term memory networks, which are extensively employed in the modelling of nonlinear dynamic systems. Recent studies have utilized algorithms that promote sparsity, extended dynamic mode decompositions, and Koopman system identification alongside neural network modelling techniques to approximate nonlinear systems.

(ii) In order to capture process dynamics in the operational region Ω_p, the RNN model is created utilising noise-free data from large open-loop simulations of eqn. (16.1). In addition to computer simulations, industrial measurements and experimental data can be leveraged to construct data sets. When real industrial measurements are affected by noise stemming from sensor variability and typical plant fluctuations, machine-learning modelling techniques like the co-teaching training algorithm and dropout technique can be employed to enhance approximation performance by mitigating the impact of noise.

(iii) A common challenge encountered in developing neural network models is overfitting. To mitigate this, a simple neural network typically comprises only one layer with a limited number of neurons. The complexity is gradually increased by adding layers and neurons until further enhancements are negligible. To ensure the model's effectiveness concerning both training and validation datasets, a considerable volume of data is utilized for validation purposes.

16.3.2 RNN-Based State Estimator Model

The RNN model is subsequently incorporated into the extended Luenberger observer described in eqn. (16.3) in the following manner:

$$\hat{x} = F_{\text{RNN}}(\hat{x},u)+K(y-h(\hat{x})) \tag{16.13}$$

Specifically, the state estimation based on the RNN model described in eqn. (16.12) is achieved through the following steps: First, the RNN model predicts the state at the next integration time step at $t = t_k + h_c$, given an initial state estimate $\hat{x}(t_k)$ at time $t = t_k$ and the manipulated input vector $u(t_k)$. Next, the state estimate at $t = t_k + h_c$ is derived by adding the second term $h_c \times K(y-h(\hat{x}))$ as outlined in eqn. (16.13). This process is repeated using the same input u once the state estimate at $t = t_k + h_c$ is determined, as u remains constant within one sampling period. After $\frac{\Delta}{h_c}$ iterations of the aforementioned procedure, the state estimate at the subsequent sampling period, $t = t_k + 1 := t_k + \Delta$, is finally obtained.

The state feedback control law $u = \Phi(\hat{x}) \in U$ constitutes an output feedback controller capable of rendering the RNN system specified in eqn. (16.12) exponentially stable when integrated with the RNN-based observer. This implies the presence of a C^1 Lyapunov function $V(x)$ such that, for all $x, \hat{x}$ in an open neighbourhood $\mathcal{D}$ surrounding the origin, the following inequalities are satisfied:

$$\hat{c}_1|x|^2 \le V(x) \le \hat{c}_2|x|^2 \tag{16.14}$$

$$\frac{\partial V(x)}{\partial x} F_{\text{RNN}}(x,\Phi(\hat{x})) \le -\hat{c}_3|x|^2 \tag{16.15}$$

$$\left|\frac{\partial V(x)}{\partial x}\right| \le \hat{c}_4|x| \tag{16.16}$$

The constants $\hat{c}_1, \hat{c}_2$, $\hat{c}_3$ and $\hat{c}_4$ are all positive values. It is worth noting that while the control law is discussed, based on the first-principles model outlined in eqn. (16.1), the control law in this context relies on the RNN model described in eqn. (16.12). Subsequently, a new closed-loop stability region Ω_p can be identified within D, where $u = \Phi(\hat{x}) \in U$ satisfies eqns. (16.14), (16.15), and (16.16). When the process model of eqns. (16.1) (16.2) is known, the state estimation using the RNN-based state estimator of eqn. (16.13) closely approximates the estimated value provided by eqn. (16.3). This is attributed to the RNN model being trained with a sufficiently small modelling error. Consequently, the system of eqn. (16.1) can be stabilized by the controller $u = \Phi(\hat{x}) \in U$ constructed using the estimated state from the RNN-based estimator, provided that the modelling error $|v| = |F(x,u) - F_{\text{RNN}}(\overline{x},u)|$ remains modest.

Considering the nonlinear system described in eqn. (16.1) with the following characteristics: given an initial state $x_0 \in \Omega_p$ and a stabilizing control law $u = \Phi(\hat{x}) \in U$ based on the estimated states derived from eqn. (16.13), the origin of the closed-loop system defined in eqn. (16.1) is made exponentially stable under $u = \Phi(\hat{x}) \in U$ for all $x, \hat{x} \in U$, if the modelling error can be constrained, specifically $|v| = |F(x,u) - F_{\text{RNN}}(x,u)| \le \gamma |x|$ for all $x \in \Omega_p$ and $u \in U$, where γ represents a positive real number satisfying $\gamma < \hat{c}_3 / \hat{c}_4$.

Verification: the nonlinear system with time derivative of V as in eqn. (16.1) is obtained as follows, based on eqns. (16.14), (16.15) and (16.16):

$$\dot{V} = \frac{\partial V(x)}{\partial x} F\left(x, \Phi(\hat{x})\right) \tag{16.17}$$

$$= \frac{\partial V(x)}{\partial x} (F_{\text{RNN}}(x,\Phi(\hat{x})) + F(x,\Phi(\hat{x})) - F_{\text{RNN}}(x,\Phi(\hat{x})))$$

$$\le -\hat{c}_3 |x|^2 + \hat{c}_4 |x| \left(F(x,\Phi(\hat{x})) - F_{\text{RNN}}(x,\Phi(\hat{x}))\right)$$

$$\le -\hat{c}_3 |x|^2 + \hat{c}_4 |x|^2$$

Therefore, if γ fulfils γ satisfies $\gamma < \hat{c}_3 / \hat{c}_4$ $\dot{V} \le -(\hat{c}_3 - \hat{c}_4 \gamma)|x|^2 \le 0$. This means that under the controller $u = \phi(\hat{x}) \in U$ with state estimates using recurrent neural network estimator model, the origin of the nonlinear system of eqn. (16.1) is made exponentially stable for every $x_0 \in \Omega_p$.

16.4 HYBRID MODEL-BASED STATE ESTIMATOR

In the hybrid model, the state estimator is devised employing the first-principles model with a feedforward neural network (FNN). The nonlinear components of eqn. (16.1) are approximated using the FNN model, while the first-principles model of the equation can be derived from fundamental principles such as mass and energy balances.

16.4.1 Feedforward Neural Network (FFNN)

In order to estimate the nonlinear components in eqn. (16.1), a feedforward neural network denoted as $F_{NN}(x)$ is constructed, where the input vector is $x = [x_1, \ldots, x_n]$ and the output vector is $y = F_{NN}(x) = [y_1, \ldots y_n]$. The feedforward neural network consists of three layers: the input layer, hidden layer, and output layer, as illustrated in Figure 16.2.

These equations are employed to calculate the hidden neurons *hj, j = 1, . . ., p*, and the outputs y_k:

$$h_j = \sigma_1 \left(\sum_{i=1}^{n} w_{ji}^{[1]} x_i + w_{j0}^{[1]} \right), \quad j = 1, 2, \ldots p \tag{16.18}$$

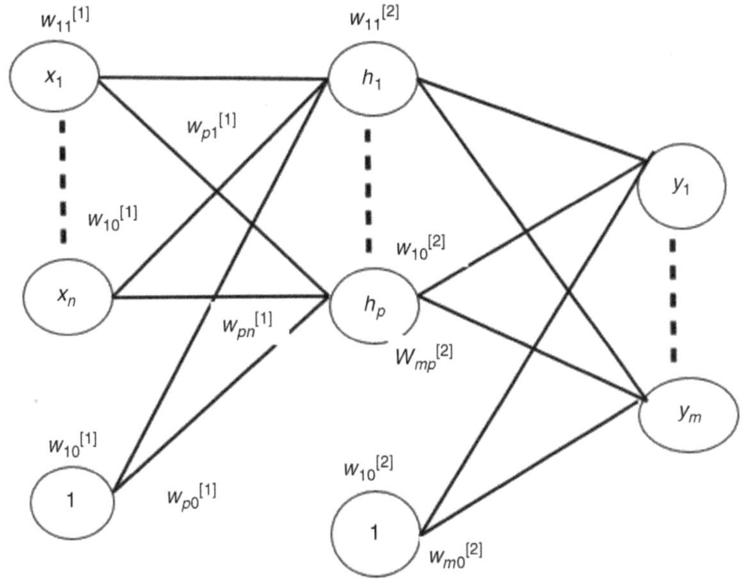

FIGURE 16.2 Three-layer feedforward neural network structure with biases.

$$y_k = \sigma_2\left(\sum_{i=1}^{p} w_{ki}^{[2]} h_i + w_{k0}^{[2]}\right), \quad k = 1,2,\ldots m \tag{16.19}$$

$w_{ji}^{[1]}$ and $w_{ki}^{[2]}$ represent the weights in the first two layers, respectively, and $w_{j0}^{[1]}$ and $w_{k0}^{[2]}$ represent the biases. The nonlinear activation functions σ_1 and σ_2 include the logistic sigmoid function $S(x) = \frac{1}{(1+e^{-x})}$ and the hyperbolic tangent function $\tanh(x) = \frac{2}{1+e^{-2x}} - 1$. The calculation of hidden neurons h_j uses the activation function σ_1 in conjunction with a linear combination of input variables x_i, while the calculation of the output variable y_k uses σ_2 in conjunction with a linear combination of hidden neurons.

Training and evaluation of FNN models datasets are necessary for their development. Thus, when creating data sets, a set of input vectors $\{x^n\}$ and matching output vectors $\{\hat{y}^n\}$ are utilized. Many techniques, such as comprehensive computer simulations and experimental data, can be used to get the data. The following loss function is minimised during training of the FNN model:

$$E(w) = \frac{1}{2}\sum_{n=1}^{N}\left|y(x^n, w) - \hat{y}^n\right|^2 \tag{16.20}$$

where N is the total quantity of training data points. The total squared error between the actual output in datasets and the output predicted by the FNN model is the loss function. The weight vectors w are produced by applying the gradient descent optimisation approach to minimize eqn. (16.20): $w^{i+1} = w^i - \eta\nabla E(w^i)$, where $\eta > 0$ represents the learning rate and i, denotes the iteration.

16.4.2 Hybrid Model-Based Estimator

Through the integration of the FNN model with the first-principles model outlined in eqn. (16.1), the hybrid model is formulated. The first-principles model is established based on fundamental physical principles, while the nonlinear components are captured by the FNN model. Below is the hybrid model for the eqn. (16.1) system:

$$\dot{x} = F_h\left(x, F_{NN}\left(x\right), u\right) \tag{16.21}$$

where F_{NN} is the FNN model used to represent the static nonlinear relationship between the inputs and outputs of the nonlinear components, and F_h stands for the hybrid model. Eqn. (16.21)'s hybrid model serves as the process model for the hybrid model-based state estimator, which also incorporates the following estimation correction term:

$$\dot{\hat{x}} = F_h\left(\hat{x}, F_{NN}\left(\hat{x}\right), u\right) + K\left(y - h\left(\hat{x}\right)\right) \tag{16.22}$$

The hybrid model for eqn. (16.1), denoted as F_h, encapsulates the dynamics of the entire nonlinear system. It is important to note that in cases where a comprehensive first-principles model is unavailable, the hybrid model is utilized. In such scenarios, feedforward neural networks can be employed to approximate certain unknown nonlinearities, such as reaction rates, leveraging the available data. The hybrid model-based state observer integrates eqn. (16.22) over one sampling period to estimate the states at the subsequent sampling time. It operates in a manner akin to the RNN-based estimator described in eqn. (16.13), utilizing as inputs the state estimate $\hat{x}\left(t_{k-1}\right)$ and the manipulated variable $u\left(t_{k-1}\right)$, alongside the latest measurement $y\left(t_k\right)$.

The integration of eqn. (16.22) is accomplished utilizing the explicit Euler method with a small integration step of h_c. Eqns. (16.6), (16.7) and (16.8) ensure closed-loop stability by stipulating that both estimators are initialized with initial conditions within Ω_p. Should they compute an estimate outside of Ω_p, this estimate is disregarded and substituted with an estimate within Ω_p.

16.5 LINEAR REGRESSION AND SUPPORT VECTOR MACHINE (SVM)

The main goal of linear regression in the context of machine learning is to model the relationships between independent and dependent variables. Let y be a collection of real numbers and x be a subset of $\mathbb{R}^d$ (d-dimensional vector) representing a set of known or labelled data. This is one way to depict linear regression.

$$L_d = \left\{x \rightarrow w,\ x + b\colon w \in \mathbb{R}^d,\ b \in \mathbb{R}\right\} \tag{16.23}$$

Here L_d are functions parametrized by $\boldsymbol{w} \in \mathbb{R}^d$ and $b \in \mathbb{R}$, $\langle \boldsymbol{w}, \boldsymbol{x} \rangle$ is the inner or dot product of $\boldsymbol{w}$ and $\boldsymbol{x}$. The loss function for $h(x)$ predictions is defined as

$$l\left(h\right) = \frac{1}{m}\sum_{i}^{m}\left(h\left(x_i\right) - y_i\right)^2 \tag{16.24}$$

A supervised machine-learning model for regression and data classification is called support vector machine (SVM). A linear prediction over mapped samples to a higher-dimensional space is modelled using SVM. Creating a high-dimensional surface, or hyperplane, that can divide data into categories is the goal of SVM models. The primary goal of support vector regression (SVR) is to locate a hyperplane where all data points lie inside the decision surfaces at a distance ϵ from the hyperplane given a decision tolerance ϵ.

$$\left|y_i - \left(w, x_i + b\right)\right| \leq \epsilon \tag{16.25}$$

The loss function of support vector regression is given by

$$l(w,x,y) = \max\left(0,\left|y - w,x\right| - \epsilon\right) \tag{16.26}$$

16.6 REGRESSION TREES AND THEIR ENSEMBLES

16.6.1 Regression Trees

Regression models are fitted to the terminal nodes, or leaves, of a regression tree. Reaching the right leaf to make the forecast is the aim. In order to make the offspring nodes 'less impure' or have less variations in their observations than the parent node, the process begins with dividing the root node. By executing the best node splits, the goal is to reduce the tree's 'impurity'.

16.6.2 Ensembles of Regression Trees

A linear prediction over mapped samples to a higher-dimensional space is modelled using SVM. Creating a high-dimensional surface, or hyperplane, that can divide data into categories is the goal of SVM models. The primary goal of support vector regression (SVR) is to locate a hyperplane where all data points lie inside the decision surfaces at a distance ϵ from the hyperplane given a decision tolerance ϵ.

16.7 FLIGHT SIMULATION DATA REPEATABILITY

The same UAV model can be used to conduct three consecutive flights in order to evaluate the repeatability of the flight simulation data. The test flights followed the same route and were consecutive. The same site and identical weather conditions are used for both flights. Plotting the flight curves together should result in an overlap as the simulated flying conditions are the same for every flight, and no deviation from these conditions is anticipated. Future research can more confidently modify the flying circumstances to include real-world situation fluctuations once the model has demonstrated its ability to produce repeatable findings. Upon preliminary examination of the data, it was found that each flight has a different time point at which data recording begins before take off. As a result, different quantities of needless 'dead time' data with values of zero for every flight aspect were generated.

Comparing graphs for repeatability is challenging, and doing so will have at least two unfavourable effects on the learning process. First off, since the learning algorithm will be searching for relationships in areas that do not impact flight performance, it will have an impact on how well it performs. Second, the data files get bigger as a result. The learning process will be delayed as a result of the algorithm taking longer to process the data. The data that caused the take-off manoeuvre to be initiated can be removed in order to resolve this problem.

All flight curves have their beginning point set at the take-off manoeuvre once the pre-take-off data has been eliminated. The output of the air-speed measurement is used to remove more pre-take-off data. The rationale is that air-speed data will increase from a constant near-zero value once the UAV starts to acquire speed, making it simple to spot for pre-processing.

16.8 FEATURE REDUCTION

Finding pertinent features that are determined to have the biggest influence on the algorithm's learning process is the aim of feature reduction. The feature reduction filter method and wrapper method are the two basic techniques. Feature reduction is carried out by the filter, which works independently of the machine-learning algorithm, using discriminating criteria. Conversely, wrapper

TABLE 16.1
UAV Weight Variations (Sample Values)

Weight (lb.)	0.95	4.75	8.55	12.35	16.15	19.95	23.75	27.55	31.35

approaches assess and choose the relevant features using the machine-learning model. Several features are used to train the regression model using the wrapper approach, and prediction tests are run to evaluate the model.

Barometric altitude, roll, pitch, yaw and airspeed data are input independently to an SVM regression model during training in order to use the wrapper reduction strategy. The same location, wind speed and temperature are all present during the flights. Predicting the UAV's weight is the algorithm's main objective. Weight is chosen as the best contender among the several parameters that can be predicted for a number of reasons. The primary explanation for this is that weight is independent of other characteristics, which means that changing the weight of an aircraft model will not change the weight of any other aircraft part or flying parameter.

This is untrue for certain parameters; for instance, altering the geometry of the air foil also modifies the coefficients of lift, drag and moment. These interactions provide needless sources of variation and can be distracting, adding to the model's complexity. The fact that the weight may be quickly and simply changed in the model's parameters file is another crucial factor in the selection process. In contrast to other characteristics, altering the density value immediately modifies the weight. This is an extremely effective and practical way to create model variations. When other model parameters are changed, other variables in the parameter file also change. The model becomes more complicated when the relationship between various factors is discovered. Additionally, because of the unknown relationships between these model parameter variables, the preparation time for the model may also grow dramatically.

Merely altering the weight is adequate to identify the basic difficulties and constraints associated with machine learning for parameter estimation. Moreover, it can offer insights into which models are better suited for testing increasingly intricate factors down the road. Initially, the regression model receives the feature data for training. A different set of input data is then used to test the model's prediction. The mean squared error (MSE) is calculated for every model in order to evaluate the accuracy of the predictions. The sum of the squared differences between the known y's and the model predictions divided by the number of instances is the mean square error, or MSE. The known y's in this instance are the known UAV weights, and the weight output is from the model prediction. Table 16.1 shows the weights used for the test.

The SVM model is trained in MATLAB after the flight data from the simulation runs is available. Randomly, the data can be divided into two categories: 89% training and 11% testing. Next, the model's MSE for each feature is calculated and contrasted with the models of other models.

The MSE of the test model and the training model are contrasted. 'High variance models' are those in the test model where yaw and airspeed show high mean MSE values. This suggests that the model's predictions for fresh data sets are not accurate when compared to similar models with smaller variance. It is also measured how much the roll and pitch models vary. Large variance and low bias characterize an over fit model. A good model must strike a compromise between low bias and low degree of volatility. There is a significant amount of volatility in any model. An adequate variance-bias balance is required for altitude. Nevertheless, it is better to reduce both in order to produce a model with a respectable level of prediction accuracy. The roll and pitch models were over fit by the SVM algorithms. According to the literature, yaw and airspeed are less volatile than pitch, roll and altitude. To illustrate, the flying system parameters of height, roll and pitch will be prioritized over yaw and airspeed if machine-learning technologies are applied.

16.9 ALTITUDE DATA

After feature reduction, pitch, roll and altitude were found to have the biggest effects on the regression model. Regression learner in MATLAB can be used to examine different features. For every trip, data partitioning is applied to the altitude input data. The learning method incorporates the time behaviour component of the data by incorporating the average altitude values over the previous 20 time steps. Four or more data point portions comprise the data. The flights that correspond to their weights provide the training's input data (see Table 16.1).

16.9.1 SVM Optimizable Regression

An SVM regression optimizable model from MATLAB's regression learner is the regression model used for the altitude data. The input data is randomly divided into testing and validation or testing before the model is trained. Twenty-five percent of the data can be preserved for model validation in this model. Determine the MSE at which the test data can be fitted by the model. Extra flights that are not included in the model's input are created independently and are not known to the regression model in order to test the model.

Plotting the known weight against the weights predicted by the model allows one to compute the difference delta between them. By deducting the minimum delta value from the maximum delta value from the flying sections, the range of the prediction is calculated from the plot. Within a given threshold value of the actual value for the first parts, the model can forecast the new UAV weight. For the later portions of the flight, the model might not be able to correctly anticipate the weight. When calculating the average of the forecasts, this has an impact on the model's overall predictability.

Hyperplanes are used in SVM regression to generate predictions. This could be the reason for the model's inability to accurately predict across all data, since the information on these flight segments may fall outside of the decision hyperplanes of the model. Ensembles of regression trees are to be investigated in order to enhance the regression model prediction in every part.

16.9.2 Ensembles of Regression Trees

For every flight data partition, the SVM model used in the previous section was unable to forecast the UAV's weight with any degree of accuracy. Ensembles of regression trees are used to increase the prediction for every part. The data is separated between training and testing/validation, same like in the prior SVM model. The model's validation uses the same 25% of the data.

The model performs better than the SVM model and fits the test data with a mean square error (MSE) significantly lower. Predictions are made to flight data that the algorithm is not familiar with in order to test the model, and the prediction results are more consistent throughout all flight data sections. The difference in actual and expected values is the maximum delta. This raises the findings' overall average. For the previous two models, a 25% holdout was used to validate the training model. Cross-validation is an alternative to holdout validation. Cross-validation's primary goal is to maximize algorithm training by determining the optimal train/validation split throughout the model-training phase. The same set of trees is trained using cross validation in the following part to assess the impact on the variation in the predictions.

16.9.3 Ensembles of Regression Trees with Cross Validation

The outcomes of regression tree ensembles using cross validation are shown in this section. Since the ensemble of the regression tree model produces consistent results across all flight data sections, it is chosen. For altitude data, three distinct models are trained and evaluated. Five, ten, and fifteen cross-validation folds, respectively, are used to validate each of the three models. The findings show that the models tend to over predict at higher folds without significantly expanding the prediction range. The roll and pitch features are treated with the same methodology in the ensuing sections.

16.10 RESULTS AND DISCUSSIONS

16.10.1 Roll Results

The roll feature findings are described in this section. The model used for regression is an ensemble of the tree optimizable model with cross validation and the same number of *f* folds, just like altitude. Roll is performed using the same data structure as the altitude forecasts.

Large delta values are displayed in all flight data sections by the model projections. In comparison to altitude, the roll feature model is not able to predict weight with any degree of accuracy. While the range among flight data sections is still large, the projected magnitude is close to the actual value. The same regression model is then used to test the pitch characteristic.

16.10.2 Pitch Results

Using the same model and data input structure as the roll and altitude features, the pitch feature is assessed separately. Once more, the regression model in use is an ensemble of trees that can be optimized for the same amount of folds and cross validation.

A more precise fit is seen, with a narrower spread of data points than with the roll feature. Similar to earlier models, the system is tested using unidentified flight data.

The data indicates that pitch and altitude are the most reliable indicators of weight. This chapter's last component builds a model by combining the training model's pitch and altitude sections. The prediction power of the model can be tested by feeding it individual pitches or altitudes.

16.10.3 Altitude Plus Pitch Results

By combining the pitch and altitude features, the final projections are produced.

Pitch and altitude features are represented by half and half of the data, respectively, as in Figure 16.3. The integrated data model makes use of the same model. An ensemble of the tree model that can be optimized using cross validation and the same number of folds is trained. After the model has been trained, the algorithm is tested using flight data that was not used during the model's training.

The test flight's forecasts for this model were carried out marginally differently than those in the parts before it. The rationale is that two distinct features, rather than just one, are being tested.

The MATLAB programme receives independent inputs from either pitch or altitude. An algorithm is trained with a portion of each flight to produce predictions. This indicates that the validation

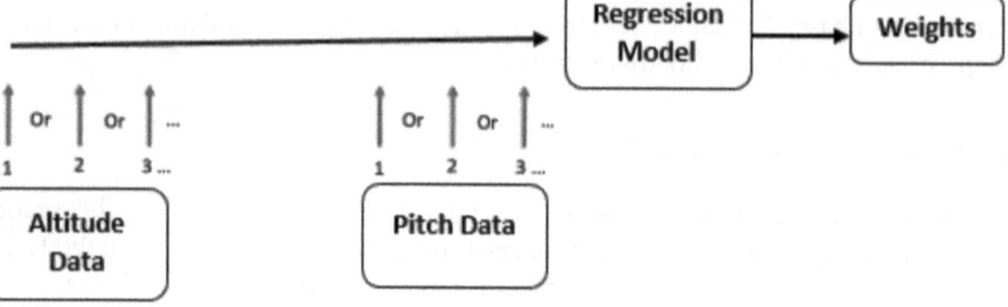

FIGURE 16.3 Altitude plus pitch ensemble of trees' optimizable cross-validation prediction.

of the pitch and altitude parts was done independently. The average results for each section were averaged over both features (pitch and altitude) in order to summarize the findings. As a result, the model can forecast the weight of the UAV independent of the feature and data segment utilized to do so (between pitch and altitude).

EXERCISES

Exercise 16.1

How can machine-learning techniques be employed for estimation tasks, and what are some common approaches used in solving estimation problems?

Exercise 16.2

Explain the concept of the Extended Luenberger Observer and discuss its significance in control systems.

Exercise 16.3

Explain the concept of the recurrent neural network (RNN)-based state estimator and discuss its significance in dynamic system modelling and control.

Exercise 16.4

Discuss how linear regression methods can be utilized for state estimation in dynamic systems, highlighting their advantages, limitations, and typical applications.

Exercise 16.5

Explain how support vector machines (SVMs) can be used for state estimation in dynamic systems, outlining their advantages, limitations and typical applications.

Exercise 16.6

Explain how regression trees and their ensembles can be used for state estimation in dynamic systems.

Exercise 16.7

Explain the concept of flight simulation and discuss the importance of data repeatability in flight simulation studies.

Exercise 16.8

Explain the concept of feature reduction in machine learning and discuss its significance in improving the efficiency and effectiveness of models.

Exercise 16.9

Explain the concept of support vector machine (SVM) optimizable regression and discuss its significance in machine-learning modelling.

Exercise 16.10

Explain the concept of ensembles of regression trees with cross-validation and discuss its significance in machine-learning modelling.

REFERENCES

1. Alhajeri, M.S., Wu, Z., Rincon, D., Albalawi, F., and Christofides, P.D., Machine-learning-based state estimation and predictive control of nonlinear processes, *Chemical Engineering Research and Design*, Vol. 167, pp. 268–280, 2021, https://doi.org/10.1016/j.cherd.2021.01.009
2. Enriquez Fernandez, A., and Everett, L.J., UAV parameter estimation through machine learning, *Proceedings of the 2009 ASEE Gulf-Southwest Annual Conference Baylor University*, American Society for Engineering Education.
3. Loh, W.-Y., Classification and regression trees, *Wiley Interdisciplinary Reviews: Data Mining and Knowledge Discovery*, Vol. 1, No. 1, pp. 14–23, 2011, http://doi.wiley.com/10.1002/widm.8
4. Lin, W., and Sontag, E.D., A smooth universal controller for nonlinear systems, *IEEE on Automatic Control*, Vol. 36, No. 12, pp.1396–1401, 1991.

Appendix A Mathematical Modelling and System Identification

A.1 INTRODUCTION

The process of mathematical modelling covers four important aspects: (i) representation, (ii) measurement data, (iii) parameter estimation and (iv) validation of the estimated models. For estimation, some *a priori* mathematical models are specified from a standard models' bank: static or dynamic, linear on nonlinear, deterministic or stochastic, continuous- or discrete-time, with constant or time-varying parameters, lumped or distributed, and even hybrid kinds of models [1–3]. The parameters are considered as random variables, since they are determined from the data, which are contaminated by the measurement noise (sensor/instrument noise) or the environmental noise (atmospheric turbulence acting on a flying aircraft or helicopter), this being of stochastic nature. The measurements (observations/data) are required for estimation purpose. The measurements generally would be noisy. For us the parameters are the coefficients that express the (strengths of) relations between the signals of interest including the states. For the signals that cannot be measured, e.g. the noise, their statistics are assumed to be known and used in the estimation algorithms. Often, one needs to estimate these statistics. The use of the so-called Cramer–Rao bounds as uncertainty bands on the estimates will provide confidence in the estimates if the bounds are very low. The real test of the estimated model is its use for control, simulation or prediction in a real practical environment.

A.2 ASPECTS OF ESTIMATION ERRORS

In parameter estimation process we need to define a certain error criterion. The optimization of this error (criterion or) cost function will lead to a set of equations, which when solved will give the estimates of the parameters of the dynamic system. Estimation being data dependant (in the form of arrays), these equations will have some form of matrices, which will be computed using the measured data. Often, one resorts to a numerical procedure to solve this set of equations.

The 'error' is defined particularly in three ways: (i) output error: the difference between the output of the model (to be) estimated and the I/O data. Here, the input to the model is the same as the system input; (ii) equation error: define $\dot{x} = Ax + Bu$, if accurate measurements of $\dot{x}, x$ (states of the system) and u (control input) are available, then equation error (EE) is defined as $(\dot{x}_m - Ax_m - Bu_m)$; and (iii) parameter error: the difference between the estimated value of a parameter and its true value; the parameter error can be obtained if the true parameter value is known, which is not the case in a real-life scenario; however, the parameter estimation algorithms (the code/script/program) can be checked/validated with simulated data, which are generated using the true parameter values of the system.

For the real data situations, statements about the error in estimated values of the parameters can be made based on some statistical properties, e.g. the estimates are unbiased, etc. Mostly, the output error method (OEM) is used and is appealing from the point of view of matching of the measured and estimated/predicted model output responses. This, of course, is a necessary but not a sufficient condition. Many of the theoretical results on parameter estimation are related to the sufficient condition aspect. Many 'goodness of fit'; model selection and validation procedures often offer practical solutions to this problem. If accurate measurements of the states and the inputs are available, the EEMs (EE methods) are very good alternative to OEM. However, such situations would not occur so frequently.

A.3 TIME-SERIES IDENTIFICATION

The time-series methods have gained considerable acceptance in the system identification literature in view of their inherent simplicity and flexibility. These techniques provide external descriptions (directly between I/O data) of systems under study and lead to parsimonious, minimum parameterization representation of the process. The accurate determination of the dynamic order of the time series models is a necessary first step in system identification, the procedure is just like the one shown in Figure 1.1 (Chapter 1).

Many statistical tests are available in the literature, which can be used to find the model order for any given process. Selection of a reliable and efficient test criterion has been generally elusive, since most criteria are sensitive to statistical properties of the process. These properties are often unknown. Validation of most of the available criteria has generally been via simulated data. However, these order determination techniques have to be used with practical systems with unknown structures and finite data. It is therefore necessary to validate any model order criterion using a wide variety of data sets from differeing dynamic systems.

A.4 TIME-SERIES MODELS

The time-series modelling is one of the specialized aspects of system identification/parameter estimation study. It addresses the problem of determining coefficients of a differential or difference equations, which can be fitted to the empirical data, or obtaining coefficients of a transfer function model of a system from its IO data. One of the main aims of time-series modelling is the use of the model for prediction of the future behaviour of the system or phenomena. One of the major applications of this approach is to understand various natural phenomena, e.g. rainfall-runoff prediction. In general, time series are a result of stochastic (random) input to some system or some inaccessible random like influence on some phenomenon, e.g. the temperature variation at some point in a room at a certain time. Hence, a time series can be considered as a stochastic phenomenon. The modelling and prediction of the seasonal time series are equally important and can be handled using extended estimation procedures.

A general linear stochastic discrete-time system/model is described here with the usual meaning for the variables:

$$\begin{aligned} x(k+1) &= \Phi_k x(k) + Bu(k) + w(k) \\ z(k) &= Hx(k) + Du(k) + v(k) \end{aligned} \tag{A.1}$$

However, for time-series modelling a canonical form (of eqn. (A.1)) known as Astrom's model is given as

$$A(q^{-1})z(k) = B(q^{-1})u(k) + C(q^{-1})e(k) \tag{A.2}$$

Here, A, B and C are polynomials in q^{-1} which is a shift operator defined as

$$q^{-n}z(k) = z(k-n) \tag{A.3}$$

For a SISO (single-input single-output) system, we have the expanded form as

$$\begin{aligned} z(k)+a_1 z(k-1)+\ldots+a_n z(k-n) &= b_0 u(k)+b_1 u(k-1)+\ldots+b_m u(k-m) \\ &\quad + e(k)+c_1 e(k-1)+\ldots+c_p e(k-p) \end{aligned} \tag{A.4}$$

Here, z is the discrete measurement sequence, u is the input sequence and e is the random noise/error sequence. We have the following equivalence:

$$\begin{aligned} A(q^{-1}) &= 1+a_1 q^{-1}+\ldots+a_n q^{-n} \\ B(q^{-1}) &= b_0+b_1 q^{-1}+\ldots+b_n q^{-m} \\ C(q^{-1}) &= 1+c_1 q^{-1}+\ldots+c_n q^{-p} \end{aligned}$$

Here, a_i, b_i and c_i are the coefficients to be estimated. We also assume here that the noise processes w and v are uncorrelated and white. In addition, we assume that the time series we deal with are stationary in the sense that first- and second-order (and higher) statistics are not dependant on time 't' explicitly. For mildly nonstationary time series, the appropriate models can be fitted to the segments of such time series.

Certain special forms are specified next. These models are called time-series models, since the observation process can be considered as a time series of data that has some dynamic characteristics, affected usually by a random process. We assume here that inputs are such that they excite the modes of the system. This means that the input contains sufficient frequencies to excite the dynamic modes of the system. This will in turn assure that in the output, there is sufficient effect of the modes and hence the information, so that from I/O time-series data, one can accurately estimate the characteristics of the process.

(a) **Astrom's model**: this is the most general linear time-series analysis model, with full form of error/noise model. Given input (u)/output (z) data, the parameters can be estimated by some iterative process, e.g. the ML method (Chapter 2). The transfer function form is given by

$$z=\frac{B(q^{-1})}{A(q^{-1})}u+\frac{C(q^{-1})}{A(q^{-1})}e \tag{A.5}$$

This model can be used to fit time-series data, which can be considered as arising out of some system phenomenon with a controlled input u and a random excitation.

(b) **Autoregressive (AR) model:** By assigning $b_i = 0$ and $c_i = 0$ in the Astrom's model, we get

$$z(k) = -a_1 z(k-1)-\ldots-a_n z(k-n) + e(k) \tag{A.6}$$

The transfer function form can be easily obtained as

$$z=\frac{1}{A(q^{-1})}e \tag{A.7}$$

Here, the output process $z(k)$ depends on its previous values (and hence the name autoregressive) and it is excited by the random signal e. It is assumed that the parameters a_i are constants such that the process z is stationary.

We can consider that $1/A(q^{-1})$ is an operator, which transforms the process e into the process z. The polynomial A determines the characteristics of the output signal z and the model is called an 'all poles' model. This is because the roots of $A(q^{-1}) = 0$ are the poles of the transfer function model. The input process e is inaccessible and immeasurable. The parameters of A can be estimated by using the least-squares method. In addition, this model is very useful for determining the spectrum of the signal z, if input process e is considered as white process noise, since the parameters of A are estimated and hence known. This method of estimation of the spectrum of a signal contrasts with the one using the Fourier transform. However, both methods are supposed to give similar spectra. It is most likely that the autoregressive spectrum will be smoother compared to the Fourier spectrum.

(c) **Moving average (MA) model**: if we put $a_i = 0$ and $b_i = 0$ in Astrom's model, we get

$$z(k) = e(k) + c_1 e(k-1) + \ldots + c_p e(k-p) \tag{A.8}$$

The process z is now a linear combination of the past and present values of the inaccessible random input process e.

The roots of $C(q^{-1}) = 0$ are the zeros of the model. The process z is called MA process and is always stationary since $A(q^{-1}) = 1$. In this form, the output signal does not regress over its past values.

(d) **Autoregressive moving average (ARMA) model:**
Letting $b_i = 0$ in Astrom's model, we obtain an ARMA model, since it contains both AR and MA parts. We emphasize here that the control input u is absent:

$$z(k) + a_1 z(k-1) + \ldots + a_n z(k-n) = e(k) + c_1 e(k-1) + \ldots + c_p e(k-p) \tag{A.9}$$

$$z = \frac{C(q^{-1})}{A(q^{-1})} e \tag{A.10}$$

So this model is zero/pole type model and has the structure of the output/input model. More complex time-series can be accurately modeled using this model.

(e) **Least-squares model:**
By letting $c_i = 0$ in the Astrom's model, we get

$$z(k) + a_1 z(k-1) + \ldots + a_n z(k-n) = b_0 u(k) + b_1 u(k-1) + \ldots + b_m u(k-m) + e(k) \tag{A.11}$$

Here, control input u is present. The model is so called since its parameters can be easily estimated by LS method. The transfer function form is

$$z = \frac{B(q^{-1})}{A(q^{-1})} u + \frac{1}{A(q^{-1})} e \tag{A.12}$$

It has an AR model for the noise part and the O/I model for the signal part. Determination of $B(q^{-1})/A(q^{-1})$ gives the transfer function model of the system. One can obtain discrete Bode diagram of the system from this pulse transfer function and then convert it to the continuous-time domain to interpret the dynamic behaviour of the system. One can use a complex curve fitting technique or bilinear/pade method.

A.5 TIME-SERIES MODEL IDENTIFICATION

The estimation of parameters of MA and ARMA can be done using the ML approach, since the unknown parameters appear in the MA part, which represents itself as unknown time series 'e'. However, parameters of AR and LS models can be estimated using LS method (Chapter 2). Assumption of the identifiability of the coefficients of the postulated models is pre-supposed.

Let the LS model be given as in eqns. (A.11) and (A.12). We define the equation error as

$$e(k) = A(q^{-1})z(k) - B(q^{-1})u(k)$$

$$r(k) = \hat{A}(q^{-1})z(k) - \hat{B}(q^{-1})u(k) \tag{A.13}$$

Eqn. (A.13) can be put in the form $z = H\beta + e$; here, $z = \{z(n+1), z(n+2), \ldots z(n+N)\}^T$

$$H = \begin{bmatrix} -z(n) & -z(n-1) & -z(1) & u(n) & u(n-1) & u(1) \\ -z(n+1) & -z(n) & -z(2) & u(n+1) & u(n) & u(2) \\ \vdots & & \vdots & \vdots & & \\ -z(N+n-1) & \cdots & -z(n) & u(N+n-1) & \cdots & u(N) \end{bmatrix} \tag{A.14}$$

N = number of total data used ; $m = n$ and $b_0 = 0$. For example, let $n = 2$ and $m = 1$, then

$$e(k) = z(k) + a_1 z(k-1) + a_2 z(k-2) - b_0 u(k) - b_1 u(k-1) \tag{A.15}$$

$$z(k) = \begin{bmatrix} -z(k-1) & -z(k-2) \end{bmatrix} \begin{bmatrix} a_1 \\ a_2 \end{bmatrix} + \begin{bmatrix} u(k) & u(k-1) \end{bmatrix} \begin{bmatrix} b_0 \\ b_2 \end{bmatrix} + e(k)$$

$$z(k+1) = \begin{bmatrix} -z(k) & -z(k-1) \end{bmatrix} \begin{bmatrix} a_1 \\ a_2 \end{bmatrix} + \begin{bmatrix} u(k+1) & u(k) \end{bmatrix} \begin{bmatrix} b_0 \\ b_2 \end{bmatrix} + e(k+1)$$

This leads to using the LS method, we get

$$\hat{\beta} = \left\{ \hat{a}_1, \hat{a}_2, \ldots, \hat{a}_n \; \vdots \; \hat{b}_1, \hat{b}_2, \ldots, \hat{b}_m \right\} = (H^T H)^{-1} H^T z \tag{A.16}$$

The parameters/coefficients of time-series models can be estimated using the system identification toolbox of MATLAB. The crucial aspect of time-series modelling is that of selection of model structure (AR, MA, ARMA or LS) and the number of coefficients for fitting this model to the time-series data.

A.6 MODEL (MODEL-ORDER) SELECTION CRITERIA

In the absence of *a priori* knowledge, any system that is generating time-series output can be represented by more popular auto-regressive (AR) or a least-squares (LS) model structure. Both represent a general nth order discrete linear time invariant system affected by random disturbance. The problem of model order determination is to assign a model dimension so that it adequately represents the unknown system. Model selection procedure involves selecting a model structure and complexity. A model structure can be ascertained based on the knowledge of the physics of the system. For certain processes, if physics is not well understood, then a black-box approach can be used. This will lead to trial and error iterative procedure, this is called sensitivity analysis. However, in many situations, some knowledge about the system or the process is always available. Then, further refinements can be done using system identification techniques. Here, we consider the modelling problem in the context of structure and order selection based on well-defined model selection criteria (MSC). We describe several such MSC arising out of different, but related principles of goodness of fit and statistical measures. The criteria are classified based on fit error, number of model parameters, whiteness of residuals and related approaches.

A.6.1 Fit Error Criteria (FEC)

We describe criteria based on the concept of fit error.

(a) **Fit error criterion (FEC1)**: one of the natural MSC is a measure of the difference between the actual response of the system and estimated response of the postulated/estimated model. Evaluate the FEC as follows:

$$FEC1 = \frac{1/N\sum_{k=1}^{N}[z_k - z_k(\hat{\beta}_1)]^2}{1/N\sum_{k=1}^{N}[z_k - z_k(\hat{\beta}_2)]^2} \tag{A.17}$$

Apply the decision rule: if FEC1<1 select the model with $\hat{\beta}_1$; if FEC1>1 select the model with $\hat{\beta}_2$. The ratio FEC can be corrected for the number (n_1, n_2) of unknown parameters in the model by replacing N by $N - n_1$ and $N - n_2$ in the numerator and the denominator of eqn. (A.19), respectively. The FEC is considered to be a subjective criterion thereby requiring subjective judgement, i.e. if FEC1≈1, then both the models would be just good, one has to prefer a model with a lesser number of coefficients (parameters).

(b) **Fit error criterion (FEC2)**: an alternative FEC, sometimes called prediction fit error (PPE) in the literature, can be used to judge the suitability of the model fit:

$$FEC2 = \frac{1/N\sum_{k=1}^{N}[z_k - z_k(\hat{\beta})]^2}{1/N\sum_{k=1}^{N} z_k^2} \tag{A.18}$$

Replacing N with N-n can correct the criterion, for the degrees of freedom. Essentially, FEC2 compares models based on the reduction of residuals to signal power ratio of successive models. Insignificant change in the value of FEC2 determines the order of the model. Essentially, one locates the knee of the curve FEC2 v/s model order. Generally, this criterion does not give sharp knee and hence again requires subjective judgment. In the parameter estimation literature (Chapters 2 and 3), this criterion is the usual fit error criterion (often used as percentage fit error – PFE = FEC2 × 100).

(c) **Residual sum of squares (RSS)**: often, sum of residuals is used for judging the model adequacy:

$$RSS = \sum_{k=1}^{N}[z_k - \hat{z}_k(\hat{\beta})]^2 \tag{A.19}$$

If any new parameter enters the model, then there should be significant reduction in RSS, otherwise it is not included in the model.

(d) **Deterministic fit error (DFE)**: for models of input/output type, this is a useful criterion. It accounts for the effects of modelling and computational errors. For the TF-type model, the deterministic fit error is given by

$$DFE = z - \frac{\hat{B}(q^{-1})}{\hat{A}(q^{-1})} u \tag{A.20}$$

Similar observations as for FEC2 can be made regarding this criterion.

The prediction error criteria (PEC) generally provide quantitative means for selecting the models that best support the measured data. The capability of a model to predict the responses of the system for a class of inputs can be judged based on these PECs given next.

(e) **Prediction error criterion 1 (PEC1)**: in this case, the data to be analyzed (measured data) are divided into two consecutive segments. The first segment of data is used in identification procedure to estimate the unknown parameters. Then, this model (parameters) is used to predict the response for the second segment and compared with it. The model that predicts this response most accurately is considered an accurate model. Again, here subjective judgement is involved since 'most accurately' is not quantified. The PEC1 can be used also as a model validation criterion. Let the identified model from the first data segment be called $M(\hat{\beta} \mid z_k, k = 1,2,\ldots,N1)$. Then prediction error time history for the second segment up to $N2$ is generated as

$$e_z(j) = z_j - \hat{z}_j\{M(\hat{\beta} \mid z_k,\ k = 1,2,\ldots,N1)\}; \quad j = N1+1,\ldots,N2 \tag{A.21}$$

Here $N > N1 + N2$.

Further quantification of $e_z(j)$ can be obtained by evaluating its power, i.e variance as

$$\sigma_{e_z}^2 = \frac{1}{N_2} \sum_{j=1}^{N_2} [e_z(j)]^2 \tag{A.22}$$

A very low value of this variance signifies a good prediction.

(f) **Prediction error criterion 2 (PEC2):** in this procedure, prediction error is estimated statistically and the criterion is a well-known Akaike's final prediction error (FPE), described next.

A.6.2 Criteria Based on Fit Error and the Number of Model Parameters

(a) **Final prediction error (FPE):** a good estimate of the prediction error for a model with n parameters is given by the final prediction error:

$$FPE = \sigma_r^2(N,\hat{\beta}) \frac{N+n+1}{N-n-1}; \quad \sigma_r^2 = \text{variance of the residuals.} \tag{A.23}$$

A minimum is sought with respect to n, the number of parameters. Absolute minimum occurs when σ_r^2 is zero. FPE includes a penalty for large model orders. This means that if n increases, the numerator increases. The penalty is paid in FPE. If n is large, then σ_r^2 will reduce, and hence a compromise is struck. For real data situations local minimum can result. This test is developed for univariate process corrupted by white noise. The penalty for degrees of freedom is greatly reduced for large N, meaning thereby that FPE is less sensitive to n, if N is large.

(b) **Akaike's information criterion (AIC/it denotes information criterion)**

Akaike refined FPE into AIC by extending the maximum likelihood principle and tool into account the parametric dimensionality of the model:

$AIC = -2\ln(\text{maximum likelihood}) + 2$ (number of independent parameters in the model)

or $AIC = -2\ln(L) + 2n$

If the two models are equally likely ($L_1 \approx L_2$), then the one with a lesser number of parameters is chosen. We see from the above expression that if the number of parameters increases, the AIC also increases, and hence the less preferable the model.For an autoregressive (AR) model of order n we get

$$AIC(n) = N\ln\sigma_r^2 + 2n \tag{A.24}$$

This is a generalized concept of FPE. For $n = 0,1,\ldots$, the value of n, for which the *AIC(n)* is minimum is adopted as true order of the model. However, *AIC* might not give consistent model order in statistical sense. We see from eqn. (A.26) that as n increases, the second term increases but due to fitting with a greater number of parameters, the first term decreases, so a compromise is struck.

These criteria, for a given model structure, may not attain unique minimum. Under weak assumptions, they are described by χ^2 distribution. It is well known that *FPE* and *AIC* are asymptotically equivalent.

(c) **Criterion autoregressive transfer function (CAT):**

Parzen and Tong advanced these CAT methods for model order determination.

(i) **Parzen (*PCAT*1)**: this criterion was advanced with a view to obtaining the best finite AR model based on a finite number of measurements used for time-series modelling. The formula for PCAT is given as

$$PCAT1(n) = 1 - \frac{\hat{\sigma}_\infty^2}{\bar{\sigma}_r^2} + \frac{n}{N}; \quad n = 0,1,\ldots \tag{A.25}$$

$\hat{\sigma}_\infty^2$ = estimate of the one-step ahead prediction error variance σ^2.
$\bar{\sigma}_r^2$ = unbiased estimate: $\frac{N}{N-1}\hat{\sigma}_r^2$

*PCAT*1 can be considered to asymptotically obtain the same order estimate as that obtained by AIC. *PCAT1* signifies the minimization of relative mean square error between nth order AR model and theoretical AR model.

(ii) **Parzen (*PCAT2*)**: a modified criterion is given by

$$PCAT2(n) = 1/N\sum_{j=1}^{n} 1/\bar{\sigma}_j^2 - 1/\bar{\sigma}_r^2 \tag{A.26}$$

Here, $PCAT2(0) = -(1+N)/N$, and minimum is sought.

(iii) A modification of *PCAT2* was proposed, since for true *AR(l)* model, *PCAT2* may prefer *AR(0)* model to *AR(l)* model. Thus, modified criterion which avoids this ambiguity, is given by,

$$MCAT(n) = 1/N\sum_{j=0}^{n}\frac{1}{\bar{\sigma}_j^2} - \frac{1}{\bar{\sigma}_r^2} \text{ and a minimum is sought} \tag{A.27}$$

It has been shown that *MCAT* and *AIC* have identical local behaviour. However, global maxima of *MCAT(n)* and *AIC(n)* do not necessarily occur at the same n.

A.6.3 Tests Based on Whiteness of Residuals

These tests are used to check whether the residuals are a white noise sequence, thereby asserting independence at different time instants. We describe two such tests.

(a) **Autocorrelation based whiteness of residuals (ACWRT)**: the test is performed as follows: Estimate the autocorrelation function $R_{rr}(\tau)$ of residual sequence *r(k)*, for lag $\tau = 1,2,\ldots,\tau_{max}$

$$\hat{R}_{rr}(\tau) = 1/N \sum_{k=\tau}^{N} r(k)\, r(k-\tau) \tag{A.28}$$

Here it is assumed that *r(k)* is a zero-mean sequence. $\hat{R}_{rr}(\tau)$ is considered an asymptotically unbiased and consistent estimate of true autocorrelation. Also, under null hypothesis, $\hat{R}_{rr}(\tau)$ for $\tau = 1,2,\ldots$ are asymptotically independent and normal with zero mean and covariance of *1/N*. Thus, they must lie in the band $\pm 1.96/\sqrt{N}$ at least for 95% of the times for the null hypothesis. Usually the normalized ratio is used: $\hat{R}_{rr}(\tau)/\hat{R}_{rr}(0)$. The autocorrelations tend to be an impulse function if the residuals are uncorrelated.

(b) **Whiteness of residuals (SWRT)**: Stoica has proposed another test to check the residual of estimation for whiteness. If a discrete time series is a white sequence, then

$$\sum_{\tau=1}^{\tau_{max}} R_{rr}^2(\tau) \le (k_j + 1.65\sqrt{2k_j})R_{rr}^2(0)/N \tag{A.29}$$

$k_j = \tau_{max} - n_j - 1;\ \tau_{max} = 20$

A.6.4 F-Ratio Statistics

The ratio test is based on the assumption of normally distributed random disturbances and requires *a priori* specifications of acceptance-rejection boundaries. Due to this, such tests should be used in conjunction with other tests:

$$F_{n_1 n_2} = \frac{V_{n_1} - V_{n_2}}{V_{n_2}} \; \frac{N - 2n_2}{2(n_2 - n_1)} \tag{A.30}$$

V_{n1} and V_{n2} are the minimum values of the loss function for a model with n_1, n_2 parameters, respectively. The random variable F for large N is asymptotically $F(n_2 - n_1, N - n_2)$ distributed. When the number of parameters is increased by 2, we have

$$F(2,100) = 3.09 \Rightarrow \text{Prob}(F > 3.09) = 0.05 \quad \text{and}$$
$$F(2,\infty) = 3.00 \Rightarrow \text{Prob}(F > 3.00) = 0.05$$

Thus, at a risk level of 5% and $N > 100$, the quantity F should be at least 3 for the corresponding reduction in loss function to be significant. A slightly different version of this criterion, where R could be any statistic computed using square of a variable, e.g. covariance of residual, etc., is given as

$$F(j) = \frac{R(0,\hat{\beta}_j) - R(0,\hat{\beta}_{j+1})}{R(0,\hat{\beta}_{j+1})}(N - n_{j+1} - 1); \quad j = 1,2,\ldots \tag{A.31}$$

In the above, $R(0)$ can signify the autocorrelation at zero lag, implying the variance of the residuals.

A.6.5 Tests Based on Process/Parameter Information

(a) **Entropy**: entropy signifies disorder in the system. This test is based on the amount of information measure of an AR process (of order n), which is characterized by the entropy. It is possible to judge the order of the given process before estimating the parameters because computation is based on the correlation matrices of different orders for assumed AR models:

$$E_n(j) = \ln \frac{N - n_j}{N - 2n_j - 1} + \ln\left|S_{j+1}\right| - \ln\left|S_j\right| \tag{A.32}$$

Here, S_j = correlation matrix with its elements as autocorrelations, $\hat{R}_{rr}(\tau)$; $\tau = 1,2,\ldots \tau_{\max}$; $|S|$ = determinant of S.

The value of n_j for which $E_n(j)$ is minimum is selected as the adequate order. This test can be regarded as a pre-estimation criterion. It has to do with the minimization of the difference in the adjacent entropies. Decrease in entropy signifies the increase of 'order' in the system and hence leads to a proper model order of the system.

(b) From the definition of information measure, it is known that the amount of uncertainty in estimates and hence dispersion are related to the inverse of the information matrix. Thus, near singularity of this matrix means large standard deviations of the parameter estimates. Near singularity could also signify that the model structure has been overly large, thereby losing the parameter identifiability property.

A.6.6 Bayesian Approach

The criteria based on this approach are as follows:

(a) **Posteriori probability (PP):** this test is based on a Bayesian-type procedure for discrimination of structure of the models. If C_j is the class of models, then the appropriateness of a class C_j to represent the given data set z is measured by the posteriori probability $P(C_j \mid z)$. A low value of $P(C_j \mid z)$ indicates that C_j is inappropriate for representing z. This test gives a consistent order selection criterion, the simplified version is given as

$$PP(n_j) = -N \ln(\sigma_r^2) - n_j \ln(\sigma_z^2/\sigma_r^2) - (n_j + 1)\ln N \tag{A.33}$$

Here, σ_z^2 = variance of the given time series. One chooses n_j that gives the largest value of PP.

(b) **B-statistic:** another consistent order determination statistic is given as

$$B(n_j) = N \ln(\sigma_r^2) + n_j \ln N \tag{A.34}$$

The model with minimum B is chosen, thus giving adequate (AR or ARMA) model with n_j coefficients.

(c) **C-statistic:** it is interesting to note that B statistic is similar to another statistic:

$$C(n_j) = N \ln(\sigma_r^2) + n_j \mathbf{h}(N) \tag{A.35}$$

where $\mathbf{h}(N)$ is any monotonically increasing function of number of data, and satisfies the following condition: $\lim_{N \to \infty} [\mathbf{h}(N)/N] \Rightarrow 0$. The decision rules based on 'C' are statistically consistent.

A.6.7 Complexity (COMP)

This criterion is based on a compromise between the whiteness of model residuals and the accuracy of estimated parameters. It must be recalled that a good predictor should incorporate all the available information (residuals being white) and one should include accuracy of the parameter estimates in the model discrimination process. The criterion is given as

$$COMP(n_j) = \frac{1}{n}\sum_{j=1}^{n_j} p_{jj}^2 - \left\{\frac{trace(P)}{n_j}\right\}^2 + \frac{2}{n_j}\sum_{j=1}^{n_j}\sum_{l=j+1}^{n_j} p_{jl}^2 + \frac{2}{n_j}\sum_{\tau=1}^{\tau_{\max}} (N-\tau)\,\mathrm{R}_{rr}(\tau) \qquad \text{(A.36)}$$

Here P is covariance matrix of estimated parameters and p_{jl}, the elements of P. Within a given structure, with large number of parameters, increased interactions (P) will tend to positively contribute to *COMP*. The residuals will tend to be white, thereby making the fourth term to decrease. Thus, *COMP* provides a trade-off between the accuracy of the estimates and whiteness of the residuals. However, the computational requirement is more than that for *AIC, B* statistic and *FPE* tests. This COMP criterion can be used for model structure as well as model order determination.

A.6.8 Pole-Zero Cancellation

For I/O type of the models, the process of cancellation of zeros with poles can provide a model with a lesser degree of complexity. In conventional method, the numerator and denominator polynomials are factored and cancellation then becomes obvious. However, subjective judgement is involved, since the cancellation might not be perfect.

A.7 MODEL SELECTION PROCEDURES

The subjective tests have been used in many applications and the main difficulty in using these has been the choice of proper levels of statistical significance. The subjective tests tend to ignore the increase of variability of estimated parameters for large model orders. It is often common to assume a 5% risk level as acceptable for F-test and whiteness tests arbitrarily. However, the whiteness test-SWR does consider the cumulative effects of autocorrelations of residuals. The pole-zero cancellations are often made visually and are again subjective. A systematic exact pole-zero cancellation is possible, but it is computationally more complex. Fit error methods are useful but again subjective, and are only necessary but not sufficient conditions.

In the objective-type tests, an extremum of a criterion function is usually sought. The final prediction error (FPE) criterion due to Akaike is based on one-step-ahead prediction and is essentially designed for white noise corrupted processes. The Akaike information criterion AIC is a generalized concept based on a mean log-likelihood function. Both the FPE and AIC depend only on residual variance and number of estimated parameters. At times, these tests yield multiple minima. The criterion autoregressive transfer function (CAT) due to Parzen has been proposed as the best finite AR model derived from finite sample data generated by AR model of infinite order. The *MCAT* is modification of *PCAT2* to account for any ambiguity, which may arise for 'true' first-order AR processes due to omission of σ_0^2 terms. Based on the experience gained, the following working rule is considered adequate for selection of the model order to fit typical experimental data:

(a) **Order determination**
  (i) evaluate entropy criterion (AR only)
  (ii) evaluate FPE
  (iii) perform F-test
  (iv) check for pole-zero cancellations (for input/output model).

(b) **Model validation**
   (i) time history prediction
   (ii) test residuals for whiteness
   (iv) cross validation

Alternatively, the readers can arrive at their own rule based on study of other criteria discussed in this chapter.

REFERENCES

1. Ljung, L., *System Identification: Theory for the User*, Prentice-Hall, Englewood Cliffs, 1987.
2. Jategaonkar, R.V., Raol J.R., and Balakrishna, S., Determination of model order for dynamical systems, *IEEE Transactions on Systems, Man and Cybernetics*, USA, Vol. SMC-12, pp. 56–62, Jan.–Feb.1982.
3. Raol, J.R., Girija, G., and Singh, J., *Modelling and Parameter Estimation of Dynamic Systems*, IET/IEfE Control Series Books, London, UK, Vol. 65, Aug. 2004.

Appendix B
Sensors, Signals and Estimation-Error Propagation

B.1 SENSORS

Several advances in electronics, computer science and engineering and optics have resulted in a massive increase in the number of types of sensors available for use in autonomous systems. The automated systems must be able to react to changes in their environment to be effective and therefore these systems must rely on sensors for information about their environment. The sensors provide good interfaces between an ill-defined and chaotic reality (not so always, but often) and a computer and a data acquisition system. Sensors can handle data only when they are within specified bounds and presented in a precise format for ease of our understanding and interpretation. Often sensors are not very simple and might have limited resolution and reliability. In complex systems, the sensors do not necessarily correspond to a *single concrete sensor*. It could be a sensor that has some pre-processing built in unit or some logic or even an intelligent processing unit (like a KF itself) built in the sensor system; and the readings from the sensors can be fused to create one *abstract sensor* [1–3].

Sensors that correspond to only one concrete sensor can be referred to as a *simple sensor* and the sensors made up of several concrete sensors can be referred to as *sensor networks* (SNWs), which are created out of many simple sensors. After fusion, the data from a sensor NW can be treated as if the data have originated from a simple sensor. The SNWs can have many different levels of complexity, several different architectures and configurations. These architectures have different requirements of data communication and computational complexity. The performance of sensors is measured in terms of: accuracy, repeatability (precision), linearity, sensitivity, resolution, reliability (minimum failures), and the range it can cover/provide. The sensor output could be a continuous signal, digital, frequency output, amplitude and pulse-modulated.

Active sensors inject/send some type of signal into their environment and then measure the interaction of the signal with the environment, e.g. radar, sonar, active far IR sensor (AFIR), an active optical seam tracker, and an active computer vision system. These sensors require an external power for their operation. The signal being measured is thus modified by the sensor to produce an output signal. These sensors are also called parametric sensors, e.g. in the thermister, the change of resistance due to change in temperature takes place. The properties of the sensor can change in response to an external signal. Passive sensors simply record energy/signals already present in the system's environment, e.g. thermometers, video cameras and infrared cameras. The input stimulus energy is converted by the sensor/transducer into output energy without the need for an additional power source or injection of energy into sensor, e.g. a thermocouple, a piezoelectric sensor, a pyroelectric detector, and passive IR sensor.

B.2 SENSOR MODELLING

The sensors have varying accuracy and might not provide one single accurate value of a physical variable. Of course, this is not the case with inherently very accurate and high precision sensors, but these sensors might be very costly. Even if a sensor could possibly deliver this type of reading, the physical values being measured generally vary statistically over a time and will be affected with measurement noises/errors. A sensor reading that covers a finite period would thus have to return a value range, rather than one number to accurately represent the environment, thus a limited accuracy is a factor in all measurement data processing. Data representations in common use have upper and lower bounds for the values that can be represented, and a maximum number of digits of accuracy, and if these limits in accuracy are not handled properly, important information can be lost or the results of calculations can be greatly in error. There is a possibility that a sensor might fail (reliability aspect) during the system's life time, or may temporarily give inaccurate readings for some reasons. No matter how well systems are designed, components will eventually deteriorate or stop working correctly or perfectly. Engineers and analysts must try to minimize the impact of such component failures (the reliability of the order of 10e-7, i.e. one failure in 10e+7 operations of that sensor) through proper design procedures and decisions. Besides normal wear and tear, the capabilities of any component are generally limited. This is especially true when the system is placed in a hostile and/or noisy environment, like nuclear reactors, battlefields, and outer space.

The mathematical models of the activities that these sensors perform can be used for characterization of the sensors, and their data. Algorithms can be developed that handle sensor data efficiently and accurately and these algorithms should be efficient: (i) they extract the maximum information from the sensors/data, (ii) avoid making wrong and absurd conclusions based on faulty data, and (iii) do not waste costly system resources. Mathematical model of sensors can be derived. One can reasonably see that a sensor can be represented as a mathematical function with two arguments: the environment and time:

$$S(E, t) = \{V(t), e(t)\} \tag{B.1}$$

Here, E is an abstract entity with several physical attributes that can be measured, and S is a function that maps environment to numeric values. These attributes can be continuous valued (temperature, velocity, range, etc.), Boolean (presence of a vehicle, temperature > given limit, etc.), or integer valued (number of aircraft, tanks, number of persons in a queue). The values read will change over time, because the system's environment also changes over time. V is a point in an n-dimensional space and is affected/surrounded by some uncertainty. Knowing everything about the sensor in use and measured signal/data, *a priori*, as much as possible would help in estimation and data-fusion process because of the following relationship between the measurement and the object's state:

measurements = sensor's math model (dynamics) times state (or parameter) + noise, or

$$z = Hx + v \tag{B.2}$$

Here, aim is to estimate state/parameter vector 'x' based on 'z', and hence any knowledge/information on H and statistics of v would help in the estimation of x: sensor type, 'range band', and measurement signal 'range'. The scale factor and bias errors can be related as

$$z \text{ (measured data)} = k\,y + b \text{ (bias)} \tag{B.3}$$

Here, 'k' is the scale factor, and 'b' is the bias.

General aspects regarding measurement preprocessing by taking into account sensor filter are (i) filter dynamics, since a low pass analogue filter might have been used to remove the high-frequency noise, (ii) sensor mounting errors (rate gyros pickup other axis signal if misaligned) and (iii) the linear accelerometer's position must be known. All these aspects will help the analyst understand performance results of any estimation algorithm and data-fusion (DF) system, or to correct the data at the outset only, like kinematic consistency for aircraft's data analysis). The analyst should know as much as possible of the entire chain as to how the data get to them from the transducer/sensors, since the screening, editing of the data, and signal processing/system identification are important functions to be performed to obtain very accurate results from the DF systems. An important point of sensor modelling is to incorporate the dominant modes of the sensor dynamics, if any, into the modelling effort for the DF systems, e.g. in state/measurement models used in the KF. A surveillance system model (SSM) could be represented as a nonlinear state-space model

$$x(t+1) = f(x(t), u(t)) \tag{B.4}$$

Here, x as the state vector consisting of the variables related to sensors, platforms, targets, and environmental conditions; and u is the decision vector representing controllable inputs (or decisions in the surveillance system, e.g. sensor allocation and platform deployment). Thus, the SSM is a stochastic model. The system output 'y' (or 'z' as in eqn. (B.3)) contains the measurements, and all control elements and decisions. The platform states and the sensor measurements are observables and are part of the system output 'y', or 'z'. The state vector x is estimated from the measurements, 'z' (of observables 'y') by the situation assessment. Basically, solving the decision-making problem is to find a policy or control law $u(t)$ that optimizes an expected utility function within the constraints of the system dynamics. Hence, the flow of the main information in such a management architecture is from the sensed environment → SSM → fusion/association → ← situation assessment → sensor management → and back to SSM. The SSM could consist of the devices like electromagnetic induction sensors (EMI), ground penetrating radars (GPR), infrared sensors (IR) mounted on autonomous ground vehicles (AGV) and air (unmanned aerial vehicles – UAV).

The targets are represented by their anticipated features within the sensor models. The IR-detected target is characterized by the size and shape of the target, since the IR sensor measures the size and shape. The sensed information is acquired in two stages: (i) target detection, and (ii) target classification, here the target is assigned a model: a static BNW or a dynamic linear (-Gaussian) model, and the BNW structure would compose of feature nodes (features of the targets) based on the BNW model of the sensor that might have taken the measurement of the target. For environment, some available models can be used or the environment states that are directly measurable are used. The senor platforms can be modelled by ordinary differential equations. For performance optimization motion planning and inner loop control should take into consideration of the sensed information/situation assessment. The sensor manager should also consider the kinematic and dynamic constraints of the sensor platform. Then the SSM can incorporate the platform models and given terrain maps. The measurements processed by the fusion process are also included in the model as and when they are available. Finally the sensor manger uses this expanded SSM to allocate the sensor resources and deploy the platforms while optimizing the desired objectives. When the model of a senor is available inference algorithms can be used to compute the unknown variables based on the observables. In fact, the measurements, the observations about the environment, and the parameters used by the sensors become the evidence to the sensor BNW. Then the inference algorithms can be used to estimate the target state. When multiple sensors are used, one should use an appropriate DF process or method to obtain improved estimation of the target states. This information is then used for situation assessment (SA) and possible sensor management. The SA algorithm would produce a picture of the tactical situation inferred from the SSM with an aim of updating the estimated state of the surveillance system in time. This is based on the inference and sensor fusion, and inferring the values of the (hidden) states. This SA picture is used by the sensor manager to make decisions for allocation of the resources and optimize the objectives.

B.3 PROPERTIES OF SIGNALS AND ESTIMATORS

We derive or measure the signals from a sensor suite mounted on a body, or placed in a remote place, viz. radars in a chosen area. These signals are processed in a state or parameter estimator to obtain filtered states or parameters of the dynamic system under test and observation.

A good estimator should possess certain properties in terms of errors in parameter estimation and/or errors in the predicted measurements or responses of the mathematical model thus determined. Since the measured data used in the estimation process are noisy, the parameter estimates (estimated parameters) can be considered to have some random nature. In fact, the estimates that we would have, are the mean of the pdf, and hence the estimation error would have some associated covariance matrices. Thus, due to the stochastic nature of the errors, one would desire that the probability of the estimate being equal to the true value, be 1. We expect an estimator to be unbiased, efficient, and consistent – not all of which might be achievable.

We briefly describe several properties of signals, matrices, estimators and estimates that would be useful in judging the properties and 'goodness of fit' of the parameter/state estimates and interpreting the results. Many of these definitions, properties, and other useful aspects are used or indicated in various chapters of this book.

Some aspects of signals required for state/parameter estimation from flight test data are mentioned here. Minimum practical set is (i) time (t); (ii) control positions δe, δa, δr (elevator, aileron, and rudder control surface deflections in radians); (iii) angular rates p, q, r (roll, pitch, yaw in rad/sec.); and (iv) linear accelerations a_x, a_y, a_z (forward, side, vertical in rad/sec^2). Other useful signals are (i) α, β; (ii) Euler angles; (iii) angular accelerations $\dot{p}$, $\dot{q}$, $\dot{r}$; (iv) ρ (air density); (v) engine parameters rpm, thrust; and (vi) configuration parameters m, inertia; fuel, loading data (C.G. aft or fore of a datum).

Most important aspects are as follows: for control surface positions the direct measurements of surface hinge angles (rudder, aileron and elevator) would be desirable. If these signals are taken at the actuator command, the measurements require the knowledge of actuator dynmics. For the angular rates the alignment of the sensors should be within 0.1 deg. For the linear accelerations, the mounting location should be known-ideal being at the C.G. The flow angles, the vanes are normally used for which the vane location must be known. If the nose boom is used, then empirically the nose (boom) length should be 2.5 to 3 times the fuselage diameter. For pressure ports, three locations are a minimum to deduce the flow angle in one-axis. The aircraft velocities can be obtained by integrating accelerometer measurements. Engine parameters (engine thrust; pressure, temperatures at various places in the propulsion system; rotational speeds; fuel flows; engine control settings) are not useful for stability and control derivatives estimation but are useful for estimating drag coefficients. The positions of flaps, landing gear up or down, tank and fuel weight are other useful parameters. Typical allowable measurement and installation errors for identification of longitudinal

TABLE B.1
Permissible Error Budget

Parameter/Signal	Allowable Errors
Mass, m	1%
Moment of inertia, Iy	2%
Position of C.G. $X_{C.G.}$	0.00 5m
Dynamic pressure, $\bar{q}$ (1/2 ρ V^2)	2%
Velocity of aircraft, V	1%
Angle of attack, α	0.06 deg
Pitch angle, θ	0.3 deg
Pitch rate, q	0.01 rad/sec
Pitch acceleration, $\dot{q}$	0.05 rad/sec^2

motion of a small aircraft are given in Table B.1. Linear accelerations range should be at least 4 g and the accuracy of 0.1 g might be tolerable. Body rates accuracy of +/-0.1 deg/s would be preferable. The control surface deflection measurements should be at least accurate to 0.1 deg. The flow angles accuracy should be better than at least 0.1 deg.

B.4 PERFORMANCE MEASURES AND METRICS FOR ESTIMATORS

Here, we briefly describe important measures that can be used for judgement of goodness of the estimators.

B.4.1 Autocorrelation

For a random signal *x(t)*, it is defined as

$$R_{xx}(\tau) = E\{x(t)x(t+\tau)\}; \ \tau \text{ is the 'time-lag'} \tag{B.5}$$

Here, E stands for a mathematical expectation operator. For stationary process, R_{xx} is dependent on τ and x only and not on 't'. Its value is maximum when $\tau = 0$, then it is the variance of the signal x (assuming the mean of the signal is removed). As the time tends to be large, if the R_{xx} shrinks, then physically it means that the nearby values of the process x are not correlated and hence not dependent on each other. Autocorrelation of the white noise/process is an impulse function. Autocorrelation of discrete-time residuals is given as

$$R_{rr}(\tau) = \frac{1}{N-\tau}\sum_{k=1}^{N-\tau} r(k)\, r(k+\tau); \quad \tau = 0,\ldots,\tau_{\max}\ (\text{discrete time lags}) \tag{B.6}$$

In order that residuals are white, the normalized values R_{rr} should lie within $\pm 1.97/\sqrt{N}$ band; only 5% of R_{rr} are allowed out of the band. This property is used for checking the performance of state/parameter estimation algorithms. In practice, about 30 to 50 autocorrelation values are obtained and checked if at least 95% of these values fall within the band. Then it is assumed that practically, these autocorrelation values are zero and hence, the residuals are white, thereby signifying that they are not 'autocorrelated'. This means that nearly complete information has been extracted out of the measurement data for parameter estimation.

B.4.2 Confidence Level

In parameter/state estimation, requirement of high confidence in the estimated parameters/states is imperative without which the results cannot be trusted. Often this information is available from the estimation results. A statistical approach and judgment are used to define the confidence interval within which the true parameters/states are assumed to lie with 95% of confidence, signifying high probability with which truth lies within the upper and lower intervals. This signifies that the estimation error, e.g. $\tilde{\beta}_{LS}$, should be within a certain interval band. In that case, one can define $P\{\, l < \beta < u \,\} = \alpha$; it means that α is the probability that β is constrained in the interval (l,u). In other words, the probability that the true value, β, is between l (the lower bound) and u (the upper bound) is 'α'. As the interval becomes smaller, the estimated value $\hat{\beta}$ can be taken, more confidently, as the value of the true parameter.

B.4.3 Consistency of Estimates

One can study the behaviour of an estimator with increased amount of data. An estimator is called asymptotically unbiased, if the bias approaches zero as the number of data tends to infinity. An asymptotically efficient estimator is obtained if the equality in CRLB is approached as the number of data tends to infinity. It is very reasonable to postulate that as the number of data used increases, the estimate tends to the true value. This property is called 'consistency'. This is a stronger property than asymptotic unbiasedness, since it is to be satisfied for single realization of estimates and not 'on the average' behaviour. It means that the strong consistency is defined in terms of the convergence of the individual realizations of the estimates and not in terms of the average properties of the estimates. Hence, all the consistent estimates are unbiased asymptotically. The convergence is required to be with probability 1 (one) and is expressed as

$$\lim_{N\to\infty} P\left\{\left|\hat{\beta}(z_1, z_2, \ldots, z_n) - \beta\right| < \delta\right\} = 1 \qquad \forall \delta > 0 \tag{B.7}$$

This means that the probability that the error in estimates (w.r.t. the true values) is less than a certain small positive value is one, as the number of data used in the estimation process tends to infinity.

B.4.4 Correlation Coefficient

$$\rho_{ij} = \frac{\text{cov}(x_i, x_j)}{\sigma_{x_i}\,\sigma_{x_j}} \qquad -1 \le \rho_{ij} \le 1 \tag{B.8}$$

Here, $\rho_{ij} = 0$ for independent variables x_i and x_j. For certainly correlated process, $\rho = 1$. Thus ρ defines the degree of correlation between two random variables. This test is used in model error method for parameter estimation. For example, in KF theory, often an assumption is made that the state error and measurement error or residuals are uncorrelated. If a variable d is dependent on several x_i, then correlation coefficient for each of x_i can be utilized to determine the degree (extent) of this correlation with d as

$$\rho(d, x_i) = \frac{\sum_{k=1}^{N} (d(k) - \underline{d})\,(x_i(k) - \underline{x}_i)}{\sqrt{\sum_{k=1}^{N} (d(k) - \underline{d})^2}\,\sqrt{\sum_{k=1}^{N} (x_i(k) - \underline{x}_i)^2}} \tag{B.9}$$

Here, 'under bar' represents mean of the variable. If $|\rho(d, x_i)|$ is nearly equal to 1, then d is considered linearly related to particular x_i. In that case, the x_i terms with higher correlation coefficient can be included in the model.

B.4.5 Covariance

It is defined as

$$\text{Cov}(x_i, x_j) = E\left\{\left[x_i - E(x_i)\right]\,\left[x_j - E(x_j)\right]\right\} \tag{B.10}$$

For the independent variables x_i and x_j, the covariance matrix is null. But if the matrix is zero, it does not mean that x_i and x_j are independent. The covariance matrix is supposed to be symmetric and positive semi-definite, by definition. However, in practice, when the estimation (iteration) proceeds the matrix may not retain these properties. Covariance matrix plays a very important role in KF time propagation and measurement data update equations. It provides theoretical prediction of the state-error variance and covariance-matching concept can be used for judging the performance/consistency of the filter (tuning). A similar concept is also used in method of model error for tuning the deterministic state estimator. The square roots of the diagonal elements of this matrix give standard deviations of the errors in estimation. It must be also emphasized here that the inverse of the covariance matrix gives the indication of the information content in the signals about the parameters. Thus, large covariance matrix signifies higher uncertainty and low information and low confidence in the state/parameter estimation results.

Several error-based criteria are defined in Appendix A.

B.4.6 Measurement Noise Covariance Matrix

This matrix for discrete-time noise, given as *R(k)*, is called the noise covariance matrix. For continuous-time measurement noise, the covariance matrix *R(t)* is called spectral density matrix.

In the limit $\Delta t \to 0$, $R(k) = \frac{R(t)}{\Delta t}$, such that the discrete noise sequence tends to the infinite valued pulses of zero duration. This ensures that the area under the impulse autocorrelation function $R_k \Delta t$ = the area R under the continuous white noise impulse autocorrelation function.

B.4.7 Parameter Estimation Error Norm (PEEN)

This is always given in percentage (%):

$$\text{PEEN} = \frac{norm(\beta - \hat{\beta})}{norm(\beta)} \times 100 \tag{B.11}$$

B.4.8 Root Sum Square Error (RSSE)

Let x_t, y_t, z_t be the true trajectories and $\hat{x}, \hat{y}, \hat{z}$ be the estimated/predicted trajectories. Then

$$RSSE(t) = \sqrt{(x_t(t) - \hat{x}(t))^2 + (y_t(t) - \hat{y}(t))^2 + (z_t(t) - \hat{z}(t))^2} \tag{B.12}$$

This is valid also for the discrete-time signals.

$$\text{Percentage RSSE}(t) = \frac{RSSE(t)}{\sqrt{x_t^2(t) + y_t^2(t) + z_t^2(t)}} \times 100 \tag{B.13}$$

The variance of residuals is evaluated as

$\sigma_r^2 = \frac{1}{N-1}\sum_{k=1}^{N}(r(k) - \underline{r})^2$. Here, $\underline{r}$ is the mean of the residuals.

B.4.9 Root Mean Square Error (RMSE)

$$RMSE(t) = \sqrt{\frac{(x_t(t) - \hat{x}(t))^2 + (y_t(t) - \hat{y}(t))^2 + (z_t(t) - \hat{z}(t))^2}{3}} \tag{B.14}$$

Percentage RMSE can also be defined.

B.5 TRANSITION MATRIX METHOD

This method is used for solving any matrix Riccati equation. Based on the development in Chapter 12, we have the following set of linear equations (for $a = Sb$):

$$\dot{b} = -f_{\hat{x}}^T b + 2H^T R^{-1} H a \quad \text{(B.15)}$$

$$\dot{a} = \frac{1}{2} Q^{-1} b + f_{\hat{x}} a \quad \text{(B.16)}$$

or, in a compact form, we have

$$\begin{bmatrix} \dot{b} \\ \dot{a} \end{bmatrix} = \begin{bmatrix} -f_{\hat{x}}^T & 2H^T R^{-1} H \\ \frac{1}{2} Q^{-1} & f_{\hat{x}} \end{bmatrix} \begin{bmatrix} b \\ a \end{bmatrix} \quad \text{(B.17)}$$

$\dot{X} = FX$; and its solution can be given as $X(t_0 + \Delta t) = \Phi(\Delta t) X(t_0)$. Here, Φ is the transition matrix given as

$$\Phi(\Delta t) = e^{F \Delta t} = \begin{bmatrix} \varphi_{bb} & \varphi_{ba} \\ \varphi_{ab} & \varphi_{aa} \end{bmatrix} \quad \text{(B.18)}$$

Since the elements of matrix F are known, the solution X can be obtained which in turn gives b and a. Thus, S can be obtained as

$$S(t_0 + \Delta t) = \left[\varphi_{ab}(\Delta t) + \varphi_{aa}(\Delta t) S(t_0) \right] \left[\varphi_{bb}(\Delta t) + \varphi_{ba}(\Delta t) S(t_0) \right]^{-1} \quad \text{(B.19)}$$

This procedure can also be used to solve the continuous-time matrix Riccati equation for the covariance propagation in the continuous-time Kalman filter.

REFERENCES

1. Raol, J.R., *Multisensor Data Fusion with MATLAB*, CRC Press, Florida, 2010.
2. Raol, J.R., *Data Fusion Mathematics – Theory and Practice*, CRC Press, Florida, 2016.
3. Raol, J.R., Girija, G., and Singh, J., *Modelling and Parameter Estimation of Dynamic Systems*, IET/IEE Control Series Books, London, UK, Vol. 65, Aug. 2004.

Appendix C
Neural Networks, Fuzzy Logic and Genetic Algorithms for Estimation

C.1 INTRODUCTION

Artificial neural networks/nets (ANNs) attempt to mimic and exploit the parallel processing capability of the human brain in order to deal with precisely the kinds of problems that the human brain itself is well adapted for. ANNs are new paradigms for computations and modelling of dynamic systems. ANN simulates a highly interconnected parallel computational structure with relatively simple individual processing elements called neurons.

Fuzzy-logic-based systems (FLS) are useful for modelling vagueness and imprecision in real-life dynamic phenomena. The real-life experience shows that some extension of the crisp logic is needed. Events or occurrences leading to fuzzy logic (FL) are (i) the light could be dim, (ii) day could be bright with a certain degree of brightness, (iii) day could be cloudy with a certain degree and (iv) weather could be warm, cold or hazy. Thus, the idea is to allow for a degree of uncertainty with the truth and falsity (1 or 0) being at the extremes of a continuous spectrum of this uncertainty. This leads to multi-valued logic to fuzzy set theory. FL has seen applications in the process control industry, home appliances and traffic. Fuzziness is based on the theory of sets, the characteristic function is generalized to take an infinite number of values between 0 and 1; mA (x) is a membership function of x on the set A and is a mapping of the UoD x on the closed interval.

Genetic algorithms are based on the nature's evolutionary aspects, and are computationally very simple, but are of iterative nature. They provide robust solutions to optimization problems in control theory. They can be easily parallelized. They can be used for modelling, parameter estimation and control.

Much of the discussion in this Appendix is based on [1–10].

C.2 ANNS

ANNs have found successful applications in image processing, pattern recognition, nonlinear curve fitting/mapping, flight data analysis, adaptive control, system identification and parameter estimation. In many such cases, the ANNs are used for prediction of the phenomena that they have learnt earlier by way of training from the known samples of data. They have a good ability to learn adaptively from the data. The specialized features that motivate the strong interest in ANNs are (i) The neural network (NW) expansion is a basis (recall Fourier expansion/orthogonal functions as a basis for expression of time-series data/periodic signals), (ii) the NW structure does extrapolation in adaptively chosen directions, (iii) the NW structure uses adaptive bases functions – the shape and location of these functions are adjusted by the observed data, (iv) NW's approximation capability is good, (v) the NW structure is repetitive in nature and has advantage in HW/SW implementations, and (vi) this repetitive structure provides resilience to failures. The regularization feature is a useful tool to effectively include only those basis functions that are essential for the approximation and is achieved implicitly by stopping rules and explicitly by putting a penalty for parameter deviation.

One can develop new generations of ANN design providing new learning-based capabilities in control and system identification. These new designs could be characterized as extensions as an aid to existing control theory, and new general control designs compatible with ANN implementation.

The inputs are weighted by a set of coefficients and the adaptive linear combiner (ALC) outputs linear combination of inputs. For training, the measured input patterns/signals and desired responses are presented to the ALC. The weight adaptation is done by LMS algorithm (least mean square, Widrow-Hoff delta rule), which minimizes the sum of squares of the linear errors over the training set. More useful ANNs have multi-layer perceptrons (MLPN) with input layer, one or at most two hidden layers and an output layer with a sigmoid (nonbinary) neuron in each layer. In this network (NW), the I/O data are real-valued signals. The output of MLPN is a differential function of the NW parameters. This NW is capable of approximating arbitrary nonlinear mapping using a given set of examples. For a fully connected MLPN, two to three layers are generally sufficient. The number of hidden layer nodes should be much less than the number of training samples. One important property of the NW is the generalization property that is a measure of how well the NW performs on the actual problem once training is completed. The generalization is influenced by the number of data samples, the complexity of the underlying problem and the NW size.

C.2.1 Feedforward Neural-Network-Based Modelling

The FFNN (feed forward neural network) is a noncyclic-type ANN with a layered topology and no feedback. FFNN is an information processing system with a large number of simple processing elements. FFNN can be thought of as a nonlinear black box (model structure but not in the conventional sense of polynomial or transfer function model), the parameters (weights) of which can be estimated by conventional optimization methods. The FFNNs are suitable for system identification, time-series modelling, parameter estimation, sensor failure detection and related applications. The chosen FFNN is first trained using the training set data and then it is used for prediction using another input set, which belongs to the same class of the data. This is the validation set. The process is similar to the one used as cross validation in the system identification literature.

C.2.2 Backpropagation Training Algorithm

The back-propagation (BPN) algorithm is a learning rule for multi-layered neural networks. The algorithm gives a prescription for adjusting the initially randomized set of synaptic weights (existing between all pairs of neurons in each successive layer of the network) so as to minimize the difference between the network's output of each input fact and the output with which the given input is known (or desired) to be associated. The back-propagation rule takes its name from the way in which the calculated error at the output layer is propagated backwards from the output layer to the nth hidden layer to the n_1 th hidden layer, and so on. Because this learning process requires us to 'know' the correct pairing of IO facts beforehand, this type of weight adjustment is called supervised learning. The FFNN with backpropagation using recursive least squares (FFNN-RLS) uses an enhanced form of the back-propagation technique. The conventional approach minimizes the mean-squared error with respect to the weights. Here, the algorithm minimizes the mean-squared error between the expected output and actual output with respect to the summation outputs and Kalman filter gain and covariance are computed in each layer to update the weights.

C.2.3 FFNN-Based Parameter Estimation

We have seen that a FFNN can be used for the nonlinear mapping of the input-output data. This suggests that it should be possible to use it for system identification/parameter estimation. The FFNN works with a black-box model structure, which cannot be physically interpreted and hence the parameters, i.e. the network weights, have no interpretation in terms of the parameters of the actual system. The parameter estimation using FFNN is done in two steps: (i) the FFNN is presented with the measured data and trained to reproduce the clean/predicted responses that are compared with the actual system's responses in the sense of minimization of the output error; and (ii) subsequently the predicted responses are perturbed in turn for each parameter that is to be estimated. This obtains a changed predicted response. Next, assume that $z = \beta x$ and the FFNN is trained to produce clean z. Then the trained FFNN is used to produce $z+\Delta z$ and $z-\Delta z$ by changing x to $x+\Delta x$ and $x-\Delta x$, and the parameter β is obtained as

$$\beta = \frac{z^+ - z^-}{x^+ - x^-} \tag{C.1}$$

This is called a Delta method. The estimates are obtained by averaging these respective parameter time histories, by removing some initial and some final points.

C.2.4 Recurrent Neural Networks (RNNs)–Parameter Estimation

The RNNs are NWs with feedback of an output to the internal states. The RNNs are more suitable for the problem of parameter estimation of linear dynamic systems, as compared to FFNNs. RNNs are dynamic neural networks that are amenable to explicit parameter estimation in state-space models. Basically, RNNs are a type of the Hopfield NWs (HNN).

The FFNNs and RNNs can be used for parameter estimation (Chapter 14). The RNNs can be used for parameter estimation of linear dynamic systems as well as nonlinear time varying systems. They also can be used for trajectory prediction/matching. Many variants of RNNs and their interrelationships have been reported. Also, the trajectory matching algorithms for these variants have been evolved, these algorithms can be used for training RNNs for nonlinear model fitting and related applications, in a similar manner as done using FFNNs.

C.3 FUZZY-LOGIC-BASED MODELLING

FL allows for a degree of uncertainty and gradation; truth (1) and falsity (0) become the extremes of a continuous spectrum of uncertainty. This leads to multi-valued logic and the fuzzy set theory. Fuzziness is the theory of sets and the characteristic function is generalized to take an infinite number of values between 0 and 1: e.g. triangular form. Fuzzy systems can model any continuous function or system. The quality of approximation depends on the quality of rules that can be formed by the experts. Fuzzy engineering is a function approximation with fuzzy systems. This approximation does not depend on words, cognitive theory or linguistic paradigm. It rests on mathematics of function approximation (FA) and statistical learning theory (SLT). Much of the mathematics is well known, and as such there is no magic in fuzzy systems. A fuzzy system is natural way to turn speech and measured action into functions that approximate the hard tasks. The words are just a tool/ladder we climb down on to get the task of function approximation. Fuzzy language is a means to the end of computing and not the goal. The membership function gives a measure of the degree to which x belongs to the set A: mA(x): X à [0,1]. Fuzzy variable low is described in terms of a set of positive integers in the range [0,100]àA = {low}. This set expresses the degree to which the temperature is considered low over the range of all possible temperatures.

The rule-based fuzzy systems can model any continuous function or system and the quality of the approximation depends on the quality of rules. These rules can be formed by the experts who have great experience in dealing with the classical systems, which are designed/developed or maintained by them. Alternatively, the artificial neural networks can be used to learn these rules from the data. The fuzzy engineering deals with function approximations. Application to washing machines might save the energy and wear and tear on clothes. This approximation actually does not depend on words, cognitive theory or linguistic paradigm. It rests on the mathematics of function approximation and statistical learning theory. Since much of this mathematics is well known, there is no magic in fuzzy systems. A fuzzy system is a natural way to turn speech and measured action into functions that approximate the hard tasks.

The basic unit of fuzzy approximation is the 'If…Then…' rule. As an example: If the wash water (in the washing machine) is dirty then add more detergent powder. Thus, fuzzy system is a set of such well-defined and composed If… Then… rules that map input sets to output sets as in the previous example. The overlapping rules define polynomials and richer functions. Each input partially fires all the rules in parallel and the system acts as an associative processor as it computes the output function. The system then combines these partially fired Then part fuzzy sets in a sum and converts this sum to a scalar or vector output. These additive fuzzy systems are proven universal approximators for rules that use fuzzy sets of any shape and are computationally simple. A fuzzy variable is one whose values can be considered labels of fuzzy sets: temperature à fuzzy variableà linguistic values such as low, medium, normal, high, very high, etc. leading to membership values (on the universe of discourse – degree C). The no. of rules could be large, say 30. For a complex process control plant, one might need 60 to 80 rules and for a small task 5 to 10 rules might be sufficient, e.g. for washing machines. Combination of two or three fuzzy conditional statements will form a fuzzy algorithm. A linguistic variable can take on values that are statements of a natural language: primary terms that are labels of fuzzy sets, such as high, low, small, medium, zero; negative NOT and connective AND and OR; hedges like very, nearly, almost; parenthesis. These primary terms may have either continuous or discrete membership functions. The continuous membership functions are defined by analytical functions.

The core of every fuzzy controller is the inference engine, which is a computation mechanism with which decision can be inferred even though the knowledge may be incomplete. This mechanism gives the linguistic controller the power to reason by being able to extrapolate knowledge and search for rules, which only partially fit any given situation for which a rule does not exist. The inference engine performs an exhaustive search of the rules in the knowledge base to determine the degree of fit each rule for a given set of causes. A number of rules contribute to the final result to a varying degree. A fuzzy propositional implication defines the relationship between the linguistic variables of a fuzzy controller:

- Given two fuzzy sets A and **B** that belong to the universe of discourse X and Y, respectively, then the fuzzy propositional implication is the following:
- R: If A then $B = A \rightarrow B == A \times B$ where $A \times B$ is the Cartesian product of the two fuzzy sets A and B.

The knowledge necessary to control a plant is usually expressed as a set of linguistic rules of the form: If (cause) then (effect). These are the rules with which new operators are trained to control a plant and they constitute the knowledge base of the system. All the rules necessary to control a plant might not be elicited, or known, and hence it is necessary to use some technique capable of inferring the control action from available rules. The fuzzy systems are suited to control nonlinear systems and multi-valued nonlinear processes. The measurements of plant variables (even it contaminated by noise) and control actions to the plant actuators are crisp. First, fuzzify the measured plant variables, then apply fuzzy algorithm (rules/inferences) and finally defuzzify the results.

For a FL-based adaptive KF, the UoD is U_r for s (.) = [0.0 0.4] and the UoD U_ψ for ψ (.) = [0.1 1.5].

Both the input and output Universe spaces have been discretized into five segments. The fuzzy sets are defined by assigning triangular MF to each of the discretized Universe. Then fuzzy implication inference leads to fuzzy output subsets. Finally, the adaptive estimation algorithm requires crisp values. A defuzzification procedure is applied using the *centre of area* method and to realize the fuzzy rule base. Defuzzification of the output arising from the fuzzy controller is done using either the centre of gravity or centre of area method. In the centre of area method, the area under the composite membership function of the output of the fuzzy controller is taken as the final output.

The basic unit of fuzzy FA is the 'If then' rule: 'If the wash water is dirty then add more detergent'. Thus, the fuzzy system is a set of 'If then' rules that maps input sets like 'dirty wash water' to output sets like 'more detergent'. The overlapping rules define polynomials/richer functions. A set of possible rules is given below:

Rule 1: if the air is cold then set the motor speed to stop
Rule 2: if the air is cool then set the motor speed to slow
Rule 3: if the air is just right then set the motor speed to medium
Rule 4: if the air is warm then set the motor speed to fast
Rule 5: if the air is hot then set the motor speed to blast

This gives the first-cut fuzzy system. More rules can be guessed and added by the experts and by learning new rules adaptively from training data sets. The ANNs can be used to learn the rules from the data. Much of the fuzzy engineering deals with tuning these rules and adding new rules or pruning old rules.

C.4 GENETIC ALGORITHMS

Genetic algorithms (GAs) are search methods inspired by nature's evolutionary systems. They can be used to obtain global and robust solutions to many optimization problems in science, engineering, economics, psychology and biology. The natural systems have evolved over millions of years; they have gone through iterations over many generations and in the process have become very robust, especially to their many different environments. Due to its strong evolutionary 'experience', the Nature offers good solutions whenever robustness is called for. Biological systems are generally more robust, efficient and flexible compared to the most sophisticated artificial systems. Genetic algorithms are based on some of the principles that govern the natural systems. Genetic algorithms are computational optimization schemes with an approach that seems rather unconventional. The algorithms solve optimization problems imitating the nature in the way it has been working for millions of years on the evolution of life forms. Inspired by the biological systems, genetic algorithms adopt the rules of natural selection and genetics to attain robustness. Acting on the premise of survival of the fittest, a population or samples of feasible solutions is combined in a manner like the combination of chromosomes in a natural genetic system. The fitter population members pass on their structures as genes in far greater measure than their lesser fit members do. As the generations evolve, the net effect is evolution of the population towards an optimum (species, solution). Genetic algorithms operate by combining the information present in different possible solutions in a way that a better solution is obtained in next/future generations.

C.4.1 Operations in a Typical Genetic Algorithm

(a) Chromosomes: they represent the encoding of information in a string of finite length and each chromosome consists of a string of bits (binary digit: 0 or 1); or it could be a symbol from a set of more than two elements. Generally, for function optimization, chromosomes are constructed from binary strings as seen from the following table:

Parameter Value	String
6	000110
34	100010

The long stretches of DNA that carry the genetic information needed to build an organism are called chromosomes. The chromosomes consist of genes. Each gene represents a unit of information and it takes different values. These values are called alleles at different locations called loci. The strings, composed of features or detectors, assume values like 0 or 1, which are located at different positions in the string. The total package or system is called the genotype or structure. The phenotype results when interaction of genotype with environment takes place.

(b) Population and fitness: genetic algorithms operate on the population of possible solutions with chromosomes. The population members are known as individuals. Every individual is assigned a fitness-value based on the objective function, or cost function. Better individuals (solutions) have higher fitness values and weaker ones have lower fitness values.

(c) Initialisation and reproduction: by randomly selecting information from the search space and encoding it, a population of possible initial solutions is created. Reproduction is a process in which individual strings are copied as per their fitness values. Thus, the strings with a greater fitness value have a higher probability of contributing one or more offsprings to the next generation.

(d) Crossover: in a crossover, a site is selected randomly along the length of the chromosomes, and each chromosome is split into two pieces at the crossover site. The new ones are formed by joining the top piece of one chromosome with the tailpiece of the other.

(e) Mutation: it is a small operation in which a bit in a string is changed at a random location. The main idea is to break monotony and add a bit of novelty. This operation would help gain information not available to the rest of the population. It lends diversity to the population.

(f) Generation: each iteration in the optimization procedure is called a generation. In each generation pairs are chosen for crossover operation, the fitness is determined, and mutation is carried out during crossover operation (during or after has a subtle distinction). With these operations performed, a new population evolves which is carried forward.

(g) Survival of the fittest: the individuals may be fitter or weaker than some other population members. So the members must be ranked as per their fitness value. In each generation, the weaker members are allowed to wither out and the ones with good fitness values take part in the genetic operations. The net result is the evolution of the population towards the global optimum.

(h) Cost function, decision variables and search space: in most practical optimization problems, the goal is to find optimal parameters to increase the production and/or to reduce the expenditure/loss. That is to get maximum profit by reorganizing the system and its parameters that affect the cost function. Since, this in effect, reflects on the cost, it is represented by the cost function. A carefully devised and convergent computational algorithm would eventually find an optimum solution to the problem. The parameters of the system that decide the cost are termed decision variables. The search space is a Euclidean space in which parameters take different values and each point in the space is a probable solution.

C.4.1.1 Steps in GA

Step 1: Create population of N samples from a chosen search space – denoting the decision variables.
Step 2: Produce series of 0s and 1s to create chromosomes – i.e. encoding the decision variables.
Step 3: Calculate the cost function values and assign fitness (values) to each member.
Step 4: Sort the members accordingly to their respective fitness values.

Step 5: Carry out crossover operation taking two chromosomes at a time.
Step 6: Mutate the chromosomes with a given probability of mutation.
Step 7: Retain the best members of the population and remove the weaker members based on their fitness values.
Step 8: Replace the old generation by the new one and repeat steps 3 to 8.

C.4.2 Stopping Strategies for Genetic Algorithms

One needs to know as to where and when to stop the genetic algorithm iterations. If the population size is fixed, then more generations might be needed for the convergence of a genetic algorithm to an optimal solution. One way is to track the fitness value for no further improvement. As the algorithmic steps progress, a situation would occur wherein we need a large number of generations to bring about a small improvement in the fitness value. One can define a predetermined number of generation/iterations to solve the problem. Also, an insignificant change in the norm of estimated parameters can be tracked for a few consecutive iterations before stopping the search. It must be possible to do an effective search if one exploits some important similarities in the coding used in genetic algorithms. Another way is to evaluate the gradient of the cost function and use the conventional approaches for assessing the quality of the estimates for their convergence to true values. It is possible to use GA with a gradient-based approach for evaluating the estimation accuracy as is done for OEM (Chapter 3). Again, here as is true with all the other parameter estimation methods, the matching of time histories of the measured data and model responses is a necessary and not sufficient condition. An increase in the number of samples would generally increase the success rate.

C.4.3 Genetic Algorithms without Coding of Parameters

Genetic algorithms become more complex because of coding the chromosomes, especially for more complex problems. In the problems of science and engineering, we come across real numbers. Thus, we need to use real numbers and still use genetic algorithms on these numbers for solving optimization problems. Major change is in the crossover and mutation operations. Averaging the two samples, i.e. the two sets of parameter values can perform the crossover operation. After the crossover, the best individual is mutated. In mutation, a small noise is added. Assume that two individuals have β_1 and β_2 as numerical values of the parameters. Then after crossover, we obtain the new individual as $(\beta_1 + \beta_2)/2$. For mutation we have $\beta_3 = \beta_1 + \varepsilon * v$, where d is a constant and v is a number chosen randomly between –1 and 1. Thus, all the GA operations can be performed by using real numbers like 4.8904, etc., without coding the samples. This feature is extremely well suited for several engineering applications: parameter estimation, control, optimization and signal processing.

REFERENCES

1. Eerhart, R.C., and Dobbins, R.W., *Neural Network PC Tools – A Practical Guide*, Academic Press, New York, 1993.
2. Raol, J.R., Girija, G., and Singh, J., *Modelling and Parameter Estimation of Dynamic Systems*. IET/IEE Control Series, Vol. 65, IET, London, 2004.
3. Raol, J.R., Aerodynamic modelling and sensor failure detection using feedforward neural networks, *Journal of Aeronautical Society of India*, Vol. 47, No. 4, pp. 193–197, Nov. 1995.
4. Hopfield, J.J., and Tank, D.W., Computing with neural circuits – A model, *Science*, Vol. 233, No. 4764, pp. 625–633, Aug. 1986.
5. Raol, J.R., Parameter estimation of state space models by recurrent neural networks. *IEE Proceedings – Control Theory and Applications*, Vol. 142, No. 2, pp. 114–118, March 1995.

6. Kosko, B., *Neural Networks and Fuzzy Systems – A Dynamic Systems Approach to Machine Intelligence*, Prentice Hall, Englewood Cliffs, 1992.
7. King, R.E., *Computational Intelligence in Control Engineering*, Marcel Dekker, New York, 1999.
8. Kashyap, S.K., and Raol, J.R., *Unification and interpretations of fuzzy set operations*. NAL PD FC 0502, FMCD, NAL, Bangalore, March 2005.
9. Goldberg, D.E., *Genetic Algorithms in Search, Optimisation and Machine Learning*, Addison-Wesley Publishing Company, Reading, MA, 1989.
10. Raol, J.R., Jalisatgi, A.M., and Jose, J., Parallel implementation of genetic and adaptive portioned random search algorithms, *Institution of Engineers, IE(India) Journal – ET*, Vol. 80, pp. 49–54, March 2000.

Index

A

Absolute, 287
Acceleration, 35, 104, 108, 164, 276
Accuracy, 14, 24, 87, 100, 126, 148, 272, 275, 280, 296
Activation, 249–250, 252, 261, 266, 290, 293
Adaptability, 190, 275–276
Adaption, 33, 89, 102, 127, 154
Additive, 43, 53, 104, 220, 244
Adjoint, 193
Advances, 274
Aerodynamic, 89, 102, 258, 273
Aeronautical, 273–274
Aeroplane, 256
Aerospace, 26, 89, 176–177, 191, 273, 284
AHFFDOM, 16
AHIF, 150, 162, 164–170
AHIFcompared, 166
AHIFDOM, 150, 154–156
AHIFFDOM, 161–162
Aircraft, 26, 45, 76, 89, 257, 273–274
Airfield, 103
Airplane, 76, 247, 273
Airship, 274
AKFDOM, 156
AKFFDOM, 161–162
Algebraic, 5, 10, 24, 45, 113
Algorithm, 29, 162, 167, 169, 220
Alpha, 275
Alternative, 68, 90, 150, 180, 212, 234, 297
Altitude, 296–299
Amplitude, 25, 79, 179
Analyse, 195
Analytically, 21, 23, 28, 88, 101, 148, 184, 190, 236, 241, 248
ANFIS, 119–121, 123, 126–127
ANN, 120, 247
Antecedent, 120
Ape, 76–77, 88–89, 101
Approaches, 1, 3, 119, 150, 181, 296, 299
Appropriate, 32, 46, 51, 54, 58, 62–63, 67, 71–104, 119–120, 181, 216, 241, 280, 283
Approximation, 10, 20, 22, 76, 77, 90, 213, 237, 264, 266, 291
Artificial, 3, 119–120, 180, 247, 273–274
Asymptotic, 141, 221
ATC, 103
ATM, 103
Atmospheric, 274
ATR, 46, 103
Attack, 84, 98
ATTAS, 26
Attitude, 104
Augmented, 119, 121, 123, 125, 127, 128, 132–133, 136, 142–143, 150, 177
Autocorrelated, 11
Automatica, 26, 75, 148–149, 191, 233, 246
Autonomous, 103–104, 275
Autopilot, 104
Aviation, 76

B

Backpropagation, 248, 250
Bandwidth, 18
Barometric, 296
Bayesian, 89, 102
Bernoulli, 47, 51, 67
Beta, 275
Bias, 8, 14, 16, 25, 37, 85, 98, 248–249, 260–262, 264, 272–273, 296
Bierman, G. J., 45, 118, 149
Binary, 103, 106–108, 123, 143
Biological, 54
Boundary, 20, 106, 193–195, 223
Brain, 104, 274
Branches, 3

C

Centroid, 103, 105–109, 111, 113, 115, 117–121, 123, 125–135, 137–139, 141–143, 145, 147–149, 151, 156
Classification, 3, 273, 287, 294, 300
Clusters, 103–104, 106–107, 123
Coefficient, 23, 38, 76, 79–80, 89, 247, 290
Complement, 120
Conjunction, 32, 274, 293
Constituents, 119
Constraints, 60, 141, 204, 242, 288, 200
Convergence, 22–23, 40, 42–43, 53, 65–66, 73–74, 76–77, 80, 82, 84, 88–90, 93, 95, 101, 141, 143, 181, 188, 195, 204, 206, 212, 220, 267, 269–270, 272–273
Coordinates, 113, 142, 161
Correlation, 18, 20–21, 79–80, 103, 150
Covariance, 7–9, 13, 17–18, 20–21, 23–25, 27–36, 38, 39, 76, 80, 82, 84, 86, 88, 90, 93, 99, 101, 105, 107, 109, 110, 112, 114, 123, 128, 132, 202, 208, 213, 220, 225, 226, 234, 235, 244, 250, 255, 260, 272, 277, 280–281, 284
Cramer-Rao, 14, 16–17, 20, 22, 25
CTA, 105–106, 108–109, 119, 121, 123, 125
CTFLKF, 123
CTFSRIF, 125–126
CTFSRIMLF, 126
CTKF, 109–112, 137
CTMLF, 110
CTSRIF, 115, 117, 125–126, 129–130, 135, 147–148
CTSRIMLF, 130
CTSRVD, 131, 147–148
CTSRVDMLF, 131
CTSRVL, 131
CTSVF, 110

CTUDF, 109, 111–112, 135, 138
CTUDSRIF, 134–135, 136–139, 147–148
CTVDSRIFM, 135
CTUDSRIFMLF, 138
Cubature, 275

D

Databases, 286
Decentralized, 49, 75, 92, 102, 112, 246
Decision, 80, 103, 105, 121, 126, 180, 283, 286–287, 294–295, 297
Deep, 274
Defence, 103, 118, 176, 191
Defuzzification, 119–120, 151, 154, 156, 178
Degree, 75, 90, 120, 149–150, 154, 177, 191, 233, 246, 296–298
Delta, 248, 250, 258, 297–298
Density, 5, 10, 14, 18, 32, 77, 84, 90, 98, 197, 247, 284
Deploy, 104, 287
Derivative, 53, 60, 71, 82, 84, 95, 141–143, 184, 187, 190, 203–204, 220, 242, 244, 250, 252, 254–256, 258, 264, 267–268, 288–289, 292
Detection, 103–108, 113, 275, 287
Determinant, 21
Deterministic, 3, 6–7, 28, 43, 84, 89, 180–181, 190, 192–193, 195, 197, 199, 201–204, 207–208, 211–213, 217, 221, 223, 225, 227, 231, 233–236, 245, 247
Deviation, 7, 20, 22, 35, 43–44, 85, 99, 110, 125, 129, 146, 193, 267, 280, 295
Diagonal, 22, 31, 34–35, 105, 109, 123, 128–129, 136, 177–178, 290
Differentiable, 58, 67, 181
Differential, 6, 12–13, 20, 27, 51, 58, 64, 67, 82, 84, 140, 181, 199, 202, 206–207, 241, 267, 274
Digital, 27–28, 32, 117, 125, 274
Dimension, 10, 51, 58, 62, 67, 71, 90, 181, 217
Discrete, 3, 6, 18, 23–24, 28–29, 32, 43, 45, 64, 73, 77, 82, 84, 93, 95, 118, 128, 143, 149, 187–188, 192, 199, 201, 206, 212–213, 216, 221, 223–226, 230, 234, 237, 239, 245
Divergence, 12, 34, 154
Doublet, 39, 43, 85, 99, 220, 244, 257–258, 267, 269–270, 272
DSP, 125
Dynamic, 3–6, 12, 17–18, 21, 24–28, 36, 39–40, 42, 46–47, 53–54, 63–64, 73, 77, 83–84, 89, 91, 96, 98, 102, 106, 113–114, 126, 139–140, 156, 176, 180, 187, 236, 244–247, 260–261, 273–276, 280, 282, 284, 286–287, 290–291, 299

E

Eigenfactor, 128, 131
Eigenvalue, 53, 88, 131, 142, 190, 206–207, 211–212, 243, 288
Eigenvector, 128–129
EKF, 3, 27–28, 33, 37–39, 43–44, 54, 58, 63, 72–73, 75–78, 80, 85–88, 90, 93, 149, 180, 191–192, 233, 246–247, 274
Embedded, 195
Ensemble, 297–298
Estimation, 3–8, 10, 12–14, 16, 18, 20–28, 30, 32–34, 36–40, 42–45, 48, 50, 54, 56, 58, 60, 62, 64, 66–68, 72, 74–80, 82, 84–90, 92, 96, 98, 106, 108, 110, 112, 114, 117, 180–181, 183–184, 188, 190, 192–194, 197–198, 200–202, 204, 206, 208, 212, 214, 216, 220–222, 224, 226, 228, 230, 232–234, 236, 240, 242, 244–250, 252–262, 264–277, 280, 282–284, 286–300
Estimator, 3, 5–9, 11, 16–17, 21, 30, 46, 77, 90, 148, 150, 193, 197, 199, 202, 208, 212, 223, 225, 244, 276–277, 284, 286, 288–289, 291–294, 299
Euclidean, 288
Euler, 64, 73, 188, 206, 245, 290, 294
Eulrich, 273
evalfis, 143, 187

F

Factorization, 14, 33–35, 39, 45, 105, 109, 118, 128, 131, 135, 149
Feedback, 54, 286–287, 289, 291
Feedforward, 3, 28, 247–248, 250, 271–272, 274, 292–294
Filtering, 5, 27–35, 37–39, 43–51, 53–55, 57, 59, 63, 65, 67–68, 71, 73, 75–77, 86, 89–90, 99, 102–105, 112, 117–118, 120, 127–129, 131–132, 134–135, 137, 139, 148, 150–151, 155, 159, 161, 163, 165, 167, 169, 171, 173, 175–177, 179–181, 191, 212, 220–221, 233, 235, 245–247, 273, 275, 280, 284–285
FIS, 119–120, 151, 177–178
FKF, 165–170
FLHIF, 150–151, 154, 156, 167, 169, 171, 173
FLHIFF, 150–151, 154, 156, 158, 167
FLHIFIS, 169
FLIR, 105, 107, 113
FLKF, 121, 123–124
FMF, 139–143, 146, 148, 180–181, 183–184, 187–188, 190
FNN, 292–294
FPGA, 125
Functions, 13, 22, 37, 39, 62, 76–77, 83, 91, 96, 119, 180–181, 183, 190, 192, 268, 288, 293–294
Fundamental, 14, 275, 292–293
Fusion, 3, 46–51, 53–55, 57, 59, 62–63, 65, 67, 71–73, 75, 90, 92, 103–105, 109, 125, 129, 132–133, 138, 139, 141, 143, 145, 147–150, 167, 171–173, 175–177, 188, 191, 221, 223, 225, 227, 230, 233–237, 239–241, 243–246
Fuzzification, 3, 119–121, 123, 125–127, 131, 139–140, 143, 148, 150–152, 154–156, 159, 161, 163, 165, 167, 169, 171–180, 187, 190–191
Fuzzifier, 120

G

Gamma, 48, 73–74, 188, 226
Gauss, 5, 19, 22, 27–28, 164–165
Gaussian, 3, 6, 12–13, 18, 21, 27–28, 32, 43, 76–80, 83, 85–91, 93, 95, 99–102, 107–108, 123, 162, 178, 192, 234, 247, 267, 275, 280–284
Genetic, 3
Gradation, 119
Grade, 3

Gradients, 22–23, 25, 248–249, 260, 266–270, 272, 287, 293
Gramians, 48, 57, 59, 63, 72–73, 92–93, 141, 143–145, 148, 150, 180, 184, 189–190, 203–204, 206–208, 212–213, 216, 225–226, 228, 231, 235, 239–240, 244–245
Guesstimate, 6, 10, 132

H

Histograms, 106
Homogeneous, 28
Hybrid, 3, 120–121, 135, 137, 234–237, 239–241, 243–245, 276, 286, 292–294
Hyperbolic, 293
Hyperplane, 294–295
Hypothesis, 284

I

Identity, 7, 9, 13, 46, 59, 103, 141, 184, 203
Images, 3, 103–109, 111, 113, 115, 117–121, 123, 125–135, 137–139, 141, 143, 145, 147–149
Impedance, 261
Indefinite, 163
Infinity, 47–50, 72–73, 150–151, 155, 159, 161–163, 165, 167, 169, 171–177, 179–181, 183, 187–192, 212, 220–221, 230, 233–235, 237, 239, 241, 243–245
Information matrix, 22
Infrared, 105, 107, 113
Integral, 76
Integrate, 287
Intelligent, 119
Intensity, 105–108, 140, 202, 208
Intermittent, 58
Iteration, 22, 79, 250, 262, 270, 293
Iterative, 5, 10, 13, 19, 24, 39

J

Jacobians, 47, 51, 58–59, 64, 68, 73, 78, 92, 183, 184, 188, 202, 206, 208, 212, 217, 241

K

Kalman, 23, 27–33, 35–39, 43–45, 75–77, 79, 83, 85, 87, 89–90, 101–103, 105–107, 109, 111, 113, 115, 117, 126, 129, 134, 137–138, 150, 156, 162, 169, 177, 180, 191–192, 234–235, 246–247, 250, 255, 260, 272–275, 277, 280–281, 284
Kinematic, 103, 118, 143, 156

L

Lagrange, 193, 197, 200
Lasers, 104
Latency, 46, 58
Likelihood, 5–7, 9, 11, 13–14, 17–19, 21, 23, 25–26, 276
Linear, 3, 5–9, 11–12, 18, 21, 23–24, 27–28, 33–35, 37, 44–45, 54, 75–76, 90, 112–113, 131, 139–140, 150, 180, 190, 212–213, 233–234, 236, 244, 246–247, 274–276, 286, 288, 293–295, 299
Linearity, 8
Linearization, 37, 39, 67, 76, 90, 181, 247
Localization, 104
Longitudinal, 43, 84–85, 99, 256–257, 267, 273
Luenberger, 54, 139, 180, 234, 286–289, 291, 299
Lyapunov, 3, 46, 53, 71, 76–77, 84, 88, 90–91, 93, 95, 99, 101, 131, 180–181, 187, 190, 202–203, 206, 234, 245, 247–249, 252–257, 259, 261, 265, 267, 269, 271–274, 289, 291

M

MAE, 287
Magnetic, 104
Markovian, 284
Mass, 25, 84, 98, 257, 292
Matlab, 3, 46–47, 53, 63, 73, 76–77, 90, 110, 112, 119, 121, 123, 127, 132, 143, 156, 164, 181, 187, 190, 192, 202, 206, 212, 256, 260, 267, 284, 296–298
Maximize, 14, 27, 297
Measurementvector, 98
Mediate, 249
Metrics, 47, 54–55, 74, 109, 111–112, 115, 123, 125–126, 129–131, 134–135, 137–139, 143, 146–147, 165–166, 176, 220, 230, 244–245, 283, 287
Minimum, 7, 9, 17, 19, 23–24, 30, 33, 165, 212, 275, 297
Model, 3, 5–6, 9–13, 21–25, 27–28, 32–33, 35, 37, 39, 43–44, 47, 51, 53, 58, 63–65, 71–72, 76–77, 84–85, 88, 90, 98–99, 106–109, 113, 120, 143, 151–152, 154, 156, 163, 232–233, 244, 247, 256–257, 267, 274–277, 282, 284, 286–299
Modes, 23, 46, 220, 244, 276–277, 280
Modulating, 119, 142
Multisensory, 46, 75, 90, 102, 127, 245

N

NASA, 26, 273
Navigation, 103, 118, 177, 284
NNKF, 123
Nodes, 112, 247, 255, 259, 295
Noises, 28, 46, 54, 108, 212, 220
Nonadaptive, 183, 190
Nondeterministic, 162
Nondimensional, 84–85, 98–99, 256, 267
Nonfuzzy, 120
Nonlinearities, 3, 5–6, 9–10, 12–13, 19–20, 22–51, 53–54, 58–59, 62–65, 67–68, 71–77, 80, 83, 88–91, 93, 96, 101–102, 141, 181, 183–184, 187–193, 195, 197, 199, 201–204, 206–208, 211, 227, 231–233, 241, 242, 244–249, 252, 255, 260–262, 266, 268, 269, 272–274, 284–286, 288–294, 300
Nonpositive, 33
Nonstationary, 176
Nonsteady, 6
Normalization, 3, 31, 48, 54, 59, 65, 67–68, 73, 82–83, 85, 95–96, 99, 141, 143, 150, 162, 167, 180–181, 184, 188, 192, 199, 203, 206, 212–213, 220–222, 224, 226, 236, 244–246, 257, 277

O

Object, 46, 103–106, 108, 115, 117–118, 148
Observations, 5, 14, 21, 46, 54, 58, 76–78, 91, 104–106, 109, 111, 114, 123, 143, 234–235, 295
Observer, 27, 46, 51, 53–54, 58–59, 62–68, 71–76, 80, 82–84, 88–90, 93, 95–96, 101, 128, 131, 139–146, 148–149, 180–181, 183–184, 187–191, 202–204, 206–208, 211–212, 217, 220–221, 223, 226, 233–234, 240–242, 244–246, 286–289, 291, 294, 299
Occlusion, 105, 113
OEM, 5–7, 21–23
Optimal, 5, 27, 45–46, 53, 58, 67, 76, 79, 90, 150, 181, 246–248, 275, 283, 297
Orbit, 37
Origin, 105, 291–292
Orthogonal, 7, 10, 114, 129, 133–134, 137–138
Orthogonalization, 34, 110

P

Parameterization, 128, 180, 193, 294
Parameters, 5–9, 11–14, 20–27, 33, 37, 39–40, 43, 76, 84–88, 98–101, 105–106, 108, 120–121, 192, 226, 268–271, 273, 280, 287, 290, 296
Particle, 77, 89, 106, 284
Perpendicular, 7
Perturbation, 23
PFE, 36, 54–55, 73–74, 123, 134, 143, 146, 162, 165–167, 169, 171–172, 174, 188, 206, 220, 244
PFEX, 110, 112, 115, 125–126, 130–131, 134–135, 138, 147
PFEY, 110, 112, 115, 125–126, 130–131, 134–135, 138, 147
Pitch, 84, 98, 256, 296–299
Positioning, 35–36, 54, 104–105, 108, 115, 123–124, 127, 138, 143–144, 161–162, 164–166, 171–175, 281–282, 284
Predictions, 6, 10, 17, 21, 22, 29, 33, 35–36, 38, 65–66, 73–74, 91, 111, 138, 178, 188–189, 193, 206–207, 211–212, 221, 249–250, 274–276, 280, 286, 287, 293, 294, 296–298, 300
Priori, 23, 27, 29–31, 33, 35, 80, 88, 91, 93, 110–114, 128, 179
Probabilistic, 5, 8, 24, 27, 284
Probability, 5, 14, 18, 77, 90, 106, 180, 277, 280, 284
Propagation, 29, 31, 33–35, 38–39, 44, 47–50, 78, 90–91, 110–112, 114, 128–129, 131–133, 136–137, 148, 178, 213, 216, 235, 239–240, 248, 260
Propeller, 273
Proximity, 107, 123
Pseudo, 7

Q

Quadratic, 11, 163, 249
Quantization, 108
Quasi, 6, 19–20, 23

R

Radar, 275, 280, 282–284
Radiation, 108
Random, 5–7, 13–14, 16, 18, 27, 32, 35, 43–44, 46, 53–54, 60, 67, 77, 85, 91, 98–99, 106, 150, 180, 184, 203–204, 236, 267, 286
Raphson, 19, 23
Recursive, 4, 26–28, 30, 33, 45, 76, 190, 195, 199, 224, 248–250, 272
Reliable, 33–34, 276, 298
Resonance, 104
Riccati, 28, 32, 51, 58, 67, 140, 145, 199, 202, 207, 241
RMS, 44, 123, 143, 167
RMSE, 280–283
RMSPE, 110, 112, 115, 123, 125–126, 130–131, 134–135, 138, 143, 146–147
RMSVE, 110, 112, 115, 123, 125–126, 131, 138, 146–147
Robotics, 53, 104, 180, 192, 221, 245, 274–275, 287
Roll, 296–298
RSSPE, 143, 161, 165–166, 168–170

S

Scalar, 32–33, 35, 43–44, 51, 67, 83, 96, 111
Seismic, 104
Sensor, 8, 28, 46–50, 53–55, 63, 72, 75–76, 90, 92, 103–104, 107–109, 118, 125, 129, 138, 150–151, 154, 156, 162–163, 167, 171–175, 221, 234–235, 241, 245, 275, 291
Sensory, 188
Sigmoidal, 249, 252, 255, 259, 262, 268, 293, 322
Sigmoidslope, 259
Signal, 6, 10–11, 16, 27, 39, 44, 46, 58, 73, 77, 89, 105, 119, 125, 143, 188, 202, 206, 208, 212, 220, 244, 248, 252, 274, 284
Singular, 112, 128–129
Sonar, 118
Spectral, 6, 10, 32, 105, 129, 197, 236
Spectrum, 18, 276
SQRT, 254, 267
SRIF, 3, 105, 111, 113–114, 117, 119, 125, 128–137, 148
SRVD, 128, 130–132, 135
SRVL, 134
Stability, 3, 26, 39, 46, 53, 58–59, 74–76, 80, 84, 88, 90–91, 93, 95, 99, 101, 105, 112–113, 128, 131, 141, 148–149, 180–181, 183–184, 189–193, 195, 197, 199, 201–203, 207–208, 211–213, 217, 221, 223, 225, 227, 231, 233–234, 242, 245–249, 253, 255–257, 259–261, 265, 267, 269–274, 287, 289–291, 294
Stabilization, 289
Stationary, 6
Std, 110–112, 123, 125–126, 129–130, 135, 138, 143, 146–147
Stochastic, 45, 47, 59, 77, 91, 104, 180, 235, 287
Sugeno, 143, 187
Supervised, 294

Surveillance, 275
SVD, 128–129, 132–133
SVF, 46, 50, 58, 63, 72, 109, 150, 154, 225–226, 230
Symptotic, 241
Synthetic, 107–108, 118–119, 123, 143

T

TGNSTD, 110, 112, 115, 125–126, 129, 134–135, 138, 143, 146–147
Timestep, 91, 93, 133, 137–138, 237
Tracking, 3, 45–46, 53–54, 58, 103–109, 111–115, 117–121, 123, 125–135, 137–139, 141–143, 145–151, 154–156, 158–159, 161, 163–177, 179, 190–191, 221, 245, 274–277, 280–285
Traffic, 103, 275
Training, 123, 247–250, 257–260, 268, 274, 287, 290–291, 293, 296–298
Trajectories, 28, 37, 162, 189, 221, 245, 276, 280–283
Transform, 8
Trapezoidal, 119, 178
Traverse, 195

U

UDF, 34, 86, 105, 109–110, 131, 135
UDKF, 35, 43
UDSRIF, 135, 137–139
Unmanned, 273
Unscented, 77, 89, 247, 273, 284
Unstable, 12, 181

V

Vagueness, 180
Variables, 5, 8–9, 18, 21–23, 25, 37, 44–45, 59–60, 67, 77, 83–85, 96, 98–99, 119–120, 150–151, 154, 162, 177–181, 184, 193, 195, 203–204, 226, 234–236, 248, 256–258, 267, 286, 293–294, 296
Variance, 16–17, 24, 27, 29, 33, 43–45, 48, 107–108, 111, 139, 181, 216, 236, 244, 275, 296
Variant, 77, 162, 281
VDSRIF, 134–135

W

Wavelets, 118
Weighted, 8, 30, 34, 77, 79, 91, 93, 280
White, 6, 10, 13, 18, 25, 27–28, 33, 43, 91, 108, 123, 234, 275
Wireless, 46, 53, 75, 221, 245

Y

yperr, 143
Ystem, 5, 76

FSC
www.fsc.org
MIX
Papier | Fördert gute Waldnutzung
FSC® C083411

Kolibri 360 GmbH
Ferdinand-Jühlke-Straße 7
99095 Erfurt, Deutschland
produktsicherheit@kolibri360.de